WOMEN, WORK AND COMPUTERIZATION: FORMING NEW ALLIANCES

IFIP TC 9/WG 9.1 International Conference on
Women, Work and Computerization
Amsterdam, The Netherlands, 27–29 April, 1988

NORTH-HOLLAND
AMSTERDAM • NEW YORK • OXFORD • TOKYO

WOMEN, WORK AND COMPUTERIZATION: FORMING NEW ALLIANCES

Proceedings of the IFIP TC 9/WG 9.1 International Conference on Women, Work and Computerization
Amsterdam, The Netherlands, 27–29 April, 1988

edited by

Kea TIJDENS
Dept. of Social Science Informatics
University of Amsterdam, The Netherlands

Mary JENNINGS
Microsyster
London, United Kingdom

Ina WAGNER
Institute for Practical Informatics
Technical University of Vienna, Austria

Margaret WEGGELAAR
Centre for Labour and Welfare Studies
University of Leiden, The Netherlands

1989

NORTH-HOLLAND
AMSTERDAM • NEW YORK • OXFORD • TOKYO

ISBN: 0 444 70548 1

Publishers:

ELSEVIER SCIENCE PUBLISHERS B.V.
P.O. Box 103
1000 AC Amsterdam
The Netherlands

Sole distributors for the U.S.A. and Canada:

ELSEVIER SCIENCE PUBLISHING COMPANY, INC.
655 Avenue of the Americas
New York, N.Y. 10010
U.S.A.

Library of Congress Cataloging-in-Publication Data

IFIP TC 9/WG 9.1 International Conference on Women, Work,
and Computerization (1988 : Amsterdam, Netherlands)
Women, work, and computerization.

Bibliography: p.
1. Women--Employment--Effect of technological innovations on--Congresses. I. Tijdens, Kea. II. Title.
HD6053.I34 1988 331.4'25 88-30937
ISBN 0-444-70548-1

PRINTED IN THE NETHERLANDS

PREFACE

This volume includes the papers and the speeches of the third conference on Women, Work and Computerization, organized by working group 9.1 on 'Computers and Work' of Technical Committee 9 'Computers and Society' of the International Federation of Information Processing (IFIP). The conference was held in The Netherlands on April 27-29, 1988.

This book is divided into five sections, following the themes of the workshops. Section one contains the papers on organizational change. Section two deals with papers on system design and in section three contains the papers on training experiences are included. Section four is on empowering strategies and the last section contains the papers on the Third World. At the end we arrive at a conclusion from the conference. Each section will start with an introduction in which one of the editors outlines the papers and the discussions in the workshops.

In Appendix I the report of the second conference in Dublin, 1986 is included.
Appendix II contains the programme of the conference.

The views expressed in this book are those of the authors. They should not be ascribed to the IFIP or to the editors.

Mary Jennings
Kea Tijdens
Ina Wagner
Margaret Weggelaar

July 1988

ACKNOWLEDGEMENTS

The first IFIP conference on Women, Work and Computerization was held in Italy, Riva del Sole in 1984. The second conference was held in Dublin, Ireland in 1986 and the third conference was held in The Netherlands on April 27-29, 1988. The first and the second conference concentrated on exploring the opportunities that the use of computers could have on women's participation. The third conference we concentrated on strategies : How to improve the position of women, working and unemployed, in this area of new technologies. The discussions about strategies were based on the knowledge of how new technologies can influence the position of women in a positive and negative way.

The women who are active in IFIP working group 9.1., took the initiatives for the first and the second conference. The initiative for the third conference was taken by the three Dutch women who participated in the Dublin conference. They formed an organizing committee of six Dutch women, who all had shown quite a lot of knowledge of computerization. The members of the organizing committee were:

* Ria Hermanussen, working at the Research Institute Branche Electronics (SOBE);
* Ellen de Ru, working at City Council of Deventer and before at the Dutch PTT;
* Loes Steensma, working at Netherlands Postal and Telecommunications Services (PTT);
* Kea Tijdens, working at the University of Amsterdam;
* Marja Visser, working at the Alida de Jongschool and afterwards at the Building- and Woodworkers-union FNV (Bouw en Houtbond FNV);
* Margaret Weggelaar, working at the University of Leiden.

NGI, the Dutch Association for Informatics is a member of IFIP. One of the working-groups within NGI is called SAIA (Social Aspects of Informatics and Automation) and is narrowly related to the IFIP-working group.9.1. They really supported this conference.
The Stichting Informatica Congressen (Foundation of Informatics Congresses, SIC) was responsible for all practical arrangements and guarantied financial support in case we did not play equal. The organizing committee was assisted by Iris Allebrandi, working at SIC and Karin Eckenhuysen-Smit, a student at the University of Leiden.

The programme committee for the conference met three times and discussed the main items for the conference. They made quite a lot of efforts in making this conference a success:

* Gro Bjerknes, Institute for Informatics, University of Oslo (Norway)
* Frank Boddendijk, Equal Opportunities Council, the Hague (Netherlands)
* Titia Bos, Dutch Trade Union Confederation, Amsterdam (Netherlands)
* Tone Bratteteig, Institute for Informatics, University of Oslo (Norway)
* Tilly Brons, Interdepartemental Commission on Information Policy, the Hague (Netherlands)
* Boel Carlsson, Federation of Civil Servants (Statstjänstemannaförbundet), Stockholm (Sweden)
* Monique Chalude, University of Brussels (Belgium)
* Fiorella de Cindio, University of Milano (Italy)
* Isabel da Costa, Reseau Femme Informatisation Bureaux, Paris (France)
* Maria Crouz-Alonzo, FAST Programme of the CEC, Brussels (Belgium)
* Ghiselaine Doniol-Shaw, G.E.D.I.S.S.T./CNRS, Paris (France)
* Ben Elkerbout, Netherlands Postal and Telecommunicatons Services (PTT), The Hague (Netherlands)

* Sabine Gensior, Wissenschafts Centrum Berlin (Federal Republic Germany)
* Judith Gregory, Department of Communication, California (United States of America)
* Rita Hellman, University of Turku (Finland)
* Mary Jennings, Microsyster, London (United Kingdom)
* Werner Korte, Büro Empirica (Federal Republic Germany)
* Thea de Poel, Foundation of Women and Informatics, Amsterdam (Netherlands)
* Cleo Presvelou, University ofWageningen (Netherlands)
* Cidney Pullman, The Labor Institute, New York (United States of America)
* Joop Ramondt, State Commission on Technology, Labour and Organization, Rotterdam (Netherlands)
* Els Rienstra, Alida de Jongschool, Utrecht (Netherlands)
* Felix van Rijn, Netherlands Society for Informatics, Amsterdam (Netherlands)
* Arie Rip, University of Twente, Enschede (Netherlands)
* Sharon Szymanski, The Labor Institute, New York (United States of America)
* Kathleen Vandermassen, Flanders Technological Foundation, Antwerp (Belgium)
* Ina Wagner, Institut für Praktische Informatik, Vienna (Austria)
* Thea Weijers, Technology and Police Foundation, Apeldoorn (Netherlands)

The plenary session of the conference was attended by roughly 370 people. We were very glad with the performance of Clio Presvelou as a conference chairperson and the speakers: Saskia van Hoek (The Netherlands), Judith Gregory (USA), Kari Thoresen (Norway), Ursula Huws (United Kingdom), Cynthia Cockburn (United Kingdom) and Boel Carlsson (Sweden).
As we realized that the use of new technology was not only a question of the Northern countries, but also of the Third World, we decided to organize a plenary meeting about problems and possibilities for women in the Third World. Four women, two researchers, a computer professional and a trade-unionist, accepted the invitation to come to the conference. The evening of Thursday April 28 was confined to their view on technological developments in their countries and its impact on the labour market position of women. We would like to thank Pilar Vasquez Rubio and Soledad Robina from Mexico, Lim Gek Kheng from Singapore and Rosalinda Pineda-Ofreneo from the Philippines for their speeches as well as Maartje van Putten from The Netherlands, who chaired this meeting.

We also would like to thank the people who made it possible for us to organize this conference, especially the Dutch Ministry of Social Affairs, the co-ordination Bureau of Emancipation, the Dutch Ministry of Economic Affairs, the Dutch Ministry of Education and Science, UNESCO, the European Economic Community, the Dutch Postal and Telecommunications Services PTT, the city of Amsterdam and IBM. They gave us moral and financial support.

We wish to thank all the participants to the conference, who have put so much effort in it and made this conference so successful.
Hope to see you all at the next conference.

Ellen de Ru,
on behalf of the organizing committee.

CONTENTS

Women, Work and Computerization: Forming New Alliances
K. Tijdens, M. Jennings, I. Wagner, and M. Weggelaar (Editors)
Elsevier Science Publishers B.V. (North-Holland)

REPORT OF THE CONFERENCE

Kea Tijdens
Department of Social Science Informatics
University of Amsterdam
The Netherlands

1. INTRODUCTION

The three Dutch women who participated in the Women, Work and Computerization conference in 1986, took the initiative for this Dutch conference. They set up a local organizing committee, supported by an international program committee. Via a network of contact persons in and outside European countries the conference was announced. A 'call for papers' was issued in September 1987 and in November, about 70 abstracts were received. The program committee decided to organize workshops where a few papers could be discussed intensively in small groups, so 30 authors were invited to write a paper.

During the preparation of the conference two ideas behind the conference were discussed. The first was that the conference should be attractive to Dutch women and men to get involved in the developments at stake. Secondly, the conference should be attractive for women and men from all over the world to deepen their knowledge and experiences. Therefore, a 'double formula' was chosen: a plenary session on the first day with five speakers from different countries, and a second and third day with small groups discussing papers. The first day was held in the meeting hall of the Free University in Amsterdam and was visited by 370 people. On the second and the third day the participants stayed in a conference-building at Noordwijkerhout, a place at the seaside about 50 kilometres from Amsterdam. More than 140 people visited this part of the conference. The conference really was a great success.

As said before, the purpose of the first day was to give general but actual information on the subject. Keynote speakers from several countries focussed on past and future developments regarding new technology and on real life examples which have already created opportunities for women or have the potential to do so. The first speaker was Saskia van Hoek from The Netherlands. At the time of the conference she was director of the Alida de Jongschool, a school for education in the field of informatics for re-entering women older than 25 years. She dealt with the question whether the strategy of separate school systems are in the benefit of women. Van Hoek gave a review of the encouraging results of vocational training experiments for female adults and hypothesizes that only a coherent strategy will be successful. Her speech is in part 3 of this book.

Judith Gregory (United States), researcher at Department of Communications of the University of California, formerly a research director of '9 to 5', the National Association of Working Women, and now a staff liaison for the Standing Committee on

Women's Employment, sketched the 'good' and the 'worse' of trade union policies concerning women and new technology in the United States. Her speech can be found in part 1 of this book.

After the lunch Kari Thoresen, who is working at the Norwegian Computing Centre in Oslo, Norway, delivered her speech about a 'female-friendly way of system designing'. She discussed different ways of system designing and argued for an alternative method: systems should be no more than just a tool for users. Her speech is in part 2 of this book.

Ursula Huws from the United Kingdom, director of Empirica London, published a lot about VDU work and about technology in the third world. She pointed the relations between technological and organizational changes like decentralization, flexibilization and an increase of sub-contracting. Unfortunately, her speech is not in this book.

Cynthia Cockburn (United Kingdom), senior researcher at the City University of London, author of such well known books as 'Brothers' and 'Male Dominance' was the last keynote speaker. She dealt with the question of how women relate to technology in history and today. Women have been said to be invisible in technology although they developped own technologies. Cultural mechnisms in society are important to understand the marginalization of women's role in technology. Her speech is included in part 4 of this book.

On the second and the third day of the conference, daily 8 workshops were held. Each of them had about 8 to 20 participants, depending on their preferences and one up to four papers were discussed. The authors had a short explanatory note on their paper. The chair(wo)men and one or two referee(s) made some statements about the papers in advance, which guarantied an intensive discussion afterwards. On the plenary meeting at the end of each day the chair(wo)men gave brief reports.
Five themes were chosen for these workshops:

* Challenges of organizational change,
* Patterns of access to and opportunities on the labour market,
* Strategies for system development,
* Training: new opportunities for women
* Empowering strategies.

Participants especially showed much interest in the workshops about System designing and about Training, as well as in one workshop about organizational change. Because of the overwhelming number of participants some of the workshops were doubled.
A scheme of the conference can be found in Appendix II.

2. THE PARTICIPANTS

148 participants were on the three-day conference and they came from 22 countries. On the past conferences people came from 13 countries in Dublin in 1986 and from 12 countries in Riva del Sole in 1984. In the table you will find the number of participants from the different countries, compared to the participants at the conferences in 1986 and 1984. The figures of 1988 and 1986 are based on the list of participants. The figures of 1984 are based on the results of a questionnaire, which was held on the conference. In Riva del Sole, 60% of the participants came from the Scandinavian countries and in Dublin this percentage was 58%. At this conference the Dutch were a majority (33%) and the participants from the Scandinavian countries were 22%. See the table at the next page.

Nationality	'88	'86	'84		'88	'86	'84		'88	'86	'84
Austria	4	(1)	(3)	Australia	-	(-)	(1)	Belgium	1	(-)	(-)
Canada	1	(-)	(-)	Denmark	10	(4)	(8)	Finland	2	(2)	(3)
France	7	(-)	(4)	FRG	9	(3)	(4)	GDR	1	(-)	(-)
Hungary	2	(-)	(-)	Iceland	1	(-)	(-)	Ireland	-	(3)	(1)
Italy	12	(6)	(11)	Japan	1	(-)	(-)	Mexico	2	(-)	(-)
Norway	8	(3)	(3)	Netherlands	49	(3)	(-)	Philippines	1	(-)	(-)
Spain	1	(-)	(-)	Sweden	13	(24)	(24)	Switzerland	-	(2)	(-)
Singapore	1	(-)	(-)	UK	12	(4)	(1)	USA	9	(1)	(1)
Yugoslavia	1	(1)	(-)								

Table 1. Nationality of the participants of the conferences in The Netherlands (1988), Ireland (1986) and Italy (1984).

Olerup et al. (1985) reports the results a questionnaire which was held under the participants at the first conference at Riva del Sole. On the third conference a similar questionnaire was distributed among the participants. 63 questionnaires were received from 94 participants in 1984. 97 questionnaires were received from 148 participants in 1988. The questionnaire was quickly made and distributed, so there very definite conclusions cannot be drawn.
In table 2 some demographic data of the questionnaire are summarized. Not so much differences were found between the results of the 1984 and the 1988 questionnaire. The age of the participants was more spread in 1988 as it was in 1984. The participants were more actively in women's movements now, compared to 1984. Education level was high in Riva del Sole (68% was a graduate) and it was even higher in Noordwijkerhout (84% was a graduate). In 1984 there were more students (22%) than in 1988.

What is your sex?		% in '88	% in '84
female	94	97%	89%
male	3	3%	11%
What is your age?		**% in '88**	**% in '84**
20-24	2	2%	2%
25-29	14	14%	21%
30-34	23	24%	21%
35-39	28	29%	30%
40-44	20	21%	)
45-49	6	6%	) 25%
50-54	3	3%	)
> 60	1	1%	) 1%

Do you have children?		% in '88	% in '84
yes	44	45%	44%
no	53	55%	56%
Are you actively in any women's organization or movement?		% in '88	% in '84
yes	51	52%	38%
no	26	27%	57%
passive	12	12%	5%
no answer	8	8%	-

What is your highest education level?		% in '88	% in '84
secondary/non-university	4	4%	10%
student	5	5%	22%
graduate (technical)	21	22%	)
graduate (non-technical)	55	57%	) 68%
graduate (techn. + non-techn.)	4	4%	)
else	5	5%	-
no answer	3	3%	-

Table 2. Findings from a questionnaire distributed at the conference Women Work and Computerization in 1988, compared to the findings of the same questionnaire distributed at the same conference in Riva del Sole in 1984.

In table 3 some data are presented about employment and occupation. In 1984 the participants raised the issue whether researchers were dominating in numbers. The results of the questionnaire showed that this was the case in 1984 and in 1988: about half of the participants were researchers and trainers (49% in 1984 and 52% in 1988). In Italy there were more system designers than four years later in The Netherlands. The same

can be said of the trade-unionists. In 1988 one-third of the participants were employed by an university. In 1988 nearly half of the participants had their university education in social and behavioral sciences or humanities, which was more than in Italy.
Comparing the results of the questionnaires in this table must be done very carefully, because in 1984 the respondents were allowed to fill in more than one alternative for these questions. The percentages are taken from the total number of answers, while in 1988 the respondents had a separate category 'more than one answer'.

What is employing organization?		% in '88	% in '84
university	32	33%	28%
research	12	12%	14%
civil service	13	13%	13%
private industry	5	5%	6%
consultancy	2	2%	17%
trade union	12	12%	24%
other	14	14%	5%
more than one	5	5%	5 out of 63
no answer	2	2%	-
What is your occupation/tasks?		% in '88	% in '84
research	38	39%	)
education and training	13	13%	) 49%
system design, programming	4	4%	12%
user	1	1%	6%
trade union activities	11	11%	27%
other	12	12%	6%
more than one	17	18%	15 out of 63
no answer	1	1%	-
If university education, what is/was your major subject?		% in '88	% in '84
social and behavioral sciences, humanities	44	45%	38%
business administration, economic sciences	9	9%	16%
law	1	1%	2%
computer science, system analysis	12	12%	23%
mathematics, physical sciences, engineering or biological sciences	7	7%	16%
philosophy	1	1%	4%
other	2	2%	-
more than one	13	13%	18 out of 63
no answer	8	8%	-

Table 3. Findings about education and occupation from a questionnaire distributed at the conference Women Work and Computerization in 1988, compared to the findings of the same questionnaire distributed at the same conference in 1984.

LITERATURE

Olerup, A., L. Schneider and E. Monod: Women, work and computerization. Opportunities and disadvantages, North Holland, Amsterdam, 1985

Women, Work and Computerization: Forming New Alliances
K. Tijdens, M. Jennings, I. Wagner, and M. Weggelaar (Editors)
Elsevier Science Publishers B.V. (North-Holland)

PART 1

CHALLENGES OF ORGANIZATIONAL CHANGE

Ina Wagner
Technical University Vienna
Austria

INTRODUCTION

The introduction of computer networks into work organizations has proven as an effective instrument for carrying through organizational change. The processes involved in such change show that automation practices vary considerably. Which path is chosen to a large extent depends on the strategies and negotiation power of the change agents involved. This makes participation of women in organizational change an issue of paramount importance.

The papers included in this part focus on observable changes of work that are detrimental to women's career opportunities as well as on positive experiences with influencing the design of organization and jobs and improving the position of women. The contributions cover a wide range of topics that can be grouped together under four headings:

* computer-based technologies in industry, with a focus on jobs in manufacturing and assembly;
* automation of clerical work;
* home-based telework;
* and changing patterns of access and opportunities on the labour market.

COMPUTER-BASED TECHNOLOGIES IN INDUSTRY

Starting out from a variety of examples observed in industries such as food-processing, shoe-making, clothing and electronics, the authors stress different aspects of organizational change and consider different points of leverage. In a case-study of the Dutch confectionary industry Eric Poutsma and Dorien Trommel analyze the impact of a largely automated packing technology on women's work. The question at stake is if and under which circumstances women are selected for key functions such as machine operators. The French group - Marie-Claude Bergouignan, Marie-José Cameleyre, Marie-Claire Rouyer - focus on women's own strategies in coping with experienced deficiencies in training, information, organization, and co-ordination between system designers and users. They point at women's difficulties to think out long term alternatives, given their neglect by French syndicalism.

AUTOMATION OF CLERICAL WORK

In contrast to the evidences from industrial sites, Martha Ormos and Maya Blameble describe two successful examples of office automation in which a redistribution of jobs and responsibilities has led to improvements of job content and work procedures. Findings according to the role of unions in organizational change are ambivalent. Whereas in the first case unions supported women's demands, success in the second example was solely due to women's own initiatives. Päivi Korvajärvi gives a less promising account of office automation practices. She perceives a wide gap between rigid organizational hierarchies and women's own ambitions and resources. Equally, Frida de Jong and Loes Steensma, using the example of the Dutch PTT, see Taylorism as the dominant paradigm at work in office automation. They put some hope though in general equality measures as well as demographic and labour market trends.

HOME-BASED TELEWORK

Two empirical studies are presented that analyze the situation of home-workers in Australia (Judith Wajcman, B.Robert, T.Tanter) and the FRG (Monika Jaeckel). The attractive side of working at home is experienced by (predominantly male) computer specialists who are in a position to convert their expert status into good pay and autonomy. Women who do wordprocessing at home, however, seem to pay a high price for whatever flexibility this arrangement offers: in terms of insufficient income and social security protection, difficulties to combine childcare and work. Still, home-based telework is perceived as meeting real needs of women, depending on whether appropriate support systems can be built up.

CHANGING PATTERNS OF ACCESS AND OPPORTUNITIES ON THE LABOUR MARKET

Papers included in this section combine discussion of organizational change with an analysis of labour market trends and women's participative behaviour. Mary Ann Madison, Vary T. Coates and Anna-Maija Lehto present statistical material on the spread of information technologies, employment trends and changes in job content in their countries, the USA and Finland. Angelika Volst and Ina Wagner argue on a more general level that computerization as an instrument of increasing organizational flexibility results in a growing polarization of jobs and career opportunities. They identify transitory jobs that are subject to far-reaching automation as the main field of political struggle over women's chances on the labour market. The polarization thesis is confirmed by Judith Gregory. Using research evidence from health and office work, she from an union perspective focuses on opportunities to reinforce tendencies towards multi-faceted jobs as well as on measures for coping with the growing precarization of work.

STRATEGIES

Awareness of the necessity to adjust and modify overall strategies to the particularities of national political backgrounds, trade union structures and jobs involved, characterized the workgroup discussions. With this consciousness of diversity in mind, several areas for strategic action as well as guiding principles were singled out:

1. Special attention should be given to the design of work organization and job content. Even small changes in the degree of job autonomy, task variety and co-operativity do matter to those concerned. To exert influence on job design was considered of particular relevance for women in "transitory jobs", as in these fields

far reaching decisions concerning women's chances for learning and advancement are at stake.

2. Broad re-skilling initiatives were considered an important vehicle for women. However, such training can only be effective in conjunction with strategies of designing jobs in which the new skills can be deployed. Moreover, it was noted that training has to bridge a widening gap between the positions in offices and factories typically held by women today and those that offer opportunities.
3. Strategic thought has to be applied to the invisible skills that many women in low-paying positions deploy. Several routes to making women's unrecognized contributions to work organizations more valued need to be tried out such as
    * to seize the opportunities that open up with management's need for co-operativity in automation projects;
    * to develop desirable patterns of meaning within a work organization concerning crucial concepts such as "flexibility" and "efficiency";
    * to experiment with organizational measures of implicit and explicit recognition of women's work;
    * and finally to convert these experiences into income-raising policies.
4. Networking was discussed on different levels
    * as a solidarity-enhancing strategy that builds upon social forms of raising consciousness;
    * as making use of and gaining access to those informal structures within companies that often play a crucial role in the redesign of organization and job content.

    Networking needs time and energy. Trade unions might make a positive contribution to improving women's work situation by providing financial and professional support for networking outside the official union channels.
5. Concern for the growing number of women in "casual" or "precarious" employment was expressed. These women are often not represented by unions and/or neglected in collective bargaining efforts. Income-raising policies as well as measures for improving the quality of work contracts were singled out as strategic priorities for this group of women.

Discussion of strategies also touched upon issues that up to date have not been dealt with sufficiently. The need to develop strategies for job design brought up the question of how to cope with the contradictions between efficiency and solidarity. Similarly, discussion of home-based telework raised the problem of how to find a balance between the need for autonomy and social security. It was felt that the contradictions inherent in women's own goals need to be actively thought out.

Women, Work and Computerization: Forming New Alliances
K. Tijdens, M. Jennings, I. Wagner, and M. Weggelaar (Editors)
Elsevier Science Publishers B.V. (North-Holland)

TRADE UNION STRATEGIES: THE UNITED STATES CONTEXT

Judith Gregory
University of California at San Diego, USA

In the United States, the level of unionization is very low, yet there are examples of creativity at the level of local unions, in new organizing campaigns, and in collective bargaining over conditions of industries undergoing restructuring such as telecommunications and health care. Union strategies which benefit women include pay equity, occupational health and safety issues such as video-display-terminal work, and training programs which include attention to the occupational positions of women. Unions with high concentrations of women members especially reflect increased activism of women in unions. Research which outlines further challenges outside traditional union practice and outside the legal boundaries of equal employment opportunity law.

1. UNIONS IN THE UNITED STATES.

A few general introductory remarks regarding the U.S. context may be helpful:

1 the United States is technology advanced, but socially backward when it comes to policies related to new technologies;
2 in this short paper, the reader will find some examples of "best cases" and some examples of "worst cases", but not necessarily "typical" cases;
3 currently, unemployment rates in the United States are very low compared to many other countries, but the greater quantity of jobs is unfortunately combined with little attention to the quality of jobs.

Unions are in a struggle to rebuild and to survive. Union staff report that it is difficult to devote resources to address technological change as a special issue, but it is essential that unions understand technological change in relation to industry restructuring and the reorganization of work, in order to develop integrated analyses and strategies. Unions today represent 17 percent of the United States workforce (approximately 16.9 million workers), down from 18 percent in 1985, while the United States unions once represented 35 percent of the workforce. In the private sector, unionization is even lower, at less than 15 percent overall unionization. Important areas of the private sector economy are virtually untouched by unionization: fewer than 3 percent of finance workers (insurance and banking workers) are union members. Rates of unionization throughout the non-governmental service sector are low: 7.2 percent in trade, 6.6 percent in services, compared to approximately 25 percent in manufacturing. Unionization is almost completely absent from the microelectronics industry. Even so, ten unions gained in membership in 1987; the Service Employees International Union experienced the greatest gain. Union victories also indirectly benefit non-union workers, for example, reflected in higher wages for office workers in industrial cities which are significantly organized.

2. WOMEN IN THE WORKFORCE AND WOMEN IN UNIONS.

In 1983, about 41 percent of union members were women -- an increase from 33 percent in 1981. The increased proportion of women union members is partly the result of men's loss of jobs in manufacturing industries as the economy has shifted towards services employment, and partly the result of the influx of women into unions where organizing has been successful. Since the mid-1970s, workplaces with 70 percent or more women have been more likely to join unions, compared to workplaces which do not have a women-intensive workforce (Freeman and Leonard, 1985). Many union gains have been in health care and public employment, major employers of women, particularly black and Hispanic women. Women are central to the transformed economy. Sixty to seventy percent of new jobs in the last decade have been filled by women, a pattern which is expected to continue into the 1990s. Key areas of employment for women in relation to technological change are:

* clerical work, including business services, which have expanded dramatically, especially temporary help services, which incorporate information technology skills;
* health care, the third largest employer in the U.S;
* "high tech" production workers--women are the majority of microelectronics production jobs, with Hispanic and Asian women, often new immigrants, holding 60 to 75 percent of these jobs;
* at the high end of informatics industry employment, women have made important gains in computer programming and design jobs, although men continue to dominate the best jobs. In 1986, 40% of computer programmers and 64% of computer operators were women, earning 81% and 73% of men's earnings, respectively;
* the telecommunications workforce is about 45 percent women--at 37.5 percent unionization. It is the most unionized sector among the "information technology" industries; since deregulation of AT&T/the Bell Companies communications industry has been subject to radical restructuring and expansion of non-union firms.[1]

Unions are virtually absent from both the high/professional end and the low-production end of the informatics/microelectronics industries

Table 1. Percentage of female workers in some occupations

Office and clerical workers	80 % women	
Health care workers		
Practical nurses	97% women	19% minority women
aides, orderlies, attendants	87% women	24% minority women
medical technicians	75% women	
"High tech" workers		
production workers	65-70% Hispanic and Asian women	
assemblers	72-77 percent women	
computer programmers	39.7% women	
computer operators	63.8% women	
Telecommunications workers	45% women	
Changes in women's position in computer jobs	1970	1980
computer scientists/systems analysts	15%	22%
computer programmers	23%	31%
computer operators	29%	59%

Source of data: from Strober and Arnold in [Women, Chips and Paper Clips, eds. Hartmann et al., NAS, Washington, D.C., 1986.

Overall, women in the U.S. earn 64.3 percent of the earnings of men. The Institute for Women's Policy Research estimates that three-fourths of the improvement (from 59 cents on a dollar previously) results from women earning more, while 25 percent of the "improvement" in women's wages results from the fact that men are learning less due to the permanent loss of jobs in high-wage industries. Job segregation exerts a powerful

1 Gregory, Kohl, Loble, Novotny, and Sacks, 1986

influence: the higher the percentage of women in an occupation, the lower the pay rate. 17 to 30 percent of the wage gap is due to overrepresentation of women in certain occupations, including data entry (92 percent women) and computer operator (64 percent women). Differences in skills and experience account for slightly less than half of the gap for college-educated women, but only 23-27 percent of the gap for non-college educated women.

3. RECENT RESEARCH ON WOMEN'S EMPLOYMENT AND TECHNOLOGICAL CHANGE.

3.1 Shifts in office clerical labour markets.

Changes in entry requirements and barriers to access or entry. Several recent studies of office computerization identify a pattern of "upskilling" or "upscaling" in which entry requirements -- educational requirements, previous work experience, computer literacy -- are raised, many lower level jobs are eliminated, and possibilities for advancement beyond the tier of computerized information-processing and customer service jobs are reduced.[1] Researchers examining the insurance industry also describe this pattern as simultaneous "upskilling" and "downwaging", as women lack the organization and power to get pay increases commensurate with increased work demands for skills, accuracy and responsibility for directly entered data and provision of service.[2] Noyelle observes that: "In many back offices (insurance, banking, public sector agencies, etc.) the threshold for the entry level has been raised, often to a 2-year community college diploma or equivalent degree" (Noyelle,1986b). The arena of education, outside of Equal Employment Opportunity law and outside of trade union practice, becomes an important arena for affirmative action in relation to shifts in job skills.
Examining the effects of technological changes in health care, Berliner predicts a polarization of employment resulting from the introduction of labour-saving technologies such as automated clinical laboratory analyzers, disposable sterile supplies, etc. "Technological change may thus ... simultaneously increase the number of both high-skilled health specific workers (for example, to operate new capital-intensive technology) and lower-skilled non-health-specific workers, reducing the numbers of mid-level hospital employees", while more and more hospital work is performed in free-standing, extra-hospital settings whose existence is facilitated by portable technologies. Given women's occupational position in the health care workforce, they are at the greatest risk of employment insecurity. This is particularly true for minority women who account for 24 percent of aides, orderlies and attendants and 19 percent of practical nurses. Generally, employers will seek workers who are "more flexible" -- able to work in varying locations and adapt to rapid changes in medical technologies (Berliner, 1984).

3.2 Contingent work.

It is estimated that one quarter of all U.S. workers is employed in the "contingent workforce.".[3] This category includes part-time and temporary industry workers as well as self-employed individuals and those who work as "contract employees." Following the deep recessionary period of 1980-82, employers sought to limit their "core" workforce of permanent, full-time employees; instead, when possible, they expanded

---

1 For example, see Murphree, 1983, on "technological discrimination"; Nelson, 1986, 1984, on spatial separation and skills; Appelbaum, 1984a, 1984b.

2 Appelbaum, 19841, 1984b; Baran, 1987, 1985; Noyelle, 1987, 1986a, 1986b, 1985; 9to5, National Association of Working Women, 1985.

3 Appelbaum, 1987; Appelbaum and Gregory, 1987

their "peripheral" workforce of contingent workers. The expansion of part-time employment is of particular concern to unions; one in ten of all workers works part-time, but one in four women works part-time. In particular, "involuntary" part-time work has exanded. Of the 2.5 million increase in part-time jobs from 1979 to 1986, only 600,000 workers report that they were looking for part-time jobs, the rest reporting that they settled for part-time work when they could not find full-time positions. The use of temporary workers, predominantly women, has grown rapidly in the 1980s. While the annual average number of temporary help services (THS) workers is (600,000?) on any given day, the industry assocation estimates that five million men and women move in and out of temporary employment each year.

Christopherson argues that contingent work is a key element of economic restructuring "which has arisen in conjunction with structural changes in the way goods and services are produced. If this is so," she writes, "then the demand for a more flexible work force can be expected to be a long term trend, not explainable solely in terms of the characteristics of the labor force nor as a cyclical fluctuation" (Christopherson, 1987). In other words, we are witnessing the evolution of a permanently impermanent workforce.

3.3 Disparate impacts of office mobility.

In an case study of "suburbanization" of computerized back office work in the San Francisco Bay Area, Nelson found a disparate racial impact in that the movement of jobs particularly hurt the employment opportunities of central city women, among whom minority women and women who are sole supporters of themselves and their families were proportionately greater Nelson found that this was not "an unfortunate side effect, but rather a calculated choice on the part of clerical employers to avoid the potential future demands of such women, in other words to break up the pre-conditions for unionization, while at the same time taking advantage of new pools of available labor among well-educated housewives to meet the higher skills levels of the computerized jobs.[1]

3.4 Recognizing different corporate strategies

Appelbaum observes that company strategies towards their employees have powerful effects which should be addressed in trade union strategies: a strategy towards "paraprofessionalization" -- towards multi-faceted jobs in which employees are viewed as responsible and given training in the fundamentals of the business (in this case, insurance) with technology to support the work -- has very different implications for job design, decision-making, and potential autonomy than does an employer strategy of deskilling and technology-as-replacement which facilitates increased use of temporary workers and subcontracted work. The fight to keep meaningful content in jobs, not to lose broad-based knowledge of the industry, is important to preservation of both quality and quantity of jobs.

4. UNION PRACTICE IN THE U.S.

What follows is only a brief outline of the state of union activity, some areas of difference between U.S. and European trade union experiences, and some longer-range problems identified in recent research.

Technological change is more likely to be "taken for granted" in the U.S. compared to five years ago. In other words, new technology is not necessarily seen as "new" anymore, but rather as a fact of modern worklife. In the mid and later 1980s, it is more

1 Nelson, 1986, 1984; research separately conducted by Baran and by Noyelle confirms the disparate impact found by Nelson

likely that the "handle" workers and their unions have on computerization in its combination with job reclassifications, expanded use of part-time or temporary workers, subcontracting of work previously done by union members, or occupational health issues.

4.1 Union practice related to women, work and computerization.

Affirmative action and equal employment opportunity principles and guide-lines have been included in collective bargaining agreements, mainly in unions with large memberships of women, often with majorities of women. The Reagan era caused uncertainties and instabilities in the civil rights laws and enforcement of these laws, hence a number of unions decided to incorporate provisions of the laws into agreements, to protect themselves and their members. Such provisions, along with pay equity or comparable worth efforts undertaken by unions mainly in the public sector, have been important vehicles for education within unions between men and women members (i.e., they necessitate consciousness-raising among the men). Still, these measures are true for some, but certainly not all unions.

4.2 New technology and video-display-terminals (VDTs).

Technological change has been firmly established on the unions' part as an occupational health and safety issue; it is strongly contested by employers and the computer industry. The Occupational Safety and Health Act (OSHA) was made law in 1970, after a decade's struggle by the labor movement. The OSHAct does not cover many broader work environment issues and potential health hazards, and does not address VDT work, job stress, or office hazards generally.

In the last five years particularly, unions have been able to include protections in bargaining agreements (beyond "letters of agreement" which preceded them), making concrete gains based on the period of raising awareness of new technologies' effects on health and the quality of work life. Agreement to attend to proper equipment design, work station design, (minimal) notification regarding introduction of new technology, some participation in technology decisions, have become routine parts of collective bargaining for many more unions. A handful of agreements provide women the right to transfer to non-VDT work during a term of pregnancy, without loss of seniority or benefits. Agreements which provide significant involvements in decisions about job design, systems design or significant influence on technology decisions are virtually non-existent in the United States. There is little challenge of traditional "management rights" or "prerogatives" in organizing how work is to be done, despite calls by some union leaders to take up this challenge. Ergonomics and "work environment" tend to be interpreted narrowly, in close relation to equipment and physical office layout.

4.3 Pay equity/comparable worth.

Several unions have won major victories in the area of pay equity or comparable worth for their women members, and also for black and Hispanic men and women. Among unions leading the fight for pay equity are the American Federation of State, County and Municipal Employees (AFSCME), the Service Employees International Union (SEIU), the Communication Workers of America (CWA), and others. While pay equity was first addressed by an industrial union, the International Union of Electronic, Electrical, Salaried, Machine and Furniture Workers (IUE), success in bargaining has come almost exclusively in public sector employment, where unionization of women is greatest. Building public support has been essential, as funding for pay equity reparations must come from taxpayers' dollars. Within pay equity efforts, the effects of technological change on women workers' jobs can be better understood, based on careful job evaluation analysis of skills used in clerical and other work.

During the AT&T/Bell Telephone company negotiations in the summer of 1986, the telephone company attempted a strategy we can describe as "pay equity in reverse." Generally, union women earn 34 percent above their non-union counterparts (the differential for men is 21 percent); among clerical workers (men and women combined), union wages are 41 percent higher, among service workers, 68 percent higher. Both unionization and a 1973 EEO consent decree which opened up non-traditional jobs for women created the basis for significant improvement of women's wages among AT&T/Bell employees. Telephone operators earn salaries equal to the average "man's wage. The company argued that telephone operators and phone company clerical workers' wages should be "bench-marked" to wage levels for private sector clerical workers in each local labor market. This "local labor market" strategy would have meant wage reductions of 20 to 40 percent

4.4 Part-time and contingent work.

Union attitudes and practice towards part-time and contingent work is widely varied and depends to a great extent on the power of the union in its local context. There is a body experience among the unions in improving part-time working conditions, pay and benefits, and opportunities to participate in training programs necessary for advancement.[1] In a very few cases, temporary workers have been unionized or "transitioned" into union bargaining units if they remain on the job beyond a certain length of time. These experiences can be drawn on more broadly to address changes in the structure of employment, as 25 percent of women in the U.S. work part-time and 75 percent of part-time workers are women.

4.5 Training programs related to technological change.

Training programs are increasingly an important area for trade unions. As unions push employers to provide information about jobs "at risk" of redundancy or other forms of technological impact, and for information about jobs which are predicted to open up for the future, such programs tend to benefit women as women are again and again concentrated in "at risk" jobs. "Joint training" efforts rely on labour-management committees. In a major pilot project now under way, the usual representative committee structures at the top have been augmented with committees representing rank-and-file workers from every department of the health care workforce. Some problems encountered in these joint efforts are the difficulty for the union and its representatives to maintain an independent voice (independent from the management), and the lack of experience in this type of bilateral decision-making, as this is not part of the U.S. labour tradition.

"Professional judgement" clauses and bargaining "ethical standards" clauses into collective agreements. In health care, nurses, doctors, and other health care workers in some cases, have bargained principles into collective agreements to protect their right to make decisions in the best interest of patients, when confronted with an onslaught of shifting federal reimbursement policies (for example, Diagnostic Rate Groups--DRGs) which mandate the shortest possible time for treatment of 470 specific diagnoses of illnesses, meaning that the hospital is only repaid for the standardized, average time of care), cost-saving pressures resulting from government policies and from increasing "privatization" of health care -- the transformation of formerly public institutions into privately owned, for profit corporations. These types of cost-benefit measures and new pressures are made possible and reinforced by information technologies.

---

1 See Appelbaum and Gregory, 1987, 1988

4.6 Computer monitoring

Computer monitoring of individual work performance is an even greater problem in the U.S. than five years ago. Estimates range from 20 percent to 40 percent of office workers have their work performance measured or monitored by computer -- somewhere between four and six million workers (Shaiken, 1987; [VDT News], May/June 1987). This practice, deeply embedded ideologically in the U.S. management tradition of Taylorism, is literally embedded in software packages and is now considered "upwardly mobile" -- professional workers are targets as well as clerical employees. There are very few union agreements which provide any protection against this practice, nor does any legislation exist today which protects against its abuses.

4.7 Systems design

Efforts to influence systems design are still not on the agenda of U.S. trade unions, and links with computer designers, programmers, and scientists are severely lacking, increasing the difficulty of developing strategies regarding systems design. These are especially important issues for women workers which have yet to be addressed in the U.S. context.

5. IMPLICATIONS AND CONCLUSIONS:

What are the implications from the context in the U.S. and what do recent research findings point to as further challenges for the development of trade union strategies?

1 mobility and changes of location of work require more coordination, more international approach, recognition of greater interdependence.
2 fighting polarization of employment needs to reinject affirmative action goals in the face of restratification and re-segmentation of the workforce.
3 in the education arena formal and informal educational resources should be connected, to assist women workers with gaining both (a) basic literacy skills necessary to survive, especially design effective assistance for immigrant women and non-college educated women who will face tougher barriers to entry into the labor force and/or advancement, mobility within the labor force, and (b) critical, analytic skills necessary to have meaningful rather than token involvement in job design and systems design, to have an independent voice in shaping the future of work. In taking on the educational area, influence on the content of training and long-range goals incorporated implicitly and explicitly intraining programs is of key importance. Training must provided continuity for women, a balance between support for short-term survival goals and longer-range development.
4 concerning research, respect for knowledge of women members on the frontlines in the workplace must guide union staff and leaders, computer scientists and designers, and academic researchers. But it also crucially important that researchers share understandings of industry shifts, changes in environment, technology, systems design, with workers. These two contexts must inform each other.

To conclude: unions must recognize the centrality of women and thoroughly re-examine their strategies and practices. Mobilization and activation of women is the best strategy of preparation for the challenges of technological change, its complexities, multi-faceted direct and indirect effects, differential effects. The mobilization of women, the life of women in the unions, will make the difference between whether trade union strategies and agreements are "paper tigers" or truly empowering for women.

REFERENCES

9to5, National Association of Working Women, Hidden Victims: Clerical Workers, Automation and the Changing Economy, National Association of Working Women, Cleveland, 1985.

Appelbaum, Eileen and Judith Gregory, Union Responses to Contingent Work: Are Win-Win Outcomes Possible?, in Women's Bureau, U.S. Department of Labor, Women and Work in the Year 2000: A Look to Contingent Work, U.S. Government Printing Office, Washington, D.C., forthcoming, 1988.

Appelbaum, Eileen, "Restructuring Work: Temporary, Part-Time, and At- Home Employment," in Hartmann et al, eds., Computer Chips and Paper Clips, Technology and Women's Employment, Volume II, National Research Council, National Academy Press, Washington, D.C., 1987.

Appelbaum, Eileen, Technological Change in the Insurance Industry 1960-90, Research Report, Department of Economics, Temple University, Philadelphia, 1984b.

Appelbaum, Eileen, Technology and the Redesign of Work in the Insurance Industry, Project Report No. 84-A22, Institute for Research on Educational Finance and Governance, School of Education, Stanford University, 1984a.

Baran, Barbara, "Office Automation and Women's Work: The Technological Transformation of the Insurance Industry," chapter in Castells, Manuel, ed., High Technology, Space and Society, Vol. 28 Urban Affairs Annual Reviews, Sage Publications, Inc., Beverly Hills, 1985.

Baran, Barbara, "The Technological Transformation of White-Collar Work: A Case Study of the Insurance Industry," in Hartmann et al, eds., Computer Chips and Paper Clips, Technology and Women's Employment, Volume II, National Research Council, National Academy Press, Washington, D.C., 1987.

Department for Professional Employees, AFL-CIO, Here Comes Tomorrow: Technological Change and Its Effects on Professional, Technical and Office Employment, Report of a conference held in 1986, DPE, AFL-CIO, Washington, D.C., 1988.

Freeman, Richard B. and Jonathan S. Leonard, Union Maids: Unions and the Female Workforce, Cambridge, MA, 1985.

Gregory, Judith and Karen Nussbaum, "Race Against Time: Automation of the Office, An Analysis of the Trends in Office Automation and the Impact on the Office Workforce," Office Technology and People, 1 (1982) pp 187-236, Elsevier Science Publishers, Amsterdam.

Gregory, Judith, "Clerical workers and new office technologies," in Office Workstations in the Home, National Research Council, National Academy Press, Washington, D.C., 1985.

Gregory, Judith, George Kohl, Leslie Loble, Louise Novotny, and Karen Sacks, Women's Wages: A Key to Preserving Middle Income Jobs, Communication Workers of America, Washington, D.C., 1986.

Gregory, Judith, "Technological Change in the Office Workplace: Implications for Organizing," in Labor and Technology: Union Responses to Changing Environments, eds. Donald Kennedy, Charles Craypo, and Mary Lehman, Department of Labor Studies, Pennsylvania State University, University Park, PA, 1982.

Gregory, Judith, "The Next Move: Organizing Women in the Office," in The Technological Woman: Interfacing with Tomorrow, ed. Jan Zimmerman, Praeger Publishers, New York, 1983.

Hartmann, Heidi I., Robert E. Kraut, and Louise A. Tilly, eds, Computer Chips and Paper Clips, Technology and Women's Employment, Volume II, National Research Council, National Academy Press, Washington, D.C., 1987.

Hartmann, Heidi I., Robert E. Kraut, and Louise A. Tilly, eds., Computer Chips and Paper Clips, Technology and Women's Employment, Volumes I, National Research Council, National Academy Press, Washington, D.C., 1986.

Jacobson, Carolyn, "Women in the Workforce," [AFL-CIO Federationist], Special Report, Washington, D.C., Spring 1986.

Labor Relations Weekly, Bureau of National Affairs, Washington, D.C. U.S. Department of Labor, Bureau of Labor Statistics, [Employment and Earnings]

Marschall, D. and J. Gregory, eds., Office Automation: Jekyll or Hyde? Highlights of the International Conference on Office Work and New Technology, Working Women Education Fund, Cleveland, 1983.

Murphree, Mary C., "Brave New Office: The Changing World of the Legal Secretary," in K.B. Sacks and D. Remy, My Troubles Are Going to Have Trouble with Me, Rutgers University Press, New Brunswick, 1984.

Murphree, Mary C., "Dealing with technological discrimination," Secretary Speakout '83: The Professional Secretary's New Identity in the Information Age, Report of a conference March 9-11, 1983, Professional Secretaries International, Boston, 1983.

National Research Council, Technology and Employment Effects, Interim report of the Panel on Technology and Women's Employment, Committee on Women's Employment and Related Social Issues, National Academy Press, Washington, D.C., 1985.

Nelson, Kristin Louise, Back Offices and Female Labor Markets: Office Suburbanization in the San Francisco Bay Area, Ph.D. dissertation, Department of Geography, University of California, Berkeley, 1984.

Nelson, Kristin, "Labor Demand, Labor Supply and the Suburbanization of Low-Wage Office Work," chapter in Scott, Allen J. and Michael Storper, Eds., Production, Work, Territory, Allen & Unwin, Winchester, MA, 1986.

Noyelle, Thierry J., "The New Technology and The New Economy: Some Implications for Equal Employment Opportunity," in Hartmann et al, eds., Computer Chips and Paper Clips, Technology and Women's Employment, Volume II, National Research Council, National Academy Press, Washington, D.C., 1987.

Noyelle, Thierry, Beyond Industrial Dualism: Market and Job Segmentation in the New Economy, Westview Press, Boulder, 1986a.

Noyelle, Thierry, "Employment Discrimination in Today's and Tomorrow's Economy," in Women's Bureau, U.S. Department of Labor, Women, Clerical Work, and Office Automation: Issues for Research, Women's Bureau and the Panel on Technology and Women's Employment, National Research Council, U.S. Government Printing Office, Washington, D.C., 1986b.

Noyelle, Thierry, New Technologies and Services: Impacts on Cities and Jobs, University of Maryland, Institute for Urban Studies Monograph Series No. 5, College Park, 1985.

Office of Technology Assessment, U.S. Congress, Automation of America's Offices, 1985-2000. OTA-CIT-287, U S. Government Printing Office, Washington D.C., 1985.

Office of Technology Assessment, U.S. Congress, New Communication Technologies: Implications for Privacy and Security (Monitoring Office Work), U.S. Government Printing Office, Washington, D.C., 1987.

Pullman, Cydney and Sharon Szymanski, The Impact of Office Technology on Clerical Worker Skills in the Banking, Insurance and Legal Industries in New York City: Implications for Training, a study for the Private Industry Council of New York City, Technology Policy Project, The Labor Institute, New York, 1986.

Werneke, Diane, Microelectronics and Women: a Literature Summary, Committee on Women's Employment and Related Social Issues, National Research Council, Washington, D.C., 1984.

Women's Bureau, U.S. Department of Labor, Women and Office Automation: Issues for the Decade Ahead, U.S. Government Printing Office, Washington D.C., 1985

Women's Bureau, U.S. Department of Labor, Women and Work in the Year 2000: A Look to Contingent Work, U.S. Government Printing Office, Washington, D.C.: forthcoming, 1988.

Women's Bureau, U.S. Department of Labor, Women, Clerical Work and Office Automation: Issues for Research, Report of a Conference Co-Sponsored by the Women's Bureau and the Panel on Technology and Women's Employment, National Research Council, U.S. Government Printing Office, Washington, D.C., 1986.

BIOGRAPHICAL NOTE

Judith Gregory worked as research director for 9to5, National Association of Working Women, from 1979 to 1984. 9to5 is an organization of women office workers which jointly created a new union for clerical workers, District 925 of the Service Employees International Union (SEIU), in 1981. While working for 9to5, Ms. Gregory wrote [Race Against Time: Automation of the Office] (1981), among other publications. From 1984-86, Ms. Gregory worked for the Department for Professional Employees of the American Federation of Labor-Congress of Industrial Organizations in Washington, D.C., where she served as staff liaison for the Standing Committee on Salaried and Professional Women. She joined the Ph.D. program in the Department of Communication, University of California-San Diego, in the fall of 1986, where she is studying advanced information technologies in service sector industries. Correspondence address: Judith Gregory, Department of Communication, D-003, University of California at San Diego, La Jolla, CA 92093, USA

Women, Work and Computerization: Forming New Alliances
K. Tijdens, M. Jennings, I. Wagner, and M. Weggelaar (Editors)
Elsevier Science Publishers B.V. (North-Holland)

NEW TECHNOLOGY AND THE EMPLOYMENT POSITION OF WOMEN IN THE CONFECTION-ERY INDUSTRY

Erik Poutsma, Dorien Trommel
The Netherlands

In this paper the impact of new technology in the Dutch sugar processing industry is described. The authors analysed the decision in the designing process concerning the organisation of the labour process of female workers. This paper deals with the problem whether new technologies and subsequent reorganisations can make a contribution to the improvement of conditions and contents of women's jobs. Contents of this paper are based on research done at Delft University of Technology and the Dutch Research Institute for Small and Medium-Sized Business.[1]

INTRODUCTION

In this contribution the influence is investigated which new technologies may have on the position of women employed in the Dutch sugar processing industry, specifically in the production of confectionery: biscuits, chocolate and sweets. There is reason for some concern that the reorganisations entailed by the introduction of new technologies may put the position of women in jeopardy. The concern springs from the clear division of work between men and women in this sector on the one hand, and the management's organisation strategy on the other.
The contents of this contribution are based on research carried out by associates of the Netherlands Economic Research Institute for Small and Medium-Sized Enterprise in Zoetermeer and Delft University of Technology, on behalf of the "Voedingsbond FNV" (Foodstuff Workers' Union, associated with the Netherlands Federation of Trade Unions), the largest trade union in the sector. Besides, it was subsidised by the Dutch Ministry of Social Affairs and Employment. Briefly formulated, the objective of the investigation was to find out whether alternative strategies might lead to the quantitative and qualitative improvement of the employment position of women in industry. To that end, the automation choices made have been analysed. Statistical material about the confectionery industry has been examined, and interviews held with experts in the field.

1 The complete research report: Meijdam, C., D. Trommel, E. Poutsma and R. Braaksma, "Vrouwenarbeid in de Zoetwarenindustrie. Mogelijkheden van technische vernieuwing: Restarbeid of Vakarbeid [Female labour in the Confectionery Industry. Prospects of technological innovation: Marginal work or professional work]" can be ordered from the EIM; telephone 079 - 41 36 34. A comparable investigation has been carried out in the butchers' sector: "Vrouwen in de Slagerij. Mogelijkheden van technische vernieu- wing: Hulpkracht of Vakvrouw" [Female labour in Butcher's Shops. The Prospects of Technical Innovation: Assistant or Professional], also available from the EIM.

Moreover, three case studies of selected establishments have been carried out. These case studies comprise interviews with the general management, personnel managers, technical executive staff, female employees, employees'representatives and trade-union executives. Some research results will be presented in this article.

The advance of micro-electronics offers new challenges to management. Its purpose is to achieve maximum control of the production process at a minimum of human effort. In many confectioners' plants the processing part of the production process is already largely automated. Currently, the sector is taking up the automation of packaging activities, traditionally mostly women's work. The current developments give rise to several questions:

* How does the introduction of new wrapping systems affect female labour?
* What choices do managements make with respect to the structure of the organisation and the tasks?
* What jobs will be allotted to women in the automated production process? Is the division of tasks and functions subject to (conscious or unconscious) discrimination between men and women?
* What task organisation is likely to open better prospects to female employees?

The study started from the principle that the labour organisation is conceived by persons according to their ideas about the desirable organisation, in particular the technical foundations and the norms and values to be adhered to. In that view, technology constitutes an input factor towards the desired structure of tasks and functions. In other words, technology is a means to realise an envisaged solution (the desired employment and production process). This means the possibility of choice. But, it will become clear that there is no deliberate conscious strategy in this respect.
We will deal with important aspects as: the prevailing opinions and attitudes with regard to "women's jobs", the interests involved, the behaviour of the participants in the reorganisation process, the division of work between men and women, and the organisational choices made on that score by management.

FEMALE LABOUR IN THE CONFECTIONERY INDUSTRY

The share of women in Dutch industry is small: 13 per cent. Almost a quarter of the women employed in industry are found in the food sector. The investigation focuses on women in the confectionery industry, in particular those employed on the work floor. The confectionery industry produces pastry, chocolate products, and sweets. Some 3,400 women are employed in bakeries (between 35 and 40 per cent of total employment). One third of these women work part-time, against only 17 per cent of the total number of employed. The chocolate and sweets industry employs over 2,600 women, making the total number of women working in the plants that are relevant to our investigation, as compared to some 14,000 men. From statistics, more than 90 per cent of the women in the confectionery industry appear to hold low-skilled jobs, most of them working as wrappers. For quite some time now, wrapping activities have been involved in a process of automation. New, advanced wrapping machines can wrap up the products automatically. Machines are now being developed which integrate wrapping and packaging and at the same time stick on information on selling and distributing, mostly by means of so-called bar codes. In that way, a link is laid between wrapping and packaging and the information needed to control the production. Packing technology combines sophisticated mechanical techniques and micro-electronics to wrap products in foil (flow-pack). The latest micro-electronically controlled machines, called PLC machines (PLC standing for Programmable Logic Control) work at very high speeds. Suppliers praise the quickness of the new machine, which allegedly replaces four conventional ones, and, operated singlehandedly, wraps up chocolates at the rate of 1,400 a minute. Obviously, that maximum speed is rarely feasible; the optimum will be considerably slower. In the bakery sector, staffs are being reduced. In the last five years, the level of employment has dropped by 12.5 per cent, that is, to below the 1973 level. The number of persons (mostly women) engaged in packaging has diminished by 22 per cent in the same period. The progressive mechanisation of the wrapping activities seems to make female jobs disappear from the sector. The work is mostly short-cycled and easy to automate. Redundancy in the chocolate and sweets sector has hit women more than men: the relative decline of the number of female workers just under 40 per cent between 1973 and 1984 is almost twice that of men (nearly 20 per cent). Apart from the decline in employment, the quality of the women's jobs also seems to deteriorate under the influence of automation. The investigation was concentrated on the qualitative aspects and on the consequences of automation as manifest within companies. The focus was on the question how to prevent that the negative effects, if any, of technological innovations should hit women workers disproportionally. Attention was given to the way the automation was carried through, to the tasks eliminated or created, to the allocation of these tasks to functions and to their division between men and women (see the diagram hereafter).

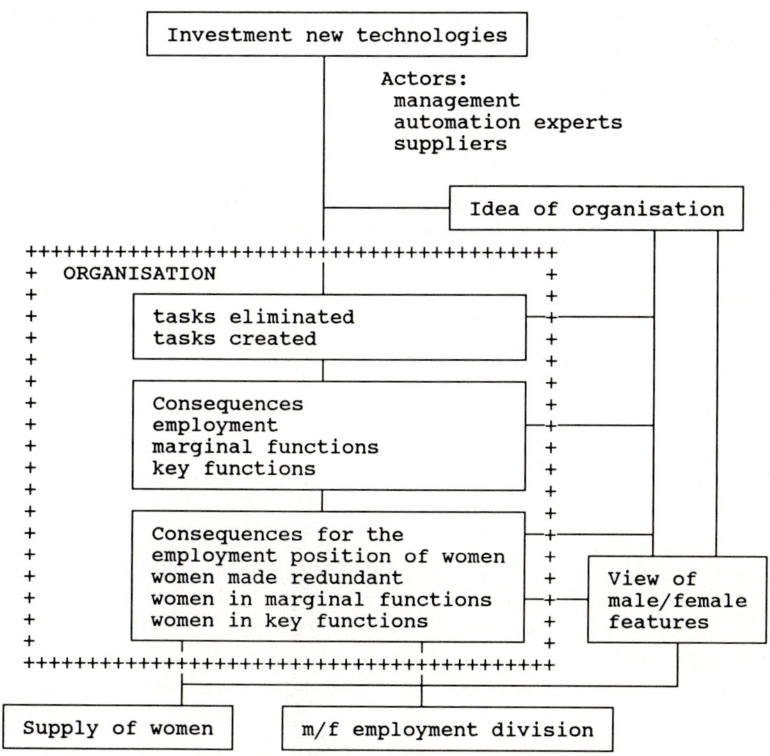

Diagram: Factors influencing job contents and division

THE REORGANISATION OF THE WORK PROCESS

Packaging is still mostly done by hand; products are picked up by hand and put in boxes and on trays. These boxes are then packed in larger cardboard containers. The development from handmade sweets to the mechanised and automated industrial production of confectionery has been accompanied by scale enlargement and concentration. The labour organisation has developed according to the Taylorist model, by which the work is broken down into the shortest routine movements possible. With the introduction of the assembly line, many jobs were reduced to monotonous short-cycle tasks, such as "placing", "shoving", putting candied fruits on cakes, picking up always the same number of biscuits, etc. Functions have been eroded, a striking example being the function of "biscuit aligner".

In an automated process the following kinds of tasks can be distinguished:

* process-controlling tasks, such as checking the performance of the machine, and checking the quality of the products;
* programming and operating tasks, operator's tasks (input of such data as velocity, total quantity of a given product, etc.), and feeding in the packing materials;
* complementary tasks such as input and output of products.

Technically, most complementary tasks could be automated, and with integrated systems frequently are. That they are often still done by hand is because production machines, wrapping machines and transport lines are frequently defective or nonintegrated. That is especially true during reorganisations and in transition periods in which new machines and production processes are being tested. Indeed, at the stage of transition on the way to a fully automated process women will be allotted marginal or conversion tasks. Such tasks, often done by temporary workers, serve to bridge the gaps between parts of the production process that are not (yet) co-ordinated. Once the automation is sufficiently advanced, many of these temporary workers will be laid off. That strategy enables managements to test their new equipment and methods without spending much on labour. As soon as this period is over and the function structure more or less stabilised, the staffing strategy will be changed. Permanent jobs are created and the demands made on employees adjusted (again).

KEY AND MARGINAL FUNCTIONS

From the investigation, the technical staff appears to play an important part in the decision-making process and in determining the conditions under which staff is selected for the new jobs. The staffing department (if any) has little say in the matter, the less so as the introduction of new machines is considered mostly a technical matter. Therefore, the organisational changes to be made are not fully recognised beforehand. Indeed, little is known about the tasks to be carried on outside the direct operation of the new machines. Actually, the tasks directly associated with the machine, in particular operator's tasks, are looked upon as the core of a new job, while complementary tasks are considered not to belong to the operator's job, and in many cases split off and given to other workers to do. There is no integral vision of the contents of jobs. In other words, there is more a rather technical division between the so called "key" task functions and "marginal" taskfunctions.[1] We found that this distinction influences the possibilities for the female workers. So, the prospects for women depend on their access to those key functions. The way manpower is recruited for key and marginal functions at the time of automation and reorganisation is decisive for their future employment position. In principle two opposite developments are possible. By one train of thought, employers will be inclined to recruit the operators of new wrapping

1 "Key" and "marginal" refer not to a social division in the first place, like concepts as "core" and "periphe- ral" jobs do.

machines to be recruited among the employees who so far have done the work by hand, that is, among the women wrappers, because suitable, instantly employable staff would be hard to find on the external labour market. However, from the investigation, the idea of recruiting among the women wrappers often had not even crossed the employers' mind. By another reasoning, only men will be recruited for operator functions, on the frequently used argument that women possess little technical ability, which allegedly is needed to operate the machines.

On the other hand a strategy of breaking down functions into small tasks might lead managements to make a sharp distinction between key and marginal functions, with little eye for integration. The implication might be that routine, short-cycle functions specifically reserved to women are constantly reproduced. In sum, two questions are decisive for the prospects of women: whether reorganisations entailed by automation will lead to the integration or to the breakdown of tasks; how, and by what criteria, staff for new jobs will be selected, either internally or externally.

The answers to both questions determine the women's chances of employment, and whether they will be admitted to marginal functions or key functions.

In the remaining sections of this contribution, both the organisation of tasks and functions and the selection of staff will be dealt with in some detail.

SELECTION FOR NEW FUNCTIONS

The selection criteria for the function of an operator of wrapping machines are different from those for the traditional wrapping function. The new machines are faster, more complicated, more expensive, and therefore more vulnerable than the old ones. For that reason, more than before great store is set by technical skills and such social qualifications as reliability and loyalty to the company. In brief, management looks for full-time, fully employable, reliable, technically trained staff to fulfil the key functions in the company.

Because automation is regarded as a purely technical matter, the technical skills needed to operate machines are apt to be overappreciated in comparison with other selection criteria. Moreover, such technical skills are often attributed to men; women are not supposed to have them. Likewise,men are excluded from certain "women's jobs" because of the prejudice that they are too rough and their fingers too thick ("less nimble"). Such stereotypes are responsible for a (renewed) stringent division of work between men and women. The operation of machines is looked upon as typically men's work,but what is understood by technical skill is mostly left unspecified. Sometimes it is understood as the quality not to panic when the machine breaks down, and to act fast and adequately to set the matter at rights. That quality is also called "feeling for the machine". The conclusion is that the requirements associated with certain functions are influenced by ideas about the capacities and qualifications of men and women.

The changes described here show a tendency in favour of male workers. For one thing because men are expected to prefer a fulltime job with complete employability. For another, women are considered less suitable for jobs requiring technical skill. Moreover the assumption is that men are less likely than women to break off their career with the company, so that their diligence and the profitability of their (re)training would be more ensured. Finally women are expected to prefer a part-time job, or at least to have less objection than men to part-time functions. The result is that men occupy the more promising jobs while women are "condemned" to marginal functions, often on the basis of flexible employment contracts for short working hours, part-time jobs, or standby arrangements. However, the investigation has also shown up some other possibilities, offering better prospects for women.

WOMEN IN OPERATOR JOBS

As pointed out earlier, theoretically two situations may spring from automation, as far as the work situation of women is concerned. One is the consolidation of the traditional division of work between men and women; the other a change in that traditional pattern. From the investigation, such a change can indeed be observed, if only on a small scale. There are women who find themselves in operator's jobs often retrained former hand wrappers or carrying out quality checks. To a certain extent this can be considered a breakthrough of the traditional division of work between men and women. Wherever women got employment as operators, there had been other factors at work, influencing the division of tasks and the selection in favour of women. If the employment position of women is to be improved, these factors must be looked at closer. First: that some women are employed as operators is partly the result of the traditional work division between men and women. That seems illogical, but can be explained as follows. The wrapping department is often separated strictly from the production department, both physically and socially. As a result, men and women are clearly separated as well. That separation tends to consolidate the opinion men and women hold of each other's jobs. Men are very loath to accept the socially lower assessed "women's jobs" in the packaging department, which they look upon as the women's domain. The employment of women as fellow-operators is, while not exactly welcomed, in a certain way acceptable to the men because it implies no direct "degradation" for them. Acceptance is made easier if women in operator functions go on to be recorded, and paid, not as "operators", but as wrappers. A strategy that moreover may be attractive to the employer because it saves cost. The expression used in such cases is that women are "given the opportunity" to operate the new wrapping machines.

The second factor is: an increasing number of women are threatened with unemployment as a result of automation. That makes women easier to recruit for the new operator functions, not only because more women are available, but also because more female workers will understand that they need to be in the market for new functions. Of course, without support that would be a rare occurrence. Evidently, modern management ideas with regard to staffing, employment conditions, and the position of women in society, can work in favour of the selection of women wrappers for operator functions, and has been observed to do so in some instances. However, in most cases management will be more inclined to choose the staff by technical and economic criteria.

The third factor we observed is the influence of a tendency towards more flexible employment of staff in key functions of the company; it has been responsible for the employ of women as operators not only of wrapping machines but also replacing men at production machines. Flexibility must be understood here to be both functional (multifunctional employment at several places within the company) and numerical (part-time and temporary work). The introduction of part-time shifts to lengthen the original working hours of the company from eight to ten or twelve hours caused one respondent company to make its production, as they put it, less "man-dependent". The labour costs for male staff are relatively higher than for female staff, partly because men tend to have longer average careers with the companies concerned. The lower wages may also induce management to consider women for new jobs.

CONCLUSIONS AND RECOMMENDATIONS

Clearly, from the above, reorganisation and restaffing as a result of automation lead to potentially better opportunities for men than for women. That is due in part to the selection based on prejudices about the qualities which male and female workers are supposed to possess. In firms where automation and reorganisation are attended by lay-offs and transfers, the position of women is mostly worsened. The research results reveal that in that case women either lose their jobs or have fewer prospects for new and higher qualified jobs. On the other hand, there are examples of women who have succeeded in securing one of the new functions. On the one hand that may be regarded as breaking through the traditional work division between men and women; on the other it reflects not so much a positive action as the strategy on the part of managements to find the cheapest, most flexible workers, namely, women. Evidently, the introduction of new technologies affects female employment in two ways.

Evaluating, we find that if a reorganisation process works out favourably for women, then that is not the result of planning. The introduction of new wrapping machines is not accompanied by a conscious strategy of management and experts. Rather, decisions tend to be step-wise, the technique being the primary concern. Mostly, the process starts with the decision to invest in new equipment. The next decision concerns the technical requirements which the new equipment has to meet. Then, a given configuration offered by one or more suppliers is selected. Finally the contents of the tasks to be carried out are derived, training programmes are drawn up, lay-offs and recruitment arranged, etc. In most cases, only at the final stage of decision making are employees and their representatives involved in the decision making, and then it is often too late to influence the nature of the planned investment. The only remaining points of discussion are the employment conditions. Every next step in the decisionmaking process limits the choice at the next stage. That is not always a conscious management strategy; it follows rather from the fact that most managers themselves lack a clear vision, let alone an elaborate plan, which would enable them right from the first to reckon with consequences becoming urgent or manifest at later stages. The unfortunate consequence is that employees and their representatives are taken by surprise by a succession of plans, proposals and decisions. Indeed, hardly if ever is there question of any purposeful positive action within the firm. That makes the employment position of women highly vulnerable. In one case, management even intimated that it was considering to retrace the steps taken towards the appointment of women as operators. Evidently, managers still allow themselves to be swayed largely by the prevailing ideas about male workers' alleged superior technical background. On that assumption, to "invest in men" is thought to be safer than to "invest in women". Besides, the suggestion was made that if the reassessment of functions entailed a rise in the wages of the operators of wrapping machines, more men might after all be found willing to do that kind of work. That would imply that men would push out women even from their traditional work area. Such a tendency would be supported by the structurally better legal and interest position of the men. The inferior qualification level of women would narrow still further their prospects of more qualified functions.

The conclusion is that improvement of the employment position of women through automation is possible, but is at this moment left to coincidences in the company. To prevent women from becoming the victim of reorganisation and see to it that positive effects are utilised, the coincidences must be shaped into a strategy. Such a strategy should specifically take into account the consequences of automation and reorganisation for employment, for the nature and contents of the new tasks, and for their division between men and women.

Attention has to be paid to the employment situation. Rationalisation and automation will induce companies to cease certain activities and to take labour-saving measures. The question is to prevent such measures from working out one-sidedly to the disadvantage of women. To that end, an employment strategy has to be developed. Agreements between the parties concerned, to be made on a central level (which in the Netherlands means by

Collective Labour Agreement), should be part of that policy. This assumes a deliberate policy of trade unions.
Attention also has to be paid to the job contents and job structures. Tasks will disappear through automation; on the other hand, new tasks will come into being, while marginal and conversion tasks will remain until a fully automated organisation system takes over. The company's employment policy should aim at improving existing functions. Integration of more responsible elements in the existing tasks is one possibility. Revolving tasks may be considered, provided the workload is not made heavier and workers are less tied to one spot on the work floor. Besides, attention should be paid to the work conditions and the possibility of participation. The division from a technical point of view between "key" and "marginal" should be avoided.
And last but not least the division of work between men and women should have special attention. Reorganisations inspired by technological developments may offer an opportunity to reconsider the existing division of tasks and work between men and women. One impediment is the traditional organisation of production, which is characterised by the fragmentation of work, men doing more technical work than women, women working part-time and in temporary functions more often than men, women getting the lower functions. If the economic returns remain the same, a policy of integration seems possible which would lead to more women in technical, higher-level, full-time jobs. Prejudices about the skills and qualities of men and women should be exposed, to which end information and training courses may be helpful. Besides, women must be better equipped by suitable training to perform the new functions, which means that a training policy is indispensable.
A plan to improve the employment position of women would include the following elements (see also the diagram hereafter).

Supervision project given • idea of organisation • dea of men/women	women's section
Attention for:	
1. Employment:	Agreements on CLA-level Target figures Disadvantages not only shifted ont o women
2. Function structure:	Agreements on company level Expansion, integration Prevention of false function requirements
3. Training possibilities:	Agreements, training institutions Matching the function level of women Prevention of the exclusion of women
4. Recruitment/selection of women:	Agreements within staffing policy Information about career prospects Exposing prejudices about women
5. Supply of women:	"Agreements" with female employees c Offering real prospects Prevention of women being ousted by men

DIAGRAM: Plan for the improvement of women's employment position

* The selection of supervision projects, that is, schemes to follow closely companies involved in a specific reorganisation, and supervise the changes.

* Imposing on reorganising companies the duty to draw up a social plan, including a separate "women's section" with the following contents:
- indication of possible implications for women;
- indication as to which women will perform which tasks;
- a scheme to widen the prospects of women; .
- a specific training scheme for women.

* The development of training courses for women. The sectoral level seems to offer the best chance of success.

* Information and advice, to make women aware of the fact that no strategy can be successful unless they themselves are convinced of the need to improve their position. "If the wrappers are to profit from progress, they must be willing to act for themselves". From the results of the investigation, there is no doubt that the women concerned can be properly motivated.

ACKNOWLEDGEMENTS

With thanks to Ro Braaksma and Caroline Meijdam for their comments on an earlier version of this contribution. We thank the referees: Birgitta Freijhagen and Rosalinda Pineda-Ofreneo for their comments on the original paper.

REFERENCES

Cockburn, C., "Brothers, Male Dominance and Technological Change", London: Photo Press, 1983

Dam, R. van en W. Trommel, "Vrouwen, industrie en automatisering. Wat gebeurt er met het werk van vrouwen?"[Women, industry and automation. What happens with the work of women?] Industriebond FNV, Amsterdam, 1986

FNV, "Aktieprogramma Arbeid en Technologie"[Action programme Labour and Technology], Amsterdam, 1986

Hemelmans, K. en L. Wijmans, "De verdwenen kopy, automatisering in de grafische industrie en de arbeidspositie van vrouwen"[The lost copy, automation in printing industry and the employment position of women], Druk en Papier, FNV, Amsterdam, 1985

Internationaal Verbond van Vrije Vakverenigingen, "Nieuwe technologie en werkgelegenheid van vrouwen" [New technology and employment of women], Brussel, 1984

Meurs, P., M. Rohling en M. Weggelaar, "Micro-elektronika en vrouwenarbeid, Deel I [Microelectronics and women's labour], een literatuurstudie naar de kwaliteit en de kwantiteit van vrouwenarbeid", Ministerie van Sociale Zaken en Werkgelegenheid, Den Haag, 1982

Meydam, C., D. Trommel, E. Poutsma, R. Braaksma, "Vrouwenarbeid in de zoetwarenindustrie, de mogelijkheden van technische vernieuwing: restarbeid of vakarbeid" [Female labour in the Confectionery Industry. Prospects of technological innovation: Marginal work or professional work], Economisch Instituut voor het Middenen Kleinbedrijf, Technische Universiteit, Zoetermeer/Delft, december 1987

Pollert, A., "From core-periphery to flexible specialisation", paper presented to: Aston/Umist labour process conference, March, 1988 (draft)

Te Elfder Ure 33, "Arbeidsproces en Technologie [Labourprocess and Technology], themanummer", Nijmegen: SUN, 1983

Te Elfder Ure 41, "Het ontwerpen van arbeidsprocessen" [The design of labourprocesses], Nijmegen: SUN, 1988

Trommel, D., E. Poutsma, C. Meydam, R. Braaksma, "Vrouwen in de slagerij. Mogelijkheden bij technische vernieuwing: hulpkracht of vakvrouw" [Female labour in Butcher's Shops. The Prospects of Technical Innovation: Assistant or Professional], Economisch Instituut voor het Middenen Kleinbedrijf, Technische Universiteit, Zoetermeer/Delft, januari 1988

Weggelaar, M. en K. de Boer, "Micro-elektronika en vrouwenarbeid [Microelectronics and women's labour], Deel II", Ministerie van Sociale Zaken en Werkgelegenheid, Den Haag, 1984

BIOGRAPHICAL NOTE

Erik Poutsma is associated with the EIM (Research Institute for Small and Medium-Sized Enterprise) as a researcher, and detailed part-time to the Delft University of Technology.
At the time of the investigation, Dorien Trommel was employed with the Delft University of Technology.
Correspondence address: EIM, Post Office Box 7001, NL 2701\AA ZOETERMEER, The Netherlands for the attention of Erik Poutsma.
Telephone 079 41 36 34.

Women, Work and Computerization: Forming New Alliances
K. Tijdens, M. Jennings, I. Wagner, and M. Weggelaar (Editors)
Elsevier Science Publishers B.V. (North-Holland)

IMPACT OF THE INTRODUCTION OF NEW TECHNOLOGIES ON WOMEN'S WORK IN SOUTH WEST FRANCE: WOMEN'S CRITICAL ASSESSMENT AND SUGGESTED STRATEGIES

Marie-Claude BERGOUIGNAN, Marie-José CAMELEYRE and Marie-Claire ROUYER
Centre de Recherche interdisciplinaire sur les Femmes
Université de BORDEAUX III, FRANCE

Ten companies located in a French region were selected for this research concerning new technologies. The researchers interviewed top managers, observed workers and the work organization and interviewed female workers. They found that the unskilled workers did not experience any improvement of their jobs and that in highly skilled jobs enrichment is unanimously acknowledged. Women cope with new technologies in response to various deficiencies, namely training, work organization, information and co-ordination between system designers and users. The coping strategies emphasize the human dimension of work.*

1. PRESENTATION OF THE RESEARCH

Ten companies located in the Aquitaine region, (South-West France), ranging from 250 to 600 workers were selected as representing an adequate sample of the various economic sectors characteristic of the area (female ratio at least equal to the regional average ratio: 30 %). Particular attention was paid to the new technologies involved in order to include production as well as tertiary activities (Robot Technology, Automated Production Systems, Computer Assisted Design and Production Management, as well as Office automation) in the following industries: food-processing, clothing, shoe-making, furniture-making and, for the tertiary sector, an insurance company.

Within each company, our investigation was conducted in three stages: interviewing of top level executives and management; observation of the various work stations held by women and men and of the work organization; interviewing of a sample exclusively chosen among the female personnel concerned with new technologies - each interview based on a questionnaire of 97 items combining closed and open questions, lasted one and a half hours and was recorded with agreement of the worker.

The collected information was treated with the combination of two methods: quantitatively with descriptive analysis using the cross-section method and multi-criteria

* The present paper is based on the results of an investigation conducted by the authors in the Aquitaine region (South-West France) over two years (1986-87) in view of a report commissioned and funded by CNRS (National Centre for Scientific research) and the Aquitaine Regional Council.

factorial analysis; qualitatively, with content analysis of the transcripts of the recorded interviews. The present paper is largely based on this "live" material, with occasional references to the statistical results, relevant to the selected topics: job enrichment and the acquisition of new skills and women's immediate and farther-reaching strategies.

2. NEW SKILLS AND JOB ENRICHMENT IN INDUSTRIAL AND CLERICAL WORK

Our statistical analysis shows a marked discrepancy between a highly positive subjective assessment and objective signs of progress: 67 % of the women interviewed think they have become better qualified - 69 % consider that their jobs now involve a certain degree of responsibility. But only 14 % have been promoted to a higher job category (from workers to group leaders, forewomen or technicians) and no significant change can be observed in a comparison of the wage structures before and after the introduction of new technologies. On the other hand the recorded comments offer a wide range of responses according to the types of tasks.

2.1.1. In industry at the lowest level

Unskilled workers have not experienced any improvement with automated machinery, but rather a deterioration of their working conditions which has not been compensated for by any substantial enrichment of jobs, as compared with their former tasks on the assembly line. Two typical cases illustrate this deterioration:

* In a food-processing plant, a packing station used to be "manned" by a male worker controlling a fork-lift truck and a woman in charge of cleaning. She comments: "With the new automation, I am alone now, I supervise the machine and I still have to do the cleaning".
  Women's alleged promotion through computerization has obviously not relieved them from ancillary tasks (the same observation was made in a recent report on the Ford plant (BORDEAUX) where the almost exclusively male workers refuse to clean their work station in the absence of available women. (BELIS-BERGOUIGNAN et al., 1986)
* In the shoe-making and clothing industries, women used to be able to repair minor failures in the material. Now they have to call in a (male) technician: "We can't touch the machines any more. Only men are allowed to tinker inside". Moreover on automated lines, the women complain about being constantly shifted from one station to another. New technologies are supposed to encourage multi-skills. For those women who used to be and still remain on the lowest rung of the production scale, they most often simply mean an addition of equally ungratifying tasks, all ranked as unskilled, whereas genuine multi-skills should imply the acquisition of a variety of skills traditionally pertaining with distinct jobs.

2.1.2. In highly skilled jobs

Enrichment is unanimously acknowledged: computerization in removing repetitive and time consuming tasks has considerably increased the interest in and enjoyment of one's work, but this gain has to be qualified with reference to the labour selection involved.
In the clothing industry, when VDU work replaced the manual "placing" operations (i.e. finding the most economical placing of patterns in order to optimize fabric use), only 3 out of the 12 women formerly employed were retained for the job. According to management "they were chosen for their quickness of mind", that is for intellectual qualities which remained unacknowledged by the slightest upgrading of the job.
In the electronics industry, two different situations have been observed. On the one hand the THOMSON plant, owing to an extensive training programme, has enabled about 200 women workers to obtain a qualification acknowledged in collective

agreements at national level.[1] Originally those women - hired as unskilled workers - were employed in the clothing and food-processing industries; they were selected for the manual ability acquired in their former jobs.

The very complex nature of the product (rocket components) no doubt accounts for THOMSON's need to provide such training; but in the SONY plant which makes audio-cassettes, i.e. a mass-produced current consumer product, only 12 out of 320 female workers have been given a new custom-made qualification as "plant technician", characterized by genuine multi-skills, including machine maintenance and adjustment. On the other hand, in the same plant the automation of new lines involving the extension of shift-work (3X8) has been detrimental to all women's working conditions, owing to the recent removal of the ban on shift-work for women in France (June 1987 legislation).

2.2. In clerical work

In our investigation the clerical sector was represented by a fully automated insurance company with 3 top-level male executives and a female staff of 200, most of whom had a 10 years' experience of data-processing and VDU work organized on taylorist lines. On the other hand the administrative and commercial departments of the industrial companies investigated (small and medium-sized firms) provided examples of non-taylorist clerical work in computerized accountancy, stock-management etc... The women's practice of computerized work was more recent: they had experienced the introductory phase of computerization.

On the whole, the wide differences in opinions clearly and unsurprisingly stem from the types of work process involved. But the issue of the recognition of new skills proves of general relevance, as was already put forward in several contributions to the 1984 and 1986 IFIP Conferences on "Women, Work and Computerization", for instance LIE, M. & RASMUSSEN, B. (1985). Women tend to identify their new skills in terms of qualities which are not taken into account in the current descriptions of qualifications exclusively concerned with tasks. Nevertheless, those "obscure" qualities have proved valuable in concrete work situations and have so far been instrumental in the proper working of computerized systems. This phenomenon is simply enhanced by new technologies, which have extended its scope to office work; it has been observed as a regular feature in the past history of industrial work: in a first stage the introduction of a new technique calls for new qualities which ensure its successful implementation; after a phasing-in period, those qualities are taken for granted and considered as normal requisites in the existing qualification scale, and consequent undervalued or ignored by employers. (ROSANVALLON, A. & TROUSSIER, J.F., 1983).

This phenomenon is more acutely felt in taylorist clerical work in which women's jobs continue to be ranked as unskilled than in jobs in which computerization has preserved a substantial degree of autonomy for the worker, as is the case with the small and medium-sized companies of our sample.

2.2.1 In non-taylorist office-work (tertiary sector of small and medium-sized companies

Women with responsible jobs find their work more interesting and rewarding; they appreciate the substantial gain in time devoted to organizing their work. Those in middle-level positions in the decision-making scale analyze their job enrichment and the personal qualities they have developed in response to certain "qualities" specific to

---

1 Those women have not been included in our sample and therefore have not been reckoned in the statistical results concerning upgrading : in our selection of plants, THOMSON (BORDEAUX) and SONY (BAYONNE) respectively employing 300 and 320 women offered the most extensive range of new technologies in industry, but the management objected to individual interviews of the women workers, while providing information on training policies, job grading and wages.

computers: the rigorous logic of computerized systems invites them to be more methodical, more accurate; reliable data are reassuring, they boost up their self-confidence and prove a great asset in planning and decision- making. Against this largely positive assessment, one notable exception must be mentioned: the "all-purpose secretary" - currently to be found in small/medium-sized firms - has to bear an increased work load: she already occupied a key position between production and stock management, whose role has been dramatically enhanced by computerization, because organization patterns have not been altered, due to a lack of foresight on the part of management. She feels overwhelmed by her new responsibilities, just about manages to cope, at the cost of overwork, overtime and feelings of guilt and frustration.

2.2.2. In fully automated office-work

The breaking down of tasks induced by automation has leveled out former distinctions between office-work and industrial work. The women working in a fully automated office have supplied the most critical and detailed statements on job content and the problematic recognition of new skills, identified by them as new qualities required by automated jobs. It is interesting to note that the same qualities have been mentioned by those office workers and the industrial workers: meticulousness, skillfulness, adaptability and quickness of mind. After listing the qualities she thought she had developed, a woman remarked: "I am sure I am better qualified now, not in the ordinary sense, but because I have enriched myself a lot".

When office workers consider the means by which their new skills could be acknowledged, a few mention a wage increase, but most of them concentrate on the specificity of VDU work. A 35-year-old woman comments: "The fact that I am called "office worker" on my pay slip gives the impression that I don't have any skills" (in French: "je n'ai pas de métier" literally: "I have no trade").

Some of them claim access to existing job categories listed in collective agreements, such as that of "technician"; others insist on the need to invent "a new name" for their job, such as "operatrice de gestion administrative", literally "administrative data managing operator". None of them has mentioned the existing term bureauticienne seemingly more gratifying than its English equivalent "automated office worker", because it does away with the term "office-worker" and the non-skilled connotation alluded to by her colleague in the above quotation. However the wording of job advertisements shows that the term bureauticienne tends to bandied about as the term secretary has been for a long time now, merely implying a knowledge of word processing which has become as common a skill as shorthand-typing and is thus equally taken for granted by employers.

Finding new names for the new jobs will be insufficient if the variety of tasks involved in office work is not seriously analyzed, which has so far never been the case, as opposed to industry in which the introduction of new machines has always been accompanied by a minute observation of the tasks performed.

Moreover the fact that VDU work and the mastery of word processing are not recognized as skills bars women's access to other jobs requiring identical skills, such as film setting. The very protectionist and self-defensive printers' union has ruled that women experienced in automated office work should be employed as unskilled workers because they have not been trained to the traditional and now obsolete type-setting methods (MAROUANI,N., 1987).

In their critical assessment of their condition and work status the women interviewed have shown a widespread, if more or less clearly expressed, consciousness of their needs. In describing how they cope, from day to day, they have disclosed a number of strategies which will now be examined.

3. WOMEN'S STRATEGIES: HOW DO THEY COPE NOW? WHAT DO THEY DEMAND?

Women's strategies have been developed in response to various deficiencies namely in training, work organization, information and co-ordination between system designers and users.

3.1. Coping strategies

Because computerization is felt as an impersonal, rigid and therefore unnatural process, the coping strategies emphasize the human dimension of work, in women's daily adjustments to routine and hazards, and in their defiance of existing communication and bargaining patterns.

3.1.1. How do they manage "to make it work"?

They first cope with organization deficiencies by accepting an extra workload, and by resorting to their own memory to make up for the incomplete or incorrect use of the computer's, or, in commercial departments, to supply a personal knowledge of clients which could never be stored in a computer's memory.

Their concern for quality prompts them to discover practical improvements and devices in order to redress the aberrations of faultily designed systems which otherwise would never function properly. In coping with hazards - which crop up more frequently in the phasing-in stage of computerization - they prove more flexible than men since they are used to dealing with multiple and varied tasks at home.

In the insurance company we visited, although the workers appreciate the gain in time, they are shocked by their being unable to cater for the individual needs of clients with whom they used to correspond with custom-made letters. The fact that they are pressed by productivity requirements prevents them from granting special attention to delicate cases. Nevertheless the more resourceful and daring ones break management orders and make an extensive use of the "self-service space" meant to be used sparingly, for exceptional cases only. In a situation in which the notion of good service seems to be ignored by management, or at least ranked far behind productivity, those workers' efforts to preserve it betray a typical woman's concern transferred from the domestic to the professional sphere.

Moreover those automated office-workers realize that their tasks can be correctly performed only because of their experience of traditional office- work, which they try to pass on to their younger colleagues for whose future they feel concerned. They intuitively perceive that professional culture cannot be restricted to the acquisition of technical and purely functional skills, and wish to preserve a former knowledge which seems doomed to obsolescence.

3.1.2. Individual versus collective attitudes

Next to those models of self-denial and conscientiousness, others do cope, but by self-protective methods. Some deliberately slow down pace, or take advantage of the system's defects.

Most women sneak out in rota for unauthorized breaks. This self-defence which can be purely individual or imply some collective strategy is prompted by the fact that they cannot humanize the system in a more open way.

They could resort to the traditional communication channels (trade-unions and "expression groups" provided by the Auroux legislation, 1981). But women look upon them as inefficient and inadequate to cope with their own needs. In France where unionization only affects 15 % of the working population, women are notably under-unionized. Among our sample, union members have acknowledged a number of personal benefits: they have become more articulate, more confident when speaking in public, better informed on work legislation. But they object to the almost exclusively male leadership and to the emphasis on theory and ideology rather than practice. Their own claims arise from actual working conditions. More generally at present the new role of unions as partners to employers' organizations is not easily accepted by women - or men either - in a country with a long history of syndicalism. Beside, women have specific, legitimate grievances against unions which have always proved reluctant to consider their particular position in the work force, made even more vulnerable by new technologies in an international economic crisis (ROGERAT, C. 1986). In our sample the claim for breaks significantly divides the members of the two largest confederations (Breaks have to be bargained for because of the legislation vacuum on VDU work): CGT (Confédération Générale du travail) demands set breaks "so that one doesn't feel spied upon when meeting a supervisor in the corridor". On the contrary CFDT (Confédération Française Démocratique du Travail) insists on free breaks which are also preferred by non-union members: "When I am tired, it is not at ten past ten", says one.

3.2. Farther-reaching strategies

In this general context of defiance and individual solutions, women find it difficult to think out long-term alternatives. However through their critical account of training and system designing, new approaches can be outlined as a first step to farther-reaching strategies.

3.2.1. Training policies

So far on-job training has prevailed in small and medium-sized firms, usually supplied by salesmen or a previously trained group leader or foreperson (GLOGAU, M. 1982). Women complain about being taught the use of a keyboard "like a cooking recipe". They regret that their perfunctory knowledge of the software prevents them from using its full resources. They also feel that a smattering of computer science would be helpful but prove reluctant to start on a course, even if it were offered. Others wish to acquire a more thorough knowledge as an asset in their claims for acknowledged qualifications, and in the long term as an opportunity for a new position in another industry or sector.
Women's concern for self-promotion is not restricted to technical acquisitions but may include a variety of more cultural interests: their demands range from a better knowledge of the product and the work process to philosophical preoccupations: they wish to be intellectually better equipped to undertake a reflection on a more convivial and human-oriented technological society, the male approach to progress tending to concentrate on tools and means rather than persons and their future (PETRELLA, R. 1988).
It has been widely verified that women's promotion depends on their initial educational level. Continuous education can help to level out inequalities if it provides both a cultural and a technical knowledge. As to the question of women's estrangement from the computer's language, and their inability to appropriate its logic, it seems to have risen from their refusal of technological values rather than from any intellectual disability. In France, at school the streaming system is based on a child's abilities in mathematics which is particularly detrimental to girls, although the power to conceptualize can be acquired through other subjects. Our own experience in training women for a university diploma in bureautique has convinced us that women can master this kind of logic after a schooling in arts subjects. When failures occur they result from a lack in initial schooling combined with a lack of motivations. Some of our most successful trainees had left school at the age of 15, but had a substantial work experience and a strong desire to acquire both technical skills and a more general culture.
This training programme has been designed with a comprehensive view of automated office work, taking into account communication data processing and not merely word

processing. A qualified bureauticienne can claim to executive status and is able to design and organize automated communication systems in offices according to the users' needs, which are currently ignored.

3.2.3. User's need and women's participation in the design process: an opportunity for new jobs?

The following account from an office worker who, without any official status, assumes the responsibility of stock management and production planning in a 200 workers' plant, encapsulates basic anti-guide-lines for the setting up of an automated office: "They made preliminary studies at the end of July. They came and asked everybody questions: how we worked, why we did this and that, and how; and off they went ... Then suddenly in October we were told: the software will be here tomorrow. You'll have to enter the data immediately. Previously we'd been told that they would bring in someone to do it. But finally when the material was there, I had to do it myself. It's only afterwards that I was asked whether I liked it; then I was allowed to speak, but not on what was important. I think that if we had talked together more, the starting period would have been easier". Her colleagues also blame the inadequacy of the software on the users' hasty consultation, but some raise doubts about the fruitfulness of talking with system designers, as long as each party speaks a different language and puts forward different pictures of work organization.
The idea that prompted the creation of our training programme was that it belonged to women to bridge that gap, by mastering the other language while preserving a woman-oriented outlook on the reality of work. The bureauticienne 's middle position does not exclude system-designers (it would be utopian to believe that women could confiscate that sector) but provides a link between them and users. She can be a permanent member of staff who, after the phasing-in period can replace the former "all-purpose secretary", with - hopefully - a different status, because of her knowledge in planning, stock-management and accountancy; or more frequently she is hired on a six months or one year contract as full-time consultant in charge of preliminary studies, negotiations with system-designers and the supervising of the phasing-in period. Large companies appreciate those skills, but - at least in our geographical area - medium-sized firm managers are not easily convinced of the long-term benefits from such an investment, both for the employees and the firm.

4. WHAT PROSPECTS FOR THE FUTURE ?

If women's training is an essential prerequisite for full access to qualified jobs and to a degree of partnership in decision-making areas vitally relevant to their work, they cannot expect any substantial progress if they merely express individual complaints and regrets. Their own future depends on the conquest of democratic practices within the workplace, whether through existing structures or newly-created ones, but always through collective strategies.

ACKNOWLEDGEMENTS

The authors are pleased to acknowledge the enlightened comments of Birgitta FREYHAGEN and Rosalinda PINEDA-OFRENEO who acted as referees in the workshop where this paper was presented.

BIBLIOGRAPHY

BELIS-BERGOUIGNAN, M.C., BORDENAVE, G., LUNG, Y., MEYER, H. & POIRIER, M., Automatisation et recomposition d'un collectif de travail: FORD (BLANQUEFORT), BORDEAUX, IERSO, Université de BORDEAUX I. 1986

GLOGAU, M., "Les machines à traitement de texte et leurs répercussions sur les personnels concernés: la productivité,enjeu social dans un grand organisme international".Mémoire de maîtrise,Ecole des Hautes Etudes en Sciences Sociales, Paris, (Unpublished thesis). 1982

LIE, M. & RASMUSSEN, B., "Office Work and Skills", in OLERUP, A., SCHNEIDER, L., MONOD, E., eds, Women, Work and computerization: Opportunities and Disadvantages, Amsterdam: North Holland, pp. 43-52. 1985

LIU, M., "Les nouvelles logiques en organisation du travail", Revue Française de Gestion, (Juillet-Août). 1983

MAROUANI, M., & CHANTAL, N., "Du travail à l'emploi: les enjeux de la mixité", Sociologie du Travail, n°2. 1987

PETRELLA, R., "L'homme et l'outil: les valeurs qui basculent", Le Monde Diplomatique, "Des sociétés malades de leurs cultures" (Janvier). 1988

ROGERAT, C., "Nouvelles Technologies, syndicats et division secuelle du travail", Les temps modernes, (mars). 1986

ROSANVALLON, A. & TROUSSIER, J.F., "Changements techniques et qualifications ouvrières", Recherches Economiques et Sociales, N.S. n°8, Paris, Documentation Française. 1983

ROUYER, M.C., ed., BARATRA, M., BELIS-BERGOUIGNAN, M.C., CAMELEYRE, M.J., ROLLAN, F., Nouvelles Technologies et travail des femmes en Aquitaine, Bordeaux: Maison des Sciences de l'Homme d'Aquitaine et Presses Universitaires de Bordeaux. 1988

BIOGRAPHICAL NOTE

Marie-Claude BERGOUIGNAN is lecturer in Economics at the University of Bordeaux I. Her research-work includes a thesis on women's domestic work and many articles on regional economy and women's work.
Marie-José CAMELEYRE works in the extra-mural Department of Continuous Education in the University of Bordeaux III. She designs training sessions for women particularly in new technologies. Her research deals with women's training and employment in the South West region of France.
Marie-Claire ROUYER is professor of English at the University of Bordeaux III and one of her research interests is in women's studies. She also works as consultant in the extra-mural Department of Continuous Education in the same university, where she supervises women's training for a Diploma in Automated Office Management.
Correspondence address: Centre de Recherche interdisciplinaire sur les Femmes, Université de BORDEAUX III, Domaine Universitaire, 33405 TALENCE-CEDEX, FRANCE

Women, Work and Computerization: Forming New Alliances
K. Tijdens, M. Jennings, I. Wagner, and M. Weggelaar (Editors)
Elsevier Science Publishers B.V. (North-Holland)

BRINGING IT ALL BACK HOME : A STUDY OF NEW TECHNOLOGY HOMEWORK IN AUSTRALIA

J. Wajcman, B. Probert and R. Tanter
School of Sociology
University of New South Wales
Sydney, Australia
Department of Anthropology and Sociology
Department of Politics
Monash University
Melbourne, Australia.

INTRODUCTION

Until recently working from home, or homework, has been regarded as an anachronism, largely confined to the traditional and declining manufacturing sectors, and peripheral to the economies of advanced industrial societies (Rubery and Wilkinson 1981). In Australia, for example research on homework has focused on the concentration of migrant women in unregulated and non-union employment conditions in clothing and textiles. However, since the mid-1960s homework has emerged as an expanding form of production in the service and high technology manufacturing sector at the same time as it has expanded in traditional industries and services. Whilst the focus of this article is on new forms of homework, such developments must be seen in the context of the persistence and growth of homework generally in Western Europe and North America as well as in Australia.

Sociological studies of the contemporary growth in computer-based homework present strikingly contradictory interpretations. Post-industrial theorists such as Toffler (1980), Jones (1982) and Gorz (1982) are optimistic about the radical changes that they believe are emerging in industrial societies as a result of the 'microelectronic revolution'. They argue that technological innovations mean less labour being expected on industrial development, and a shift from manufacturing to service-based economies. Production in this account no longer requires personnel to be concentrated at the place of work. With the proliferation of domestic computers, employees will no longer need to come together in large scale units of production.

In this scenario the home as workspace liberates people from the discipline and alienation of industrial production. Homework offers the freedom of self-regulated work and a re-integration of work and personal life. Moreover, an expansion of homeworking will allegedly lead to much more sharing of paid and unpaid domestic labour, as men and women spend more time at home together. Toffler for example, characterizes computer-based homework as closing an historical circle which originated in the cottage of early capitalist industrialization. The circle is finally completed by the shift of work from the factory back to home, to the 'electronic cottages' of today's post-industrial society.

Whereas the post-industrial theorists see homework as part of a positive future, for others it evokes the ugly spectre of 'sweated' self-exploitation piecework. Some writers see homework as part of a more general trend towards casualisation of the labour force. The focus of their concern is the persistence and expansion of a wide range of forms of casualised labour, from homework to on-site labour-only sub-contracting (Rubery and Wilkinson 1981; Offe 1985: TNC 1985).

Research into new forms of high-technology homework typically approaches it with a set of assumptions derived from the study of traditional homeworkers. Huws (1984), for example, who studied female computer programmers, expected new technology homework to be more widespread because it is a method of production favoured by employers seeking to resist competition and protect profits. Homework can reduce employers' costs in a number of ways: wage costs are reduced by isolating workers and preventing unionization; on-costs and fixed-costs such as award entitlements are avoided; productivity can be increased by adopting the system of payment by result. It is this payment system which also provides the chief mechanism by which the output of homeworkers who are not directly supervised can be controlled (Allen and Wolkowitz 1986:29-31). In the case of computer-based homework it is assumed that, in addition to the simple control exerted by such a payment system, the computer itself can enhance employer control, by monitoring the key-strokes of home-workers, for example.

Other people in the labour market are much less vulnerable. Highly qualified professionals enter the labour market as an advantage. Their qualifications restrict access and limit competition. The choices which Toffler's utopian homeworkers can make are in fact ONLY open to those with power in the labour market, and control over work stems from labour market position. Power derives from either an ability to organize to restrict labour supply, or from historically variable shortages of particular skills.

We chose to do a study of new-technology homework as a case study to contrast two very different forms of work involving new computer technology. It is an exploratory study of two groups of new technology homeworkers, but deals with a paradigmatic case for the future of work. We interviewed 13 word-processors working from home, twelve of whom were women; and 26 computer programmers working at home, 8 women and 18 men (Wajcman and Probert 1988). At the time of writing we are completing a follow-up study a year after the original interviews.

The first group in the sample, computer programmers (in fact, this group included programmers, system analysts, and systems designers) was chosen since it appeared well placed to exploit the advantages of homework so glowingly described by Toffler. The second group, word processors was chosen because such workers seem, on the surface, vulnerable to many of the pressures faced by traditional manufacturing homeworkers. We also interviewed several data-entry shop proprietors, and individuals working in the information sector, including union officials and managers.

WORD-PROCESSING

In Australia, there are a significant number of women doing word-processing work at home. Home-based word-processing adequately meets the needs of small firms, students and professionals, offering the same service as traditional home-based typing. In cases, work is provided by the spouse's business.

In our survey we identified three different employment statuses among homeworking word-processors. There are homeworkers who are 'employees of own company' and those who are employees of a firm and borrow equipment from that firm to perform work on this last category was of little significance.

Our research found out that, like typists, home-based word-processors are overwhelmingly married women with children. They tend to live in the suburbs with most of their work coming from their local area, from local newspaper advertisements and by word-of-mouth. Unlike typists, however, word-processors require some capital since the equipment required to operate a word-processing business from home includes a personal computer or a dedicated word-processing machine, printer and software.

The word-processors we interviewed were mainly married women in their 30s, half of whom had a tertiary education. (This contrasts sharply with traditional manufacturing homeworkers, who are generally poorly educated and of ethnic origin.) The principal reason they gave for working at home was child-care, with some wanting to take care of their young children and others unable to find appropriate child-care. However combining work and child-care proved difficult, often leading the women to work at night and the weekend. One woman mentioned that although she worked at home to fit in with child-care, she eventually had to pay someone to mind her child in order to get the work done. Even when the children are older it can be difficult for women to combine part-time paid work and their responsibility for children.

In fact, at the time of the 1988 interview, most of the word-processors were no longer doing paid work at home, either because of the pressure of child-rearing and full-time parenting, or because the financial returns were inadequate. Often, child-minding interfered with remunerative work and vice-versa. It seemed that only those able to treat the work as full-time business were able to survive the pressures of competition from other word-processors and children.

> It's the children. It's the fact that nobody realizes that you're trying to work. It's the interweaving of work and normal life, because everyone else doesn't do that. So trying to answer calls with a baby screaming isn't on. You end up pushing the children to one side, and the benefits of being the one caring for them are lost.

Women also find it difficult to control their hours of work because the workload fluctuated widely, with deadlines that often required intense periods of work, day and night. Almost all of the word-processors were paid by piece rates, hourly rates, or a combination of both. Under the piece-rate system, payment depends on the number of words or number of pages (so many words per page) produced, with the average rate (in 1986) being $2 per page. The hourly ($12-$15) is usually applied to work requiring proof-reading, alteration or re-drafting. When we asked the word-processors how they came to get their rates, we were told:

> It is difficult but I try to check what the industry is charging people in the same field ..... I just ask around what people are paying ..... I in variably find that I am lagging behind.

Three quarters of the word-processors we interviewed reported earning less since they began to work at home, this being the result of several factors, such as charging below market rates and irregularity in the supply of work. Furthermore they told us that their charges did not take account of the hidden costs of working at home such as electricity, equipment maintenance, and office furniture. As far as this group of homeworkers is concerned, the evidence seems to support the pessimists' view of new technology homework as just another form of sweet labour.

PROGRAMMERS

In looking for new technology homeworkers in the information sector, the largest group we discovered were computer programmers. The existence of computer programming is clearly linked to the more general growth of contracting in this area of work. Very few of the programmers in our sample were employees. The most common status of programmers was 'employee of own company', closely followed by 'self-employed contractor'. Although these people could be described as new technology homeworkers, their pattern of work is quite different from that of traditional homeworkers and word-processors in that they did not work at home all the time. Visits to client sites by programmers varied in duration and frequency, depending on the type of work, and wether or not problems could be solved over the phone. Overall two-thirds of the programmers we spoke to regularly visited sites two or more times a week.

Homework is usually thought of as being done in the home, with occasional trips out, to deliver work, For example. Most of the work done by our word-processors fits this model, and like traditional homeworkers, many of them feel isolated as a result. Taking the programmers and word-processors together, about one third of them complained that they feel isolated and that they found it hard to motivate themselves to work. But isolation and lack of motivation were particular problems for women, with over half of them mentioning the latter, and almost as many saying that they missed the intellectual and social contact derived from working away from home. Programmers were much more likely to know other programmers and to communicate with them than were word-processors to know other word-processors.It is their concentration in word-processing work that makes these women isolated.

The programmers in our sample, on the other hand, did not work which fits neither the model of full-time homework nor full-time site work. And it is precisely the fact that they can combine working from home with working on site that they like. It is the programmers' privilege to be able to enjoy the best aspects of each. So, while Toffler idealizes a totally home-based working pattern, the evidence from our study suggests that it entails significant social costs.

Programmers appreciate their flexible and varied working patterns, but the very long hours they worked militated against any significant change to the balance between work and leisure, or work and family life. When we asked both programmers and work-processors if and working from home had changed their attitude to work, we found strong evidence of reinforced rather than transformed gender differences. Most of the sample said their attitude to work had changed, with over half saying they were more work-centred. But this was especially so for the men. Around a quarter of the sample, all of them were women, said they were less work-centred and more family-centred. Similarly, increased gender differences are reflected in the fact that nearly half of the programmers thought they had become more workaholic, while most of the word-processors were less workaholic. David, a programmer, talked about this problem:

> I would say quite honestly that the prime consideration in my life is my work....and that has caused all sort of problems, personal problems. And I am the to admit that i don't spend enough time doing the normal things that people do. I tend to think, live and breathe my work, umm [notices something coming up the computer screen] I mean to stay away from a computer for any length of time is like withdrawing from heroin..[Wife comments:] I am not kidding. you can have one of the most important discussions with him if you are anywhere within visual sight of that thing he will be keeping one eye on it the whole time...[he says:] I am noticing a

> problem here now on the screen..[Wife:] We can be talking about a life and death situation, and he has got one eye on the screen.

The way in which women actually approach the technical work itself is also an issue. It may be there are gender differences in the cognitive style of computer programmers. This intriguing line of thought was raised by one woman programmer:

> I definitely agree that [women] are more logical, better organized. To do a proper design job, you've got to be incredibly logical. You should also be a good lateral thinker. That's something I am very good at.. the design area is a very important part of the job - to look at all the different ways it can be done, and then find the best. I think women tend to be better at this.

Many systems analysts and designers emphasized that a large part of their time was taken up in discussion with other software specialists and end-users, before any design or programming work was possible. Women's empathic skills come into play here, according to the same woman:

> The other thing it requires is understanding what people mean, not what they say. The user will say 'What I want is this', but what he really means is something else. It's being able to interpret that and get behind what he is saying. I think women are better at that then men are; they're a lot more subtle in a lot of ways

Half of our programmers found that they earned more working at home, and many pointed to the lower overheads in running a business from home. Needless to say these earnings are dependent upon a continuous supply of work, particularly since they receive no holiday pay or other employee entitlements. Equipment costs for programmers varied from $8,000 to $30,000, and their hourly rates of pay varied considerably. The most common payment system for programmers is a combination of hourly (daily) rates and a 'fixed price' for the job. Here again we see that the programmers have greater control over their earnings than the worth-processors on piece rates. In the 1988 re-interviews, most full-time programmers were making at least $50,000 a year.

In addition to this equipment, most of these professionalized homeworkers (and, for that matter, the word-processors as well) said they had an office in their home. Not surprisingly then, their houses suggested a fairly affluent and middle-class lifestyle. The people who lived in Sidney, for example, came from gentrified inner suburbs or wealthy eastern ones. One programmer, when we asked why he worked at home, replied from the comfort of his fully equipped kitchen, with a gesture towards his courtyard garden and showy poinsettias, 'Who wouldn't want to work here?'

The experience of working at home as a computer programmer that emerges from our research would appear to confirm the optimistic view of the technologically-based reorganization of work. But it is important to recognize the specific characteristics of the very different occupational groups we studied, and the reasons why their experiences are not susceptible to generalization. Looking at programmers, what are the conditions for their successful homeworking? First, they have the considerable financial resources necessary to purchase very expensive equipment, and second, they have a skill which is very much in demand. Their strength in the labour market is a product of the well-documented chronic shortage of programmers and systems analysts in Australia (Daniel 1986). (Although we must remember that the group we are here calling 'programmers' is heterogeneous and has diverse market power. Demand at any one time for base-level programmers, for instance, may be quite different than that for, say, project managers.) The labour process literature has been vitally interested in the question of struggles for control over work, but it has tended to study mechanisms of control that are internal to the work place. For many groups, however, their ability to control their work depends critically on their positions in the labour market. The skills of programmers at present give them unusual market strength. Programmers may well prefer self-employment and contracting to employee status because of their ability to exploit the skill shortage in their area. The strength of their bargaining position more than compensates for the loss of industrial protection attaching to employee status.

CONCLUSION

Our study of new technology homeworkers in Australia has necessarily been exploratory in character, given the lack of empirical research in this area. The three studies of most comparative value are the UK study by Huws (1984) and the American studies by Olson and Primps (1984), and Pratt (1984). However, developments in this field have been so rapid that research carried out in the early 1980s may now be dated - that is, developments in the cost, power and capacity of microcomputers and communications between types of computer have changed the work that it is now possible to do from home. Whereas in these studies as sample were mainly employees working on remote terminals connected to Mainframe owned by their employer, our sample differs in that they are overwhelmingly self-employed contractors who own microcomputer.

When we look at rates of pay for programmers, the two comparative studies (of women and men) show that female programmers are worse off than male programmers. And Huws' study, which is specifically about women found out that pay and conditions are especially unfavorable. In our study, while programmers on the whole were guarded about their exact income, both women and men were clearly in a high income bracket. Although a few programmers did experience a fall in earnings since working at home, there was no difference between women and men in the likelihood of this occurring.

It may be that our women programmers are not as badly off as Huws' in terms of pay, working conditions and irregularity of work, because the Australian labour demand for programmers is still high. Moreover, it seems that most of Huws' sample were carrying out maintenance work on computer systems, work which in Australia is regarded as appropriate only for junior or beginning programmers. However, because of the shortage of programmers, even maintenance work on contract can command high hourly rates. Further comment could only be speculative given the absence of appropriate comparative studies.

The comparative difficulties are furthered by a widespread tendency to homogenize what are actually diverse types of computing work. Kraft and Dubnoff's (1986) study of computer workers in the US distinguished programmers according to the type of technical function and the degree of involvement with management tasks. In our on-going review of these Australian programmers, we are trying to ascertain whether any of the sample are involved with home-based managerial work and how compatible homework is with the managerial dimensions of computing work. There is no doubt we are witnessing major changes in the way work is organized in Australia, as elsewhere. Much of the literature tends to explain these changes simply as the result of technological development. What our study tries to show is that these broad changes always occur in specific contexts. Our comparison of computer programmers and word-processors point to several crucial factors. it reveals the importance of examining the nature of the work itself - what is actually being produced, the skill involved in doing it, and the age and health of the industry within which it is located.

Even a study of new technology homework fails to reveal simple trends. Just the two groups of workers that we studied showed the importance of labour market segmentation in determining their contrasting experience of work. Labour market segmentation must itself be understood as both internally dependent on skill and externally shaped by gender inequalities. We have seen that working at home means very different things for word-processors and computer workers and for women and men. For women word-processors, new technology homework still reflects their labour market vulnerability - vulnerability that stems from the domestic division of labour. That same unaltered domestic division of labour undercuts the otherwise strong market position of female programmers. It is for male computer programmers who possess skills which are in short supply that new technology homework presents an unambiguously attractive choice. But this hardly warrants the general enthusiasm for 'electronic cottages' that characterizes so much of the literature about the future of work.

REFERENCES

Allen, S., and Wolkowitz, C., The Control of Women's Labour : The Case of Homeworking, FEMINIST REVIEW 22, (1986) pp. 25-51.

Daniel, A., Work in the Computer Industry, ADVANCE, 2 (1986) pp. 9-10

Gorz, A., FAREWELL TO THE WORKING CLASS (Pluto, London, 1982).

Huws, U., THE NEW HOMEMAKERS - NEW TECHNOLOGY AND THE CHANGING OF WHITE COLLAR WORK (Low Pay Unit Pamphlet No. 28, London 1984).

Jones, B., SLEEPERS WAKE (Oxford University Press, Melbourne, 1982).

Kraft, P. and Dubnoff, S., Job Content, Fragmentation and Control in Computer Software Work, INDUSTRIAL RELATIONS 25,2 (1986) pp. 184-196.

Offe, C., DISORGANISED CAPITALISM (Polity Press, Oxford, 1985).

Olson, M. and Primps, S., Working at Home with Computers, JOURNAL OF SOCIAL ISSUES 40,3 (1984) pp. 97-112.

Pratt, J., Home Teleworking, TECHNOLOGY FORECASTING AND SOCIAL CHANGE 25, (1984) pp. 1-14.

Rubery, J., and Wilkinson, F., (ed.) THE DYNAMIC OF LABOUR MARKET SEGMENTATION (Academic Press, London, 1981)

TNC Workers Research, ANTI-UNION EMPLOYMENT PRACTICES (Sydney, June 1985).

Toffler, A., THE THIRD WAVE (Collins/Pan, London, 1980).

Wajcman, J. and Probert, B., New Technology Outwork, in Willis, E., (ed.) TECHNOLOGY AND THE LABOUR PROCESS: AUSTRALASIAN CASE STUDIES (Allen and Unwin, Sydney, 1988).

Women, Work and Computerization: Forming New Alliances
K. Tijdens, M. Jennings, I. Wagner, and M. Weggelaar (Editors)
Elsevier Science Publishers B.V. (North-Holland)

HOME BASED TELEWORK - SELECTED HIGHLIGHTS FROM A CASE STUDY IN THE FEDERAL REPUBLIC OF GERMANY*

Monika Jaeckel
Deutsches Jugendinstitut
Federal Republic of Germany

This paper reports on a qualitative study of female and male tele-home-workers in the FRG. At present in West- Germany most home-based tele-work is done on a free-lance and self-employed basis for reasons of achieving more autonomy and skills than at regular work sites, and having more flexibility to combine work and family interests. Women working with their own pc at home report using a selfteaching approach to computers and feeling more confident and comfortable about developing their skills with new technology at home than in corporate settings. The need for new social security systems and for self-help networks of tele-homeworkers is emphasized by the study.
In Germany the most common form of home-based telework at present is on a free-lance or self-employed basis. The transfer of jobs "on-line" from offices into the homes has not -despite frequent forecasts- taken place on a large scale. However, a growing group of both men and women are using the new computer technology for more self-determined, flexible and qualified job conditions by going into free-lance and self- employment, operating from home.

1. THE QUALIFICATION POTENTIAL OF HOME BASED TELE-WORK

Both the men and the women in our sample rate their choice of home-based telework (the majority had regular jobs previous to going into home-based computer work) as a step up the ladder, a possibility to develop more skills and to have more responsibility and autonomy than in the usually highly hierarchical structures of the corporate world. The women often describe their decision to leave their - in most cases - clerical jobs as a kind of "protest" against the dequalification trends in this sector.
Only a minority would quit their homebased telework if offered employment outside the home. 90% rate their work as very satisfactory. Two thirds of the women report that homebased telework has raised their job-confidence and self-esteem. Two thirds of the women and three fourths of the men believe that home-based computer work has

* This study was conducted by Gisela Erler, Monika Jaeckel and Jürgen Sass from the research group on family policy of the Deutsches Jugendinstitut (German Youth Institute) in 1986 as part of a European study on women and telework, co-ordinated by Jean Tansey and Ross Moran from the Irish Foundation for Human Development in Dublin. The main trends found in the German case study were confirmed by the results of the European comparative study. This research was funded by the European Foundation for the Improvement of Living and Working Conditions in Dublin, Ireland.

increased their professional chances. Both rate their work as interesting, qualified and challenging, albeit involving a high level of stress.

Having a personal computer at home allows for more room for unsupervised experimentation and self-teaching, which, especially in the case of women has proved an important way of access to computer literacy. The women described their experience with computer work in large corporations as highly hierarchical and intimidating, mainly involving taking orders at the bottom of the line and being discouraged to ask questions or to learn more about the range of application possibilities of the computer.

In most cases, "jumping into" task-solving and trying out the computer at home, was the way the women in our sample described the way they got acquainted with the potential of their computers. Computer manuals and computer courses were rated by the majority of the women as of little help. If they were made use of at all, usually at a stage where they felt already fairly computer literate.

Although both the men and the women describe their home-based telework as higher qualified than their previous jobs, men and women enter home-based telework at and reach different levels of qualification. The majority of the women stay in the field generally described as desk-top-publishing, where as the men go into program-development, consulting and training. That the majority of self-employed women in telework in Germany limit their range of computer use, is especially striking in view of the great experience, special potential and talents women have in the field of communication and information referral in bridging the gap between computer specialists and users both in work sites and at home.

2. HOME BASED TELEWORK AND FLEXIBILITY

The women in our sample also choose home-based telework as a more flexible way to combine work and family life. They prefer to be able to be at home when their children need them, this applying both for small children as well as for school-age children. The picture often painted, of mothers sitting at a personal computer while their child sits next to them, playing with its toys, however, is more the exception than the rule. The main times during the day (and at night) used for telework are when the child is sleeping, in childcare or at school, or out playing, or taken care of by the father or grandmother.

The greater flexibility provided by home-based telework to adjust work and family rhythms are rated positively. Flexibility, however, also can make high demands on self-discipline and having both work and family life in the same space can involve a great deal of strain, in some cases leading to serious conflicts and aggravating the relationship to the spouse and/or the children.

The greatest problems are reported by single parenting mothers whose home-based telework provides the sole financial basis for the family. For women without children home-based telework is described as allowing for more flexible work rhythms and more "colourful" life-styles than regular full-time employment on a nine to five basis.

Working hours in home-based telework are irregular, often involving work in the evenings or at night and on week-ends. The majority of the women in our sample, however, even when describing themselves as full-time workers, state that their average working time is around 6 hours per day. Peak time, however, can involve working up to 12, 16, 18 hours a day or in some cases all night long.

The majority of the women in our sample described their experience with home-based telework as having involved a crisis (physical exhaustion or relationship crisis), resulting either in breaking off computer home-working or in introducing distinct measures to keep a better balance between work and personal life. (I.e.: organizing childcare hiring household help, introducing work-free "family days" etc.)

The men in home-based telework do not do more housework than men working outside the home. However they provide a reliable back-up system of childcare, in emergencies

or when other childcare arrangements break down. They perceive themselves as participating more in family life and in the majority of cases their relationship to the child/children has improved by working at home. They have little or no problems in separating their work and family space in the home, their housewife wives often supporting them in shielding them from demands on their attention.
Social isolation and a deterioration of communication and social skills is a problem felt by the men in the sample, while the women stress more that they miss the possibility of job-oriented exchange and communication with colleagues.. Other issues concern health problems, a lack of adequate illness and old-age insurance, high pressure when deadlines are short termed and a distinct reduction of time for personal disposal.
The level of income differs greatly between the men and the women. Whereas the income level of the women roughly corresponds or is slightly higher than the average range of female wages in the clerical sector, ranging between DM 1500 and DM 2500, the monthly income of the men starts at DM 2500 and goes up to DM 10.000.

3. CHALLENGES OF ORGANIZATIONAL CHANGE

The implications of the results of our study for the society at large are summarized in the following points:

* This study (like many others) emphasizes the need for a historical reconsideration of the social welfare system in Germany, for new forms of social security to include the growing number of groups left out of a system, solely designed around labour-market participation. As we know from other studies we have conducted, home-based telework is in line with a cultural preference for self- employment among large groups of the labour force in Germany. The need to address the problem of new insurance systems for this kind of self-employment on a low or average income level is expressed especially by the women in our sample, as their level of income is a great deal lower than that of their male counterparts.
* This study calls for special attention to be paid to the barriers involved in restricting women to a limited scope of computer utilization. Employment counseling for women should raise the awareness and confidence for the expertise and potential of women in the field of the application of soft-ware, putting the computer to work in new fields, creating new professions and job openings.

As it seems, there is still a great deal of computer mystification at work, defining computer intelligence mainly in terms of the command of several computer languages or the ability to develop soft-ware. The whole range of adapting available soft-ware to specific tasks is one of the main areas of computer use now and in the future and a field where women have talents and know-how, due to their traditional familiarity with the organization and structures (including the important area of informal structures that make systems work or not) of clerical work. Counseling and training are fields where women's talents could be very successful, not only in teaching and counseling courses for women. What women need is a stronger focus on the part of public opinion and discussion on their abilities rather than on so- called "female" computer-fear or computer-abstinence, more publication of female role-models in this area.

* This study confirms the need for specific computer training programs for women, relating to female strategies of learning. Women need training programs with a low profile in hierarchical and school-like structures, allowing for a great amount of task-oriented experimentation and self- teaching, communication and co-counseling. In addition to beginners' courses, special courses for women who have learned to work with computers "autodidactically", directed towards their specific concerns and interests in further developing and increasing their knowledge should be available.
* In the corporate world more attention should be paid to creating more room for women already working with computers to explore the possibilities and to discover and develop their creative potentials in applying this new technology. This involves pri-

marily a restructuring of the traditionally very hierarchical structures of firms, a process many innovative companies are already involved in, as a very relevant basis of their success on the market.

* It is also implied by the results of our study that further formalization of the qualification requirements in the field of computer-work is not necessarily propitious to advancing women in this field, since one of the ways women find access to telework is autodidactically, and since there is not an equal amount of women-specific training courses available at present.

* The need to create new forms of networking and professional communication between women involved in home-based telework is not only made clear by the results of our study, but also by a network of women working with computers at home, induced by our study in Munich, which has been operating since last year.

* A final comment concerns the tendency of men involved in telework to deteriorate in their social skills. For women this problem did not arise in our study. The generally higher interest and priority for relationships in female lives apparently counteracts this effect. The men in our sample -especially if they were single- reported problems in their social interaction, due to growing intolerance of ambiguity and expectations of yes/no structures in human communication and in their relationships or to a general restriction of interests and vocabulary involved in prolonged and intense computer interaction. The men in our study generally reported a reduction of their social life and contacts. What this means for the culture of future society and for the quality of life as computers become a growing part of reality and what measures could be taken to counteract possible effects of social erosion by a male "computer-zombie" culture should be part of social policy considerations accompanying the introduction of new technologies into society.

4. THE SAMPLE

This study was conducted in the fall of 1986 in the larger region of Munich, one of the areas in Germany, where hightech industry is most advanced and concentrated. The sample consists of 30 female and 25 male home-based teleworkers, who returned a standardized questionnaire, 15 women and 13 men, who were interviewed in a one and a half hour long qualitative oral interview at home and two group discussions, which were attended by 16 women and 6 men. The respondents were found on a random basis by adds that we ran in two local Munich newspapers and one national computer magazine.

19 of the women and 17 of the men were married, 7 women and 6 men were unmarried, 4 women and 2 men separated. 25 of the women and 12 of the men had children. 23 of the women and 6 of the men were in the age group between 26 - 35, 7 of the women and 16 of the men were between 36 and 55 years old. 19 of the women and 7 of the men had qualifications as clerical workers, 11 of the women and 18 of the men had higher education. 20 of the women and 14 of the men were working on a freelance or self-employed basis, 23 women and 22 men worked full-time and 7 women and 3 men worked part-time. A mixture of home and office-based work was the case with 6 women and 11 men.

BIOGRAPHICAL NOTE

Monika Jaeckel, 1949, sociologist, researcher at Deutsches Jugendinstitut, is active in various projects in the field of family policy, including mother's centres, flexible working hour schedules, tele-home-worker network. Correspondence address: Freibadstr. 30, 8000 München 90, FRG, Tel.: O89-623 O6-253

Women, Work and Computerization: Forming New Alliances
K. Tijdens, M. Jennings, I. Wagner, and M. Weggelaar (Editors)
Elsevier Science Publishers B.V. (North-Holland)

STRATEGIES OF INTRODUCTION OF OFFICE AUTOMATION IN MIXED AND TRADITIONALLY FEMALE JOBS WITHIN TWO WORK UNITS OF THE FRENCH ELECTRICITY AND GAS PUBLIC UTILITY COMPANIES

Marta Ormos and Maya Blameble
France

A survey was conducted between 1985 and 1987 on two work units which are highly contrasted for type of activity and extension the structure of the organizational setting. The first unit is purely administrative and handles retirement pay, and disablement, old age and death pensions. The second is a technical and scientific research unit and covers research activities related to the construction and operation of nuclear power stations.
In the first unit, given the type of activity, half of the work force is female, but the majority of women are concentrated at the bottom of the pyramid. The second unit is composed of two distinct population categories: research engineers, and for typing and assistance to intellectual production, typist and secretaries. There are only a few women in the first category whereas the second is almost exclusive feminine.
Prior to the introduction of office based business and scientific data processing, centralized data processing was the prevailing mode in the two units. Word processors were introduced at the same time as the office based business data processing. The strategies of introduction of office automatization are analyzed in these two organizational settings which resulted from the interaction of different collective actors: managers, unions, and workers (male and female) in their specific occupations. These strategies differed in the two units. The nature of the activity and the organizational work structure appear as the main factors influencing the ways in which office based business data processing and word processing were incorporated in the units. In the case of word processing, the use of dedicated machines was not imposed in any of the units. In the case of business data processing, women and men were involved to an equal extent, but women, being mainly at the bottom of the hierarchy, were affected in a different way. This paper will relate these findings and discuss identification of the factors which have positive and negative effects on women's control over the applications of office automation.

1. INTRODUCTION

The study presented here analyzes the relationships between success in the introduction of office based data processing and the managerial strategies in two work units which vary considerably in type of activity, organizational structure, personnel make-up and

mode of decision making. It focuses in particular on the effects of redistribution of jobs and responsibilities within each unit and the growth of polyvalence of women workers. The study was conducted between 1985 and 1987 at Eléctricité de France and Gaz de France (henceforth their acronym E.D.F.- G.D.F.), two public utilities having a common personnel division and an identical status. The two companies employ a total of 152.000 people.
The first unit employs 340 individuals and handles retirement pay, disablement, old age and death pensions. The second unit deals with research and development of thermic and nuclear energy production and employs 430 people.
The findings of our study are drawn from on site observation and interviews with 127 individuals (42 from the first unit and 85 from the second) from all categories of personnel: managers, supervisors, clerical workers, as well as union representatives.

2. THE ADMINISTRATIVE UNIT

Prior to the introduction of office based data processing, the first, purely administrative unit was characterized by its extreme overspecialization. This took the form of a bureaucratic type of organization[1] with a pyramidal hierarchical structure. The unit was split into smaller and smaller parts, the basic work cell being composed of a group of 4 people carrying out the most specialized jobs. Work relations were purely vertical. Each subdivision corresponded to a hierarchical level. The hierarchical incumbent assigned work, and had a role of control and responsibility for job execution. Because of partitioning of responsibility, cell members had almost none, not even the responsibility for their own work since this was checked systematically before it left the cell.
Traditionally this unit hired its personnel on the lowest level of the hierarchy (in the national E.D.F.- G.D.F. wage scale). Career advancement was guarantied by on-the-job training and by seniority. The middle management and the supervisors thus came mainly from the rank and file. Most of the upper management on the other hand had university degrees.
As is common in this sector of activity, the percentage of women was high (47%). However, a finer analysis of the different hierarchical and pay levels shows that the distribution of personnel in the different levels remained extremely skewed: management was 18% women, half of whom had university degrees, whereas only a quarter of the men were graduates. The majority of women in this unit (37%) had supervisory functions or comparable positions in therms of salary. The clerical staff was 56% feminine.

2.1. Relocation of the Unit and Unsuccessful Computerization

The relocation of this unit to a French province in 1972 had a dual consequence.First of all relocation resulted in a lowering of the average age of the unit (68% of the personnel were now between 30 and 49). Older people tend to prefer to stay in Paris, even if they must change jobs. On the other hand, promotion in the hierarchy was used as a powerful incentive for individuals to move. Its long-term effect is to block normal career advancement possibilities for a time, since the age pyramid was far from matching the optimal shape for career advancement.
Secondly, the move away from the central Parisian office was the impetus for the unit to acquire its own computer experts at a period of time when computerization was still highly centralized for all the units in E.D.F.-G.D.F. The result was the emergence, within the unit, of two groups of social actors who rapidly developed antagonistic relations since they had conflicting interests, and since there was no go-between structure to co-ordinate them.

---

1 Positions have a fixed profile, connected to the contractual wage scale.

On one hand the group of computer experts made studies, tested and wanted to continue innovation. On the other, the group of user-staff wanted an efficient, reliable, durable product that made their jobs easier.
This conflictual situation lasted from 1973-1976, at which time the management called for a halt in computerization,i.e. a freeze in applications and innovation while conducting a review of the place and role of computerization in the unit. This measure succeeded in discontenting both parties, and no date was given for starting the review process. At the same time as the moratorium on computerization, a purely administrative reform was imposed on this unit by the management. It generated massive social conflict because of its authoritarianism and absence of prior concentration with the personnel.

2.2. A new Strategy: Consultation with the Personnel on System Design

However in 1982 the situation reversed itself and the management and the unions[1] in the unit agreed upon starting up changes in computerization and changing obsolete job practices. Two fundamental objectives were defined: the improvement of customer service and improvement in working conditions.
Two simultaneous plans were launched: a general audit of the unit (and not only a computer audit) carried out by a consultant firm and secondly a vast consultation campaign where the personnel in its entirety was asked to take part in reflections on how to improve work procedures and the organizational setting.
Aside from the participation of all the social actors, this new plan was innovative because it took economic and social factors into account from its beginnings. The simultaneous development of the organizational structure of system design and a training program integrated the requirements of provision services for costumers as well as long-term human resource management. This new plan entered into its trial phase in 1985 at the start of this study.

In the new organizational structure, the base cell became a larger group covering specialized jobs. The goal was the creation of a single job profile, and thus a rapid augmentation of polyvalent personnel made possible by office based data processing. Since office based data processing was also expected to provide greater autonomy for user staff and give them responsibility for their jobs, numerous supervisory position were transformed into adviser, trainer or co-ordinator positions.
The main features of this reform are:

1 A progressive procedure from the old structure to the new one. In the trial period only two teams (each composed of 10 individuals having different specialities) started to work on office computers and terminals.
2 These teams were entirely composed of volunteers.
3 The procedure, the specifications i.e. the menus were developed on the advice of the operating workers.
4 There were no deadlines set up for the trial period.
5 In the new work process polyvalence was emphasized.
6 The workers' tasks were enriched and not impoverished in the new work setting.
7 Management and the unions agreed that productivity gains would be used for reduction of the work force but only by "natural" attrition (retirement)[2], and it would benefit the consumer by offering them new products.

---

1 There is no closed shop in France. Four unions having different political coloration represent all the employees from operatives to managerial positions. Membership is not compulsory at any hierarchical level.
2 The statute of E.D.F.-G.D.F. guarantees life-time employment but employees can be strongly urged to leave their posts and go elsewhere.

The local and regional based training sessions ensured high attendance (41% and 25%) on the part of both women and men, since it was realized that married women with children are usually penalized by training courses requiring long absences. In this unit the percentage of women attending the training courses was much higher than national statistics for the companies.

These training courses covered varied domains.[1] Some were oriented towards general culture (economy, company operation, etc) whereas others were directly related to individual jobs. Different levels of computer training courses were planned for the clerks to allow them to gain greater mastery than simple computer operation.

The reorganization of office based data processing was carried out smoothly and successfully in the 18 months following the setting up of the first volunteer teams. In the interviews,the number of women and men stating they were satisfied in their jobs was exceptionally high.

2.3. Women's Benefits from a Polyvalent job Profile

However it appears that the introduction of office based data processing was more beneficial to women than to men. First, since there are more women than men at the bottom of the hierarchy, they benefit from greater polyvalence. Since men are more numerous in supervisory jobs, they lost their ascendancy and their role switched from an authoritarian to an advisory one, in the new procedure. This implies that they could only exert a-posteriori control and not a-priori control as was previously the case. Work satisfaction also seemed to derive from the mastery of new technology for both women and men.

Secondly, the new polyvalent jobs were not upgraded. Thirdly, polyvalence is a feature in the opening up of greater career opportunities for the workers. Work experience in these jobs will enable them to apply for higher posts either inside or outside the department in other work units of E.D.F.-G.D.F. Inside the department, polyvalence and a-posteriori control introduced a greater continuum between white collar employees and the supervisory level. Applications to posts outside should be more effective, due to their better qualifications.

Career opportunities are formally the same for women as for men. In fact on the basis of a linear extrapolation, expectations for women are lower.Women had slower career advancement in the past at all professional levels in the company; and secondly women on this job level had less geographical mobility than men (promotion can be linked to acceptance of geographical mobility).

3. THE TECHNICAL AND SCIENTIFIC RESEARCH UNIT

The second unit is a technical and scientific research unit. It deals with research activities related to the construction and operation of thermic and nuclear power stations (70% of the electricity in France is furnished by nuclear power stations).

Centralized data processing has existed for many years in the research unit because of the high degree of sophistication of the work carried out which required data processing.

In 1984 this work unit also left Paris and was relocated in a province. Prior to the relocation, the unit already possessed a hundred terminals and approximately 40 word processors as well as Computer Assisted Drafting equipment.

The introduction of the first two types of material was done by calling for volunteers and created no problems since it corresponded to workers' demands.

---

1 E.D.F.-G.D.F. devotes 10.6% of its wage bill to training.

Out of the 430 employees, 60% are research engineers of whom 10% are women, 11% are technicians and in supervisory positions, 25% are women.15% are clerical employees, of which 1/5 are women; 14% are secretaries and typists.
The research staff is extremely carefully selected on intellectual bases: the research engineers are Ph.d's from the best schools and are assisted by highly competent technicians. The administrative staff scarcely have more than a B.A.. The secretarial staff was recruited out of high schools.

3.1 Relocation of the Unit and Two Innovations on Trial

When the unit moved, a new building was built to suit its specific needs. During our study two innovations were in progress:

* a network of cables built into the floor connecting each office in the new building, and linked to the national network. Terminals use national and local computers for scientific data processing. Microcomputers are also used by the researchers and by the clerical and secretarial staffs. During the study, a project was announced to interconnect all the personnel and not only the engineers working on scientific data. Secretaries could thus start a letter in one office and correct it in another and complete it if necessary with scientific data in a third. Interconnections of secretarial staff and engineers as well as the interconnection of different secretarial offices was one of the most interesting features of this work setting project.
* the prototype of a French multi-function terminal able to simulate IBM and BULL terminals (double compatibility). The aim of this computer architecture was not to respond to current needs but rather to set up the most advanced and flexible physical structure to respond to future progress in office computerization.

The necessity of team-work based on highly specialized competencies in different fields and co-ordinate by a hierarchy characterizes this work unit. Assignment of positions in the hierarchy is closely related to knowledge, but knowledge of different kinds are taken into account along with other criteria. Highly specialized individuals as well as co-ordinators in hierarchical positions can have enormous prestige. In this case institutional authority cannot be too overbearing and the general mood of work relations is exceptionally permissive.
Women concentrated in secretarial tasks are apart, but the general permissiveness in the work ethic has extended to them as well. Before the work unit moved to the new buildings,secretaries and typists could use word processors or not, as they liked. People interested were given a three-day training course. Course content was on operation of the word processor. A highly detailed instruction manual was provided with each processor. Trainees quickly evaluated the course as too short. But in the months following the course, all the women involved described this minimal exposure as having been a source of motivation and involvement and in the final analysis, enrichment. The training course favoured familiarization with the equipment and mastery of word processing as well as reinforced their abilities to take initiatives and solve problems.

Typists' job on a national scale have two classifications and women who chose to use word processors were upgraded. The salary incentive and the general atmosphere of the department working on advanced technology led to rapid expansion of the use of word processors. The typists' pool was disbanded when the employees received the word processors and the typists were assigned to different secretariats.

3.2. The Management's Strategy for Office Data Processing

When the department moved to the new building (1984) where interconnections existed physically, the management devised a specific plan for the clerical and secretarial staffs.

It created a small "organization and communication team" to be in charge of three missions: marketing the machines (explaining their possibilities) each time the machines were purchased or leased only when an explicit request was made by the operating teams, giving technical assistance when needed (enhanced on-the-job training) and elaboration of organizational schemes, i.e. program projects which would implement the interconnected network.
The originality of this strategy developed by the management of this unit was to have dissociated the methods used in the scientific data processing from those in administrative data processing. Nevertheless because of the aim to associate the two, the decision makers in the research unit felt that the interconnection and office data processing would provide new and specific means for changing work methods.
This very different approach adopted by the management in handling scientific data processing as opposed to office based administrative data processing and typing stems not only from the fact that the problems are very different but also they concerned two typical groups of people having little or no non-business communication.

3.3. Self-Organization of the Secretariats

The relaxed atmosphere in research development was beneficial to the secretarial staff with respect to their work organization. Ever since the introduction of word processors, secretaries and typists have been in a constant process of reorganization of their own work on a less hierarchical basis. This reorganization arises from their own initiative without any intervention of outside hierarchy or their chief. It is a "de facto" reorganization with no formal redefinition of the posts, which must be agreed upon by the management and the unions.
In this research and development unit, the trend towards a telescoping of secretaries' and typist' jobs observed elsewhere has been pushed further here. There are several reasons for this evolution:

* he general trend towards a reduction in secretarial staffs due to productivity gains resulting from new equipment, which calls for greater polyvalence
* the constraints linked to working on a screen make non-stop typing over a period of hours problematic and create greater bonds of solidarity among women
* in addition, training and above all self instruction and hands-on experience with the equipment have made secretaries and typists more able to appropriate their instruments of labour and to have a pleasant and creative relationship to their tools.
* the introductory phase, involving uncertainty of all kinds, allowed these women to create an enclave of freedom, initiative, to acquire personal mastery of skill, and to demonstrate their ability to solve problems.
* since the acquisition and the taking of initiatives in the appropriation of machine operations has been positive, it motivates these women to take initiatives in organizing work within the secretariat.

This last point should be stressed since this new ability to take an active part in the organization of work and the development of creative practices with the instruments of labour will raise new questions in terms of human resource management in the near future.

3.4. Freedom of Self-Organization as a Regulation Process

For the time being the management has not expressed any opposition to trends towards self organization in the secretariat. In fact the whole department is subjected to on-going modernization and a reduction of manpower. The management tries to reduce manpower by not replacing job levers in particular low-qualified administrative jobs of all kinds inside and outside the secretariat.

But some of the levers' functions remain and must be taken on mainly by the secretariats. For example twenty photocopy machines were replaced by a laser printer, two telecopier were introduced, etc. Since this process is gradual, there is no reason to intervene in the secretariat's efforts to take over these functions. They have been "left to their own devices" to decide who will do what and when.

3.5. Polyvalence and Career

In fact, not only have low qualified administrative job opportunities been suppressed, another change in the personnel can be observed in the traditionally computerized administrative sector. Thanks to office computer applications, the employees can supervise their own duties on-line and take on duties which cover a broader range than previously. Clerks have become autonomous and polyvalent. In the past, this supervision of tasks performed manually was traditionally given to employees promoted from the rank and file (without degrees). This relegation of duties to a lower level in the hierarchy due to office based data processing endangers internal promotion to supervisory posts, since from now on higher hierarchical posts will require a university degree.
Nevertheless, polyvalence is favoured by the introduction of office data processing and is considered to be more desirable by the management; currently it is a source of satisfaction both for the personnel in the secretariats as for the administrative staff. The enrichment of job content and the gain in autonomy are entirely gratifying. This stage is however too recent for the personnel to have had the time to reflect on the long term consequences. In terms of career prospects, no promising avenues have opened up. In the administrative sector itself, the "natural" career prospects for non-university personnel vanish. It would appear that in the future, the passage from the employees, secretarial level to a higher one can only take place by long-term training leading to a recognized diploma.
Furthermore, the gap between the scientific staff (research engineers and technicals) and the secretarial staff is such that it is currently impossible to envisage that clerical employees or secretaries could take over functions of assistants to engineers.

4. UNION STRATEGIES

In these two work settings, office based computerization has brought an improvement in work procedures and in job contents, since work satisfaction and the functioning of work relations have also been improved. In the first (mixed) unit, the unions have supported employees' demands. Employees were invited to express their points of view directly in workshops but also through their representatives, in the joint advisory committee of the unit. Women's work conditions were not emphasized as such, but the collective strategy was above all beneficial to them. In the second research unit, union representatives on the unit's joint advisory committee were consulted on the principles but were not involved in the practice of system design. Women's strategy in the secretariats has neither been sustained or restrained by the unions. Nevertheless, job enrichment strategies and the transformation of hierarchical relationships inside the secretariats is a collective move on the part of the women concerned. But this move has no formalized official expression. Moreover it is likely that some of the unions would not support this trend if it develops outside this unit since their policy fits with a very rigid definition and hierarchization of post content [1] For the time being, women can expect no benefits from the unions' involvement in their organizational schemes in this unit.
Unions were very effective in obtaining the upgrading of typists' posts working on WPs on the national level within E.D.F.-G.D.F.

1 This is the case for the C.G.T. (Confédération Général du Travail).

Career development in these units will as of now mostly be a matter of individual strategy. Nevertheless, the role of the unions is very important in demanding the creation of adequate training sessions especially those leading to promotion. There are disproportionate more of these sessions currently in the technical than in the administrative field. E.D.F.-G.D.F. started a new administrative session in December 1986, which lasted until Jan 1,1988, full time, full pay. This was a positive action session reserved for women mainly from the secretariat, with the aim of bringing them up to the managerial level.

ACKNOWLEDGEMENT

The authors wish to thank referee Cecelia Bergström (Sweden) for her helpful comments.

BIOGRAPHICAL NOTE

ORMOS, Marta, Ph D in sociology (university of Paris X, 1975) works in the Personnel and Staff Relations Division of the French National Electricity and Gas Companies. She was born in Hungary and educated in Hungary, Belgium and France. She has broad experience in the field of worker promotion and further education in industry. She has taught Sociology at the National Centre of Arts and Trades and at the Business Institute of the University of Paris (Dauphine). In recent years she has studied and written about women's employment in traditionally all-male technical jobs in different types of organizational structures. She is the author of La Montée: La Promotion sociale dans une grande enterprise publique: E.D.F.-G.D.F. (the Climb: Social Promotion in a Large Public Firm: E.D.F.-G.D.F.), Editions ERES, 1982, and has published articles in several scientific journals. She is currently conducting research on issues related to new technology, organizational change and career development with a specific focus on women.

BLAMEBLE, Maya has a Masters in Psychology (University of Paris VII) and first worked as a psychologist in Senegal. After training in staff management, she has conducted research on the employment of women in particular in industry and in the tertiary sector, with emphasis on the impact of new technologies on jobs.

Women, Work and Computerization: Forming New Alliances
K. Tijdens, M. Jennings, I. Wagner, and M. Weggelaar (Editors)
Elsevier Science Publishers B.V. (North-Holland)

INFLUENCING THE CONTENT OF WOMEN'S WORK IN AUTOMATED OFFICES

Juliet WEBSTER
Research Centre for the Social Sciences,
University of Edinburgh,
Edinburgh, Scotland.

This paper gives an account of research conducted in West Yorkshire, England, between 1980 and 1985 in order to assess the impact of the word processor upon secretarial jobs. It argues that the presence and extent of the technical division of labour rather than of the new technology is the significant determinant of task fragmentation, deskilling and work intensification. In order to understand and influence the impact of office technology, therefore, we have to focus on the characteristics of office jobs *prior* to the introduction of the technology. Given that not all office jobs have been rationalised, there is scope for the re-design of such work, and the paper makes arguments for the widening of tasks and broadening of responsibility, discretion and autonomy. It recognises, however, that there are limitations on this process arising from the position of women in the labour market, the current cursory nature of training for office work, and the poor levels of unionisation among office workers - factors which diminish the bargaining position of office workers.

OFFICE AUTOMATION IN THE LITERATURE

Since microelectronic technology first became a matter of major public debate in the mid 1970s, there has been an enormous literature on all aspects of the topic. Initial responses dealt with the most identifiable or controversial forms of new technology, for example, robotics in the factory and word processors in the office, and were highly influential over the tenor of the debate on future patterns of work associated with advanced automation. It is salient to look at some of these arguments, despite the fact that they date back mainly to the late 1970s and early 1980s, and refer to forms of microelectronic technolgies which are of course constantly being superceded. Their basic propositions have still not really been adequately examined in the light of empirical developments, and many assessments of women's prospects in situations of technological change appear confused by these conflicting approaches and the lack of an underlying perspective on the introduction of microelectronic technology.

From the abundance of material concerned with the introduction of word processing (WP), two themes have emerged which are of particular relevance for a consideration of labour process change. The first describes and isolates the benefits which WPs bring, and deals with the specifics of organisational detail which may ensure a 'smooth' transition to office automation, while the second addresses the question of WP critically, regarding it as indicative of a reassertion of capitalist control over the labour process. Within the first approach it has been commonly argued that the adoption of microelectronic technology in general is crucial for the increased competitiveness and survival of the UK economy by providing new and cheaper goods and services (Department of Industry 1978; Central Policy Review Staff 1978; Advisory Council for Applied Research and Development 1980).

The benefits for office workers themselves of working with such machinery are spelt out by the WP manufacturers, who publish lavish and glossy promotional material. In the 'office of the future' - the so-called 'paperless office' - elegant young secretaries sit in air-conditioned splendour amongst pot plants, smiling into visual display units. The working environment is portrayed as being cleaner, lighter and more comfortable, and the work itself as being more skilled and satisfying for the office worker. Overall, WP is heralded as a new "skilled profession" in which the operators themselves form part of "an integrated team, each in control of, and responsible for, their work area" (Word Processing Now magazine, November 1978:26). The implication is that, freed to more responsible and intellectual tasks, the office worker will become less of a machine operator and more an administrator of information, and may even be able to progress through the office hierarchy into the realms of management.

Such an optimistic viewpoint seems to be the result of a conception of planned technological change as a balanced structural process and of a belief in the evolution of a new 'post-industrial' or 'information' society. However, it massively plays down the possibility of mass unemployment or increasing social and class divisions. Furthermore, as Liff points out, there are few examples of this perspective in action despite its apparent feasibility (1988:13).

If by arguing in favour of the liberating, enriching effects of word processing on the labour process, this perspective owes its inspiration to the work of Daniel Bell, by the same token the radical critical approach which developed in response to this is demonstrably the product of Bell's own philosophical adversary, Harry Braverman. It sets out to contest the notion that jobs destroyed by the microelectronics industry will be replaced by new occupations in new industries, and that degraded and routine work will be consigned to machines, releasing human beings to 'cleaner', more creative and more fulfilling work. It poses instead a view of the development and implementation of microelectronics as a reflection of the interests of capital. In the context of an ailing capitalist economy, new technology is regarded as a deliberate mechanism whereby employers cut labour costs and achieve greater productivity in an attempt to solve the crisis of falling rates of profit and maintain the rate of capital accumulation. The entire rationale behind the new technology is to increase 'the trend towards the real subordination of labour to capital in the form of machinery':

> It is within this overall context that we must view the introduction of word processors and other microelectronically based office equipment into offices - not as part of a technology which is autonomous, driving itself forward by its own momentum, but related to and influenced by and crucially part of capital's strategy to continually reproduce itself (Barker and Downing 1980:85).

This perspective regards the introduction of WP technology as a process which heralds the implementation of techniques more commonly associated with the assembly line. The principal features of automated office work reflect a taylorist strategy and are those which Braverman himself regards as typical of labour processes in the monopoly capitalist era: fragmentation, degradation, deskilling, intensification and the pacing of work.

According to this argument, the bedrock of WP, and the fundamental means by which typing labour processes are transformed at all levels, is its principle of task fragmentation:

> While there has been a tendency historically for the secretarial worker's job to develop as a completely manual function, many secretarial workers find they have auxiliary tasks to perform, such as filing, the responsibility for an individual's mail and making coffee. The object of word processing systems is to further fragment the secretarial job. High levels of efficiency and cost effectiveness are best achieved by stripping away all possible distractions from the actual job of typing (CSE Microelectronics Group 1980:48).

The secretarial worker, it is argued, is now devoted entirely to the single activity of keyboarding text into the machine and no longer sees her work through to its completed and printed state. Moreover, since documents can be stored in the machine's memory, she can theoretically work continuously at her screen without ever needing to leave her seat, to look them up in a file, for example (Barker and Downing 1980). Her confinement is thus absolute. However, not only is the office worker now portrayed as a mere 'inputter' of text, but the material she types itself apparently becomes fragmented. Standard text can be stored in the machine's memory. All that remains to the typist is to recall the material onto the screen and type in the variable information - a name and address, date, amount of money, or an additional legal clause, for example.

The technical and physical characteristics of the machinery are held to be crucial agents in the deskilling of typists. The capabilities of the machine supposedly not only restrict the degree of manual dexterity required of the operator by automating many of the physical functions of typing formerly performed manually, but also, and consequently, they eliminate the typist's control over her machine and her working patterns in general. We are told that the machine can produce perfect copy independently of the typist's skill - her speed, accuracy and ability to align paper - and that she needs little conceptual skill or initiative in laying out her work since the 'creative' element of typing is now performed automatically by the use of layout programmes. The operator is reduced to mere button-pressing, the simplest knack, which can be carried out unthinkingly and automatically. Confining the typist to keyboarding, eliminating all 'distractions' from this activity, and creating a typing process which is simple in the extreme to perform, is designed to heighten the intensity of labour and thus increase output. As well, therefore, as the mind-numbing monotony and loss of skill which the operator of a WP apparently suffers, she is under 'greater stress than ever before to hit keys as fast as possible, to do the most routine, tedious tasks at the greatest possible speed' (Harman 1979:9).

In their attempts to counter the technological determinism which permeates the 'optimistic' approach to office automation, radical critics place great emphasis on its capitalistic character. Suggestions that new technology in the office leads inevitably to the taylorisation of secretarial work and that it proceeds with a juggernaut-like logic to rationalise entire office labour processes, make the technology appear just as self-propelling as it does in optimistic accounts. Critics fail to situate the introduction of new technology within the context of existing patterns of work organisation in offices, and they also fail to distinguish between different kinds of office jobs. As I shall show, what is an apparently plausible approach in fact becomes untenable in the light of these omissions. The centrality of what Marx identifies as the technical division of labour for an understanding of the implementation of technology is demonstrated in Capital Volume 1, where he shows that the crucial departure for the capitalist labour process is not the introduction of machinery into the workplace, but the fragmentation and simplification of tasks and the consequent creation of the 'detail worker'.

WP IN ENGLISH OFFICES

To investigate the effects of WP, I conducted eight case studies of organisations in Bradford, West Yorkshire, representing a cross-section of local business: a small structural engineering firm, a university department, a mail order catalogue company, a firm of solicitors, a large knitting yarn manufacturer and a replacement engine parts supplier. My research indicated that, far from WP having one uniform impact on all categories of typing-related office work, several different labour processes result, from highly integrated jobs containing a variety of whole tasks to tightly fragmented ones made up of partial operations. The distinction between secretarial and typing work, and also the different degrees of the technical division of labour applied to the latter category, were crucial for an understanding of change:

> The distinction between secretarial and typing work is crucial to analysis of the likely effects of WP in offices ... female typists - essentially machine operators - have existed as long as the machines themselves. As specialist machine operators, typists can be managed like factory shopfloor workers; this may not be true for secretaries, who perform a range of other duties in addition to typing (Arnold et al1982:60).

The technical division of labour in the office emerged as the factor causing work repetition, monotony and lack of social contact with authors. The following description of working in the building society shows how tasks can be simplified without recourse to automation, and that WP is not always responsible for such fragmentation:

> It was just copy typing all day. We had these cheque letters and we just typed the name and address of the person and the roll number, and that was all you did all the time. Everything you did, you just did the name of the person, and the address. It was just standard letters all the time.

The rationalisation of office work and the associated fragmentation of typing tasks are thus not new phenomena brought about by the introduction of microelectronic technology. As Delgado (1979) shows, the typing pool dates back to at least the 1880s. In fact, rather than organising work around fragmented operations, it seems that WP, like former office techniques, simply capitalises on processes already set in motion. In my research, I was able to identify different degrees to which the different office jobs were subdivided, and these remained fundamentally unaffected by the introduction of office automation. Some jobs, in particular the secretarial jobs, were inherently varied, involving a diversity of tasks, mobility and human contact. Purely typing jobs had less variety because they were confined to essentially one activity, while typing pool/WP centre work was the most monotonous of all, being physically restricted in addition to its limited task range, and generally involving only the simplest and most tedious categories of typing.

The study also revealed an immense diversity in the extent of skill deployed, and since the feature common to all these labour processes was the presence of WP, this variation could only be explained by reference to the technical division of labour in each case. Moreover, just as Braverman has been accused of constructing an over-romanticised conception of the skilled labourer as an ideal-type artisan/mechanic (Beechey 1982; Elger 1982), so too have current writers glorified the former position of some office workers, typists in particular. The ability to manipulate the typewriter has been confounded with overall control over the labour process; consequently, in idealising the nature of typing skill, it has, as Thompson has recognised, been possible to overestimate its destruction (1983:97). The general tenor of the labour process literature is that skill is largely based on knowledge, the unity of conception and execution, and the exercise of control by the workforce, and that the deskilling process involves these general skills being reduced to job-specific ones. However, one is forced to question the extent to which pool typists in particular ever exercised overall control based on knowledge and judgement, given their long standing situation of isolation from the source and destiny of their work. Surely their typing skills have always been 'job specific'. Many writers on WP appear to be under the illusion that such a situation is new to automated offices. In fact, it would seem that there are categories of office workers who have long been subject to the 'mental deskilling' associated with fragmented work. Crompton and Jones's (1984) work on the automation of clerical work concludes similarly that deskilling is associated more with broad, long-term processes of rationalisation of office work than being a direct consequence of the introduction of new technology. Indeed, the Bradford Word Processing Study Group itself admits that the typists upon whom it focused (those working in the first large WP installation at the local authority) were, even before the advent of WP, 'little more than living word processors' (1977:12).

Accompanying the deskilling trend attributed to WP by its critics, is the drive by management towards ever-increasing working intensity and the elimination of the non-productive 'pores' in the working day. First, it is argued, WP is used to impose increasing levels of work on dramatically depleted workforces. Secondly, in deskilling the office worker and removing from her the control which enables her to regulate her working patterns, it is used to pace her, eliminate porosity and consequently heighten the effort she expends. Thirdly, an assumption is made that increased productivity has crucial implications for the intensity at which a typist works and therefore for the quality of her working day. The benefits to management imply deleterious effects for the worker and the two are written about as if they are interchangeable. In fact, a distinction has be to made between what Baldamus (1961) refers to as objective 'work realities' and subjective 'work deprivations'.

My studies suggest that the phenomenon of work intensification is in fact far more complex than critics of the new technology suggest. The two most crucial determinants of the level of effort exercised by typists are 1) structural, lying in the degree to which the division of labour operates and the consequent extent of confinement of the typist to a single subdivided operation; and 2) normative, relying on managerial pressure on workers to exert themselves. WP machinery *per se* plays no significant role in this process. While, as Marx himself noted, in machinofacture the worker follows the movements of the machine, which has a self-regulating moving force, this cannot be claimed of the WP operator. Her work station, when switched on, has no fixed and unvarying rhythm to which she is subordinated. In that sense she still makes use of it, rather than it making use of her, to borrow Marx's inversion. This was clearly demonstrated by the ability of all the women I interviewed to stop and start work at any time, and particularly to take brief breaks between each item or batch of work. Those whose work was varied had porosity actually built into their working day, in the movement from one task to the next. Furthermore, despite increased productivity brought about by WP, no one was exposed to greater subjective immiseration since WP obviated the retyping of repeated material.

Claims that WP confines and paces the worker also call up images of typists as isolated individual operating units within an automated work system. Yet, as McNally (1979) has argued, one of the most isolating aspects of factory work is the prohibition of conversation by the noise and speed of the machinery, and this is not a feature of operating a WP. It is perhaps more relevant to regard porosity as lacking rather than being lost. Since the evidence suggests that WP is insignificant for the pace at which the operator worked, what must be borne in mind in every case is whether or not porosity is a feature of that particular job before its introduction. Only where a flow system of work prevails, whereby instead of moving about the workplace to the different tools, to fetch raw materials such as stationery and to return or discuss work with bosses, the typist remains fixed while the materials flow to her, does office work begin to assume some of the characteristics of the assembly line.

Pessimistic approaches to office automation find their conclusion in the notion that underlying the processes of fragmentation, deskilling and intensification of typing is the transference of control over the labour process into the hands of management. This idea follows the tradition of labour process theory, particularly that which is strongly rooted in Braverman (Zimbalist 1979; Cooley 1980), which treats such processes as the means by which management secures dominance over the labour process. It is implied that control is derived automatically and completely through the use of the machinery. Some of the case studies which I conducted led me to question whether a deliberate rationalisation strategy was being pursued at all by management, while other firms clearly adopted alternative practices to keep labour under control and costs down which did not involve the taylorisation or fragmentation of the work.

In the first group, there were companies which were positively inefficient in their use of office automation, using WPs simply as replacement typewriters, not training staff to use the machinery and in one case actually leaving the WP standing idle while typists bashed away on typewriters. Other companies, meanwhile, clearly had no need of work rationalisation strategies. Female office workers in the north of England are very low paid; this, rather than Taylorism, often seemed to be the means by which labour costs were kept down. The recession and the constant threat of unemployment, felt especially severely in this region, act as a powerful impulsion on women workers to accept their situation and affords individual capitals control over them which need not necessarily be exercised on a day-to-day level.

Furthermore, even where Taylorism was management's preferred strategy, this was a function of the work organisation and the approach of typing supervisors, and operated totally independently of office automation. Direct control strategies were generally much more in evidence in financial institutions such as building societies and insurance companies, suggesting that here valorisation was the prime managerial imperative, less urgent a drive in other, particularly public sector, organisations. For example, at the building society, every second of the working day - tasks, movements, time spent on tasks, even physical space - was governed by management and here Taylorism was at its most forceful. Strict rules and regulations governed all conduct at work:

> We weren't allowed to go and make a cup of tea (..). So you couldn't eat at your desk or drink. You couldn't smoke. You had to go into this, like a common room, like at school (..). You could stop and talk if there was nobody watching, but you had this partition, so somebody would soon see. But me and this other girl who I was next to, we had this little hole in ours, so it was alright. If anybody saw you, you would get into trouble. It was very strict.

Aichholzer et al (1988) identify a range of managerial strategies for the introduction of new technologies and the organisation of work, which they associate with particular industrial sectors such as finance and retailing, and with the different labour market positions of workers in these industries. My own work suggests similarly that the degree and type of control exercised by management over the office labour process is mediated by the kind of organisation and its requirements, its historical use of control strategies, the position of female office workers in the labour market (including the level of wages that they can command) in particular industries and geographical areas (Yorkshire having lower than average wage levels across all sectors but particularly in the private sector), and the state of the economy in general.

In this paper I am suggesting that there is no inevitable work intensification trend accompanying the introduction of new office technology; the uneven impact of new technologies has also been noted by Littek (1986) in his study of technical and commercial workers in Germany and Daniel (1987) in his comprehensive survey of technical change in a range of workplaces. If this is the case then there are, at least in theory, opportunities for influencing the content of women's office jobs in ways which secure improvements in their daily experience of work, and utilise their human skills and abilities. Central to 'good' job design based on these criteria, is the development of task variety within an office job. This in its turn lends a job all the features which make it interesting and acceptable for the office worker. Variety gives her choices about how to organise her work and when to perform different tasks, responsibility for a complete work process or item which she sees through from beginning to end and the context of which she understands, the ability to pace herself in the movement from one task to the next, in general using her own discretion and having control over her work. She is also generally afforded some measure of physical mobility and social contact with other workers - features which are absent from jobs in which the typist is tied to a desk or in a typing pool, and which are crucial for injecting interest into jobs.

The consultation which occasionally takes place between bosses and office workers - generally secretaries - about the kinds of machinery to buy, may present opportunities for making arguments for the retention/expansion of those features of the secretarial labour process which make it interesting and demand human intellect. This also means office workers being alert to managerial attempts to fragment their jobs and equipping themselves with arguments about why such measures are not necessarily as efficient as it would seem (employee discontent, absenteeism, turnover, ignorance and consequent inability to work properly). In Britain, there have been valuable attempts to develop partipative approaches to office system design, providing evidence against the notion that 'semi-skilled' workers are incapable of contributing to this process (Mumford 1983; Murray 1987).

However, where new machinery is already present or jobs are already divided (in a WP pool, for example), there is less concrete opportunity for intervention, and no model of preferable forms of work organisation. Women office workers have few if any channels through which to influence the quality of their jobs: they are poorly unionised, occupy an increasingly peripheral position in the labour market and are in fact in danger of becoming completely marginalised as keyboard skills become more widespread and typing less the domain of typists. This poor bargaining position is exacerbated by the now minimal training for office work - one more channel through which the character of office jobs might be influenced is closed.

The prospects, then, for redesigning office work in the interests of office workers seem gloomy, based as they are on slim chances of consultation during the office automation process. What this paper has attempted to show, however, is that there are theoretical area for intervention, that automated office work is not hopelessly fragmented and beyond reconstruction. To identify this fact is an important first step; what remains is to develop the loopholes that exist in management's design of jobs and to find ways of building strategies for improving the daily experience of women in offices.

ACKNOWLEDGEMENTS

I should like to thank Cecila Bergström, who refereed this paper in the Women, Work and Computerisation workshop on 29 April 1988, and who made many helpful and stimulating comments.

BIBLIOGRAPHY

Advisory Council for Applied Research and Development (ACARD) Cabinet Office (1980) Information Technology London, HMSO.

Aichholzer, G., Flecker, J. and Schienstock, G. (1988) Introducing New Technology: Does Uncertainty Undermine Managerial Control Strategies? Paper presented at the 6th Annual ASTON/UMIST Conference on the Organisation and Control of the Labour Process, Birmingham, March.

Arnold, E. et al (1982) Microelectronics and Women's Employment in Britain, Science Policy Research Unit Women and Technology Series, Occasional Paper Series No 17, University of Sussex.

Baldamus, W. (1961) Efficiency and Effort: an Analysis of Industrial Administration,, London, Tavistock.

Barker, J. and Downing, H. (1980) 'Word Processing and the Transformation of the Patriarchal Relations of Control in the Office', Capital and Class, No 10.

Beechey, V. (1982) 'The Sexual Division of Labour and the Labour Process: A Critical

Assessment of Braverman', in Wood, S. (ed)

Bradford WP Study Group/NALGO Action (1977) Word Processing: A Report on the Introduction of the Automated WP Systems in Bradford Metropolitan District Council, Unpublished.

Braverman, H. (1974) Labor and Monopoly Capital, London, Monthly Review Press.

Central Policy Review Staff (1978) Social and Employment Implications of Microelectronics, London, HMSO.

Cooley, M. (1980) Architect or Bee? Slough, Langley Tecnhnical Services.

Crompton, R. and Jones, G. (1984) White-Collar Proletariat - Deskilling and Gender in Clerical Work, London, Macmillan.

CSE Microelectronics Group (1980) Microelectronics: Capitalist Technology and the Working Class, London, CSE Books.

Daniel, W. W. (1987) Workplace Industrial Relations and Technical Change, London, Frances Pinter.

Delgado, A. (1979) The Enormous File: A Social History of the Office, London, John Murray.

Department of Industry (1978) Microelectronics - the New Technology, London, DOI Pamphlet.

Elger, T. (1982) 'Braverman, Capital Accumulation and Deskilling', in Wood, S. (ed)

Harman, C. (1979) Is A Machine After Your Job? New Technology and the Struggle for Socialism, London, Socialist Worker's Party pamphlet.

Liff, S. (1988) Gender, Office Work and Technological Change, Paper presented to PICT Workshop, Bath University, 25-27 March.

Littek, W. (1986) 'Rationalisation, Technical Change and Employee Reactions', in Purcell, K. et al (eds)

Marx, K. (1954) Capital, Volume 1 London, Lawrence and Wishart.

McNally, F. (1979) Women for Hire: A Study of the Female Office Worker, London, Macmillan.

Mumford, E. (1983) Designing Secretaries, Working Paper, Manchester Business School.

Murray, F. (1987) Women, Trade Union Technology Bargaining, and the Need for a New Strategy, Human Centred Office System Project working paper, Sheffield City Polytechnic.

Purcell, K. et al (eds) (1986) The Changing Experience of Employment: Restructuring and Recession, London, Macmillan. Thompson, P. (1983)

The Nature of Work: an Introduction to Debates on the Labour Process, London, Macmillan.

Wood, S. (ed) (1982) The Degradation of Work? Skill, Deskilling and the Labour Process, London, Hutchinson.

Zimbalist, A. (ed) (1979) Case Studies on the Labor Process, London, Monthly Review Press.

BIOGRAPHICAL NOTE

Juliet Webster studied at the University of Bradford, where she completed a PhD thesis on the impact of microelectronic technology on secretarial work. She is now a research fellow on the Programme on Information and Communication Technologies at the University of Edinburgh, where she is carrying out a research programme on the integration of factory and office automation and the transformation of administrative labour processes.

Correspondence address: Juliet WEBSTER, Research Centre for the Social Sciences, University of Edinburgh, 56 George Square, Edinburgh EH8 9JU, Scotland.

Women, Work and Computerization: Forming New Alliances
K. Tijdens, M. Jennings, I. Wagner, and M. Weggelaar (Editors)
Elsevier Science Publishers B.V. (North-Holland)

POLARIZED PATTERNS: EFFECTS OF COMPUTERIZATION ON WOMEN'S CAREER OPPORTUNITIES

Angelika Volst and Ina Wagner
Institut für Praktische Informatik
Technische Universität Wien
Austria

This paper combines an analysis of the role of computers in reinforcing the flexibility of work organizations with a discussion of changes in women's participative behaviour on the labour market. The main argument put forward is that both, employers strategies and women's patterns of access to the labour market, result in an increasing polarization of career opportunities for women. Computerization leads to a precarization or even elimination of many "transitory" jobs, thus widening the gap between stable employment in qualified positions that require intellectual and interactive flexibility and casual employment in areas of low-skilled, labour-intensive work. Strategies are spelled out that take account of the increasingly unequal life and work situations of women on both "ends" of the polarized spectrum.

1. INTRODUCTION

Recent research on women's opportunities on the labour market draws attention to questions of inequality among working women. There appears to be a widening gap between those women who enjoy the benefits of stable employment offering career opportunities as well as full protection by collective bargaining procedures and social insurance legislation; and those who have to cope with instability and precarity in their working lives. This does not only include high risks of getting unemployed and insufficient social insurance protection but also levels of income that do not guarantee proper living conditions. This polarization between types of employment open to women is aggravated by their unequal chances of simultaneously managing a career and parenthood. Women who have access to stable employment are more likely in a position to pay for child care and domestic work (or share part of it with their partners); whereas women who find themselves confined to unstable low paying employment conditions often also have to cope with small family budgets and have less opportunity to alleviate the burden of homemaking. Two factors mainly contribute to this polarization process: changes in the organization of work for which computer-based technologies are instrumental as well as changes in women's career orientations. This paper uses a twofold approach:

* First, the developments that reinforce the inequality among working women are examined, with special attention to the role of computerization in this process;
* secondly, strategies for improving women's career opportunities on "both sides" of the emerging polarized pattern are discussed.

2. ORGANIZATIONAL FLEXIBILITY - THE ROLE OF COMPUTERS

At present, a transition from one dominant type of work organization to a new organizational paradigm can be observed. In the period from 1900-1970 the "big business" prevailed - in industry, in banks and insurances,in public administration, based on a high degree of bureaucratic control as well as a progressive "linearization" of work processes. Growing difficulties of such organizations to cope with rapidly varying, unstable environments make employers seek for more flexible ways of organizing work that allow for a higher degree of interactive complexity, capable of absorbing risk and uncertainty (Perrow 1986). Computer-based technologies support this process of organizational adaptation.

The thrive for increased organizational flexibility is at the core of most advanced automation projects. They aim at achieving a better tuning of processes between organizational units. This is achieved by establishing integrated data- networks that allow to include information generated in different parts of the organization into the day-to-day informal practices of workers. This computer-based tuning of processes and decisions increases the interactive complexity of organizations, while broadening also the basis for management planning and labour control. Computer-based integration is only at its beginning. A FRG survey from 1986, for example, shows that a quarter of industrial enterprises had installed computers in three or four functional areas - office, production - related services, manufacturing, transport or assembly; but only 9% had successfully introduced one network (Nuber et al 1987). Still, case-studies identify some basic organizational strategies for coping with increasing levels of interactive complexity. They point to a central dilemma of automation: There are limits to standardization as a means of solving problems of organizational complexity. Even computer-based procedures require a considerable degree of competent human intervention and interpersonal cooperation for processing system failures, unplanned variation and unintended consequences (Berger/Offe 1980). Thus, the main organizational challenge to be met in automation projects is to find the appropriate mixture of standardization and technical integration with personal input and social integration. Managers pursue different strategies to cope with this organizational problem (Kolm et al 1988):

* Controlled intuition: In order to master the amount of uncertainty inherent in the day-to-day functionings of a work organization, management seeks to make efficient use of workers' tacit knowledge and experiences in problem- solving. Space for workers' intuition is deliberately built into many computer-based procedures. At the same time computerization provides a powerful tool of unobtrusive control of the ways intuition is brought into the work process.
* Distributed intelligence: The increasing number of interfaces between organizational units calls for the establishment of decentralized emergency-handling structures. Hence, automation frequently goes along with strategies such as creating organizational overlap, combining different types of expert knowledge or strengthening the opportunity for informal cooperation.
* Cooperation and coalition: To bring about organizational change management actively seeks to establish cooperative networks on all levels of the organization. Such networks serve two purposes: to guarantee the transfer of knowledge necessary to establish computer-based systems; to minimize the conflicts that necessarily go along with shifts in organizational power structures.
* Standardization: Tayloristic methods of reducing complexity are mainly pursued in those areas of work within an organization for which the number of interfaces can be minimized. At these workstations human input is severely reduced and largely determined by the computer system.
* Left-over pockets of labour intensive work: These are preserved or even extended in areas of work where automation is either technically not feasible or economically not profitable.

According to these strategies, one can find - often in one and the same organization - a highly polarized pattern of chances for learning, advancement and job security. Within this pattern three types of jobs can be distinguished:

* QUALIFIED POSITIONS that combine controlled intuition, distributed intelligence and co-operativity (e.g.highly sensitive production tasks, sales operations in industrial firms, client interaction in banks and insurances, work in edp - organization and other support departments) - these usually offer the kind of stability of employment that although involving job mobility allows to build up a career.
* TRANSITORY JOBS that are subject to extensive automation (e.g. back office work including typing, routine commercial operations in industry, mass-processing in banks and insurances, routine production tasks) - dependent on managements automation strategies and union influence, some of these jobs become more diversified and integrated into complex task patterns; others become increasingly routinized and will be eliminated in the long run.
* Pockets of LOW-SKILLED LABOUR-INTENSIVE WORK (e.g. data inputting, many types of assembly-work, personal services in catering, cleaning, supermarkets, laundries) - in these areas a growing precarization of employment can be observed.

In the course of diffusing organizational flexibility "transitory jobs" have proven to be particularly vulnerable. They often are in the focus of political struggles between unions and management over questions of job design. Issues at stake are the restrictivity of jobs (in terms of number of tasks involved, degree of standardization, location), worker's inclusion in (or exclusion from) electronic networks and access to data, depth of training offered, work loads and extent of electronic control. The solutions that are negotiated or enforced will determine whether women in "transitory jobs" will join the increasing number of those who stream into the pockets of low-skilled labour intensive work that exerts a high psychological toll in terms of risk and precarity.
Research indicates that these "forgotten areas" of women's work are growing. A German study (Möller 1986) shows that in some sectors, above all retail and miscellaneous services (including catering), the percentage of workers in precarious employment situations varies between 50 and 100, with women making up for the major part of these employees. Austrian data (Goldberg 1986) single out part-time employment in low- skill commodity - related services as well as in low-skill person - related services as the main areas in which female employment is increasing. The growth of jobs in this area depends on employers' choice between investment in costly technology or use of low salaried labour.
At the same time the career opportunities of women in computer-based organizations broaden. An increasing number of qualified clerical workers (as well as secretaries) join men in the struggle for resources such as edp-knowledge, personal computers, access to data networks and specialized training in this highly competitive field.
In order to get at a fuller understanding of the widening of the gap between gainfully employed women, it is also necessary to analyse changes in women's own participation strategies on the labour market.

3. PATTERNS OF PARTICIPATION IN THE LABOUR MARKET

Women's own strategies of participating in the labour market have undergone considerable changes. There are three developments in particular, that contribute to these changes:

* women's participative behaviour indicates a growing resistance against economic crises;

** more women are willing and able to invest time and energy into a market career in high-skill positions;
*** women make systematic use of their "trained flexibility" on the labour market.

Women's participation on the labour market has grown steadily through the last two decades; contrary to all expectations and ideology, in spite of recurrent economic crises. It still holds true that the occupational biographies in particular of married women with children is "dromedar shaped". In Austria, for instance, the frequency of career breaks has even increased, due to improved maternal leave arrangements; but periods of home - time are becoming markedly shorter. At the same time unemployment rates in the female labour force are increasing. Margaret Maruani (1986) interprets this contradictory phenomenon of simultaneously growing participation and unemployment as two sides of the same coin - a demonstration of women's willing to stay in the labour market.
This development casts doubts on the image of women as an industrial reserve army: Women do get unemployed but are less easily discouraged from looking for a new job. But also employers' strategies in times of economic crisis seem to change from simply laying workers off to partly absorbing market fluctuations by shifting to part-time work arrangements (Willms-Herget 1987). Women in many areas may still be the prime victims of economic crises; but the mechanisms and reactions are changing on both sides.

Both tendencies together account for the fact that, like in the case of Austria, part-time arrangements did not only grow in low-skilled areas like cleaning, catering and simple clerical work, but as well in qualified clerical occupations, the semi-professions and the professions (Goldberg 1986). Empirical data on participation reveal the pattern of a patchwork career (Sichtermann 1987) for a growing number of women, characterized by high job mobility and periods of non- participation. The price to be paid by women with patchwork careers is precarity, including a considerable depreciation of incomes. Miller (1987) argues that years of home-time account for a large part of the gender-differential in earnings.

At the same time, the number of women who are in a position and willing to adapt their lives to the exigencies of a full time career in qualified positions is constantly growing. Numerous studies on women managers (e.g.Kanter 1987, Sessar- Karpp 1986, Mc Broom 1985) or women in technical occupations (engineering: Kvande/Rasmussen 1987, craft work: Wagner 1986) document the career paths of women seeking entrance into hitherto closed male arenas. Our own research on automation in the service sector adds to this evidence the example of secretaries or women in skilled clerical positions who informally or as project leaders grow into key participants in organizational change (Volst/Wagner 1988 a and b). This influx of women into responsible positions within work organizations does not necessarily imply that they make it to the top. Also, estimates of how fast these opportunities for women are growing vary considerably. In addition, there are contradictory findings about the price to be paid by women who enter a highly competitive field. Still, it seems that in particular employers operating under tough economic conditions find it increasingly costly to discriminate against highly qualified women.
The conclusion to be drawn from this pattern is that there exists a certain overlap of employers' strategies to enhance organizational flexibility and women's patterns of access to the labour market.

This coincidence is far from accidental. It can be interpreted in terms of women's "trained flexibility". This denotes the fact that women historically have been forced to acquire a considerable amount of flexibility. This is well documented: Working women in general have had to accept much looser connexions between training and type of work than men who tend to occupy the strongly certified positions (Milkman 1983);

women traditionally have to cope with the double obligations of a market career and family life; they are usually the ones who carry the burden of shrinking family incomes in times of economic crisis.

It appears that women now make more extensive use of their acquired flexibility on the labour market while the demand for flexibility is increasing. We should be aware however, that different types of work draw upon different aspects of a person's flexibility. In situations of precarious employment a high premium is set on the ability to cope with variable working - time arrangements, high risks of getting unemployed and insufficient social insurance protection. Success in the more qualified areas of work that we described as a result of computerization and organizational change is based on a different type of flexibility. In these areas the social and intellectual mastery of non-routine situations and high interactive complexity gain importance. This gives women who are willing to strive for highly qualified positions more opportunities.

TYPE OF EMPLOYMENT	ARENAS OF INTER-VENTION	PARTICIPATION - STRATEGIES
Type I QUALIFIED POSITIONS stable employment	competitive field for individual women; strategies on plant-level	+investment of time & energy in high formal qualifications and a full-time career +high intellectual and interactive flexibility
Type II TRANSITORY JOBS growing precarization	main area of political struggle between unions and management; strategies on plant and union/branch level	
Type III POCKETS OF LOW-SKILLED LABOUR-INTENSIVE WORK precarious employ-ment	forgotten areas of women's work; strategies on a more general social-policy level	+lack of formal qualifications +unshared family responsibilities +trained and enforced flexibility in coping with precarity +patchwork-careering

4. STRATEGIES

The arguments presented so far can be interpreted in favour of designing strategies that take account of the highly unequal life and work situations of women.

Low-skilled labour-intensive work: For women at the lower end of the polarized pattern strategies that aim at reducing the degree of precarity involved in a patchwork career seem of paramount importance. There are four points of leverage for such strategies:

* income-raising policies,
* increasing opportunities for self-regulation of working time,
* improvement of social legislation protection,
* upgrading (rather than adaptation) of skills.

Contrary to union practices in some countries (e.g.in Austria), extra effort should be made to raise wages in the lowest income groups (rather than negotiating for linear increments). Incomes that do not guarantee proper living conditions are typical of most precarious employment situations. Low income jobs are the main source of poverty traps for the unemployed. In Austria, 40% of all women receiving unemployment benefits and 50% of those receiving "Notstandshilfe" had to cope with incomes below a poverty margin of AS 4.000,- (BMAS,1987). At the same time, low earnings prevent many single mothers from seeking employment, since the additional costs involved (childcare, transport, clothes) often exceed the gain in income. In the long-run, a policy of raising the low incomes might even result in employers reconsidering extreme forms of marginal, part-time arrangements as too expensive to pay off economically.

A second set of urgent measures concerns the ways working- time arrangements are settled. In Austria, employers' willingness to concede workers' co-determination in fixing working hours has visibly weakened (Stemberger 1987). At least 4000 employees, in particular in the retail business, are forced into open arrangements (with neither a fixed length of working-time nor a stable allocation over the week), allowing employers to dispose over their workers' time at short notice. The problems involved in particular for women in organizing work and family life, makes opportunities for increased self-regulation of working-time arrangements a paramount target.

Findings on the effects of training on women in precarious employment are ambiguous. Empirical evidence suggests that lack of formal qualifications (such as a final school certificate) acts as an increasingly significant barrier to finding employment; also in low-skilled areas of commodity - related and personal services (e.g.Gottsleben 1987). Further training, on the other hand, does not necessarily increase opportunities on the labour market.

This may explain for the reluctance of many low-skilled women to participate in training courses. Unless training offers certification of skills that are in high demand it is of limited value in a labour-market that does not repay training in terms of income and more interesting or less strenuous work.

Transitory jobs: The effects of training are particularly ambiguous in the case of those women in "transitory jobs" that are directly affected by computerization. Users of computer- based technologies among women (working as data-processors, typists, routine clerical workers, cashiers etc) on average have higher formal qualifications than non-users and participate more frequently in training activities (BIBB/IAB 1987). But the type of training offered is usually restricted to short-term adaptations to the new technologies that do not result in job-upgrading nor in promotion. To be integrated into the evolving areas of qualified work these women would need to bridge a widening qualification gap. They also would need more time during the regular workday for experimenting with the computer and for taking initiatives in the redesign of their jobs.

In the case of transitory jobs, special attention has to be given to strategies on the level of system design that influence the division of labour. The dynamics of automation projects suggest the existence of a considerable "grey area" of negotiable design questions. Degrees of computer-control of work tasks vary; so does the scope of tasks involved in a certain job and their nature (Volst/Wagner 1988). Integration into mixed work teams appears to increase the chances for learning and advancement, while working in pool arrangements deteriorates them (Gottschall et al 1984). These possible gains in opportunity may seem moderate, still they do matter for the women concerned. More and more highly qualified women play an active role either in project groups that monitor and to a certain degree influence the design of systems within a firm or as external consultants in technical and organizational questions. These women need to be made more aware of the opportunities to shape the "grey area" in the design of many transitory jobs to the benefit of less qualified women within the organization.

Qualified positions: While precarity-reducing strategies are of prime importance for women on the lower end of the polarized pattern, strategies that aim at easing the way into key roles in the process of organizational adaptation are more relevant for those on the upper end. Some of these strategies can be derived from the patterns of behaviour of successful women. In automation projects within the service sector these are to be found mainly in qualified clerical positions with managerial functions or, in some cases, also in secretarial positions. A particular success - generating pattern can be observed. In many cases computerization triggers off a process of self-selection into key positions. Managers or project leaders encourage individual initiatives in integrating the computer into one's own work. Images of "floating opportunities" are created which an active employee has to seize and make efficient use of. Those workers who - frequently beyond the official working time - actively seek for creative solutions of non-routine problems and are available in emergency situations, gain crucial terrains within the organization (Volst/Wagner 1987, Flecker/Volst 1988). Apart from this general willingness to perform beyond established norms and routines, there are certain specific success-generating strategies:

* As the need for controlled intuition increases with interaction complexity, social skills gain in significance relative to process-specific skills and specialized knowledge. Such skills that e.g.secretaries have traditionally drawn upon in organizing an office, become more visible and prestigious as soon as they are associated with computers. Employers' awareness of women's potential contributions seem to grow with the adoption of computer-based networks; not least because it fits the traditional image of women's specific work "habitus".
* The principle of distributed intelligence leads to less tightly coupled forms of organization and reinforces the development of local problem-solving capacities. This offers new opportunities for women who are usually not well established within the formal hierarchy of the organization. This tendency is confirmed by the findings of Kvande/Rasmussen (1987) that women engineers find much better conditions for career development in dynamic than in static organizations.
* Management's need for cooperation and coalition leads to the establishment of informal structures that ease organizational change. Automation is monitored by schemes such as a project group, an end-user service centre or the establishment of edp-responsibilities within departments. The benefits to be drawn from participation in such structures are manifold. It gives access to a more intimate knowledge of the system and its present and future potentials. Frequently, direct involvement in the introductory process opens up additional resources in terms of numbers of terminals at one's disposal or a personal computer of one's own. Finally, these informal structures often are converted into formal ones - new departments with specific functions for the company's headquarters. Being integrated into informal pattern of cooperation and coalition in these cases means moving with shifting power centres (Volst/Wagner 1988).
* Training is also a key variable for women in stable employment and qualified positions. The scope of training investments necessary for promotion rises

dramatically in computer-based organizations. Under the heading "further training for professionalization - not as compensation" Baethge/Oberbeck (1986) give the example of a typical qualified training course that involves 200-00 hours of preparatory work plus seven months of full-time learning, usually outside one's home town.

From this short listing of areas for strategic action, it becomes clear that the growing inequality among working women calls for the design of specific strategies that meet unequal needs. Summing up, the distinction between precarity-reducing strategies and measures that enhance women's roles in organizational change seems important. If it is possible at all to single out a key problem group for action, we would at present choose women in "transitory" computer-based jobs; such as routine clerical work, typing, data-inputting or certain types of assembly work. Women in these jobs need access to high quality training programmes offering a visible, certified upgrading of skills. Only participation in such extensive further training can safeguard these women from being pushed into areas of precarious employment.

BIBLIOGRAPHY

BAETHGE M., OBERBECK H.: Zukunft der Angestellten. Neuw Technologien und berufliche Perspektiven in Büro und Verwaltung (The future of white-collar work. New technologies and career perspectives in office occupations). Frankfurt 1986.

BERGER U., OFFE C:: Das Rationalisierungsdilemma der Angestelltenarbeit (The rationalization dilemma in white-collar work). In: Geschichte und Gesellschaft, Special Issue No 7 (1980), 39-59.

BIBB/IAB - Bundesinstitut für Berufsbildung/Institut für Arbeitsmarkt-und Berufsforschung (Hg). Neue Technologien: Verbreitungsgrad, Qualifikation und Arbeitsbedingungen (New technologies: Diffusion, skills and working conditions), Berlin/Nrnberg 1988.

BMAS - Bundesministerium für Arbeit und Soziales, Frauenreferat. Tendenzen der Frauenbeschäftigung. Gleichbehandlung ist das Ziel (Trends in female employment. Equal handling is the goal), Nr.13, Wien 1986.

FLECKER J., VOLST A.: Kontrolle der Kontrolle - Formen und Folgen der Steuerung betrieblicher Rationalisierungsmaanahmen (Control of controlling - modes and effects of monitoring rationalization processes with firms). In: Österreichische Zeitschrift für Soziologie 1 (1988).

GOLDBERG CH.: Wechselwirkungen und Effekte von Tertiarisierungsprozessen, "Bildungsboom" und Krisenauswirkungen auf die Beschäftigungschancen von Frauen, (Interactive effects of tertiarization, education "boom" and economic crisis on female employment), Institute for Advanced Studies, Wien 1986.

GOTTSLEBEN V.: Randgruppen in der zertifizierten Arbeitsgesellschaft? Zur abnehmenden Bedeutung der nicht formal Qualifizierten (NFQ) am Arbeitsmarkt (Groups at the periphery of a formally qualified workforce? The diminishing access of those without formal qualifications). In: Mitteilungen aus der Arbeitsmarkt- und Berufsforschung 20/1 (1987), 1-4.

KANTER R.M.: Men and Women of The Corporation, New York 1977.

KOLM P., VOLST A., WAGNER I.: Konflikt und Innovation in computerunterstützten Organisationen (Conflict and innovation in computer-based organizations), Schriftenreihe der österreichischen Computergesellschaft, Bd 41, Wien 1988.

KVANDE E., RASMUSSEN B.: Women's and Men's Organizations: On Female Engineers. Contributions to the Fourth GASAT Conference, Vol.2, Ann Arbor Michigan 1987, 160-167.

MARUANI M.: "Homme sans famille cherche femme sans travail": les esquives de la sociologie face à la montée de l'emploi féminin, (Man without family seeks woman without work: sociology's escapes vis à vis the growing labour market participation of women).In: Socius 2/3, 1986, 16-23.

GOTTSCHALL K., MICKLER O., NEUBERT J.: Auswirkungen technisch-organisatorischer Veränderungen auf Routinetätigkeiten in den Verwaltungen der Privatwirtschaft (Effects of technical-organizational change on routine clerical tasks in the private sector), Institut für Sozialforschung, Göttingen 1984.

Mc BROOM P.: The Third Sex: The New Professional Women, New York 1985.

MILKMAN R.: Female Factory Labor and Industrial Structure: Control and Conflict over "Woman's Place" in Auto and Electrical Manufacturing. In: Politics & Society 12/2 (1983), 159-204.

MILLER P.: The Wage Effect of the Occupational Segregation of Women in Britain. In: The Economic Journal 97 (1987), 885-896.

MÖLLER C.: Ungeschützte Beschäftigungsverhältnisse. (Unprotected employment) 23. Deutscher Soziologentag, Hamburg 29.9.-2.10.1986.

NUBER Ch., SCHULTZ-WILD R., FISCHER-KRIPPENDORF R., REHBERG F.: EDV-Einsatz und computergestützte Integration in Fertigung und Verwaltung von Industriebetrieben (Edp and computerbased integration in production and administration of industrial firms), Institut für Sozialforschung. München 1987.

PERROW Ch.: Complex Organizations. A Critical Essay. 3rd Edition, New York 1986.

PULLMAN C., SZYMANSKI S.: The Impact of Office Technology on Clerical Worker Skills in the Banking, Insurance and Legal Industries in New York City: Implications for Training. The Labour Institute, New ork 1986.

SESSAR-KARPP E.: Beruflicher Aufstieg von Frauen im öffentlichen Dienst (Careers of women within public administration) In: L.Unterkircher, I.Wagner (eds).Die andere Hälfte der Gesellschaft, (The other half of society), Wien 1987, 184-193.

SICHTERMANN B.: Frauen Arbeit, (Women's Work), Berlin 1987.

STEMBERGER G.: Working Time in Austria: Reduction and Flexibilization under the Auspices of Social Partnership. Int. Symposium on Working Time, Brussels, April 2-3, 1987.

VOLST A., WAGNER I.: Organisatorische Innovationen beim Einsatz computerunterstützter Techniken. Symposium "Technikbewältigung - ein Vergleich zwischen Frankreich und Österreich" (Organizational innovation and computerization. Coping with technology - a comparisation between France and Austria), Wien 11.-12.12.1987.

VOLST A., WAGNER I.: Inequality in the Automated Office. The Impact of Computers on the Division of Labour. In: International Sociology, June 1988.

VOLST A., WAGNER I.: Arbeitsorganisation und Technikeinsatz. (Work organization and the use of technology) Project report, Institute for Work and Technology, Wien 1988.

WAGNER I.: Equal Talents - Unequal Measures. Skilled Women Workers in Austria's Metal Industry. In: European Journal of Engineering Education 11/3 (1986), 321-329.

WILLMS-HERGET A.: La situation des femmes sur le marché du travail en RFA. (Women's labour market situation in the FRG) M.Maruani, E.Reynaud (eds): France-Allemagne: Débats sur l'emploi, (France-Germany: A debate on employment) Paris 1987, 230-242.

Women, Work and Computerization: Forming New Alliances
K. Tijdens, M. Jennings, I. Wagner, and M. Weggelaar (Editors)
Elsevier Science Publishers B.V. (North-Holland)

NEW TECHNOLOGY AND GENDERED DIVISION OF LABOUR

Päivi Korvajärvi
Work Research Center
University of Tampere
Tampere, Finland

The labour market in Finland have very strict vertical and horizontal segregation according to the gender. An analysis based on empirical data on clerical work shows that in spite of the high speed of the introduction of new technology the gendered division of labour seems to remain intact. The hierarchic organizational structures are the strongest obstacle for female clerical workers to make use of new technology to improve their position in the organization. It is suggested that the focus of research should be less in technology and its impacts as such and more in the gender system and its cultural features in the structures of wage labour.

1. INTRODUCTION

There is considerable disagreement among researchers on the question of how the position of female white-collar workers is and will be affected by the process of office automation; some of the empirical results available even suggest completely opposite trends in development. The reason for these differences can be found in the theories and perspectives adopted in studying the consequences of new technology for the position of women and the gender division of labour.

Many studies deal with the shopfloor workers as if they were the objects, the victims, who themselves do not take any active role in the process of the introduction of the new technology. But in fact, female clerical workers are prepared and willing to bear responsibility for the development of their work. There can be no doubt that both in terms of occupational skills and in terms of work orientation clerical women have the resources to improve their position, but these resources are not effectively utilized at the organizational level.

In this article I have three main concerns: First, I shall discuss the different theoretical perspectives through which researchers have approached the problem- area and consider their relevance to describing the position of women. Second, I shall present empirical research results from white-collar workplaces in Finland. And third, I am concerned with the tenacity of the gendered division of labour and with the chances and resources of women to overthrow this structure.

In the empirical part of this study, my specific concern is with Finnish society its structures and culture and Finnish female white-collar workers at their workplaces. It is reasonable to argue that in international comparison Finnish women in general have a relatively good and equal position both on the job and in society at large. In Finland and the other Nordic countries women are to a large extent integrated into the labour market,

and responsibility for the care of children, the elderly, the ill and the disabled is mostly assumed by the welfare state (e.g. Siim 1987). Today, women represent almost half (47 % in 1980) of the gainfully employed population in Finland. Over 80 % of women participate in the wage labour in Finland. It is worth to note that the great majority of them have full-time job (88 %). Only one of ten Finnish women is a housewife.
However, within the labour market there is a very rigid horizontal and vertical division of labour: there are men's jobs on the one hand and women's jobs on the other, and at the same time women are rarely engaged in managerial tasks, in spite of their high level of education. Almost half of the gainfully employed women are so-called lower white-collar employees, and in this category more than 70 % are women (Position of Women 1984).

2. PATTERNS OF ANALYZING WORK ORGANIZATION

In studies concerned with the consequences of office automation, new technology is often regarded as some sort of a determining variable: it it assumed that the application of new technology has certain consequences for the content of work and for the position of women in that particular context, that new technology is a cause which has certain effects leading in a certain direction. Technological determinism seems to be at least an implicit starting-point in a number of studies. However, psychological and sociological studies of work have clearly disproved these assumptions, and suggest that the relationship is much more complicated (Heiskanen 1984, Rantalaiho 1977). In addition, it has been shown that there is a very close connection between the content of work and the organization of work, which seems to be a significant factor in improving the position of women (Gunnarsson and Ressner 1983).

Studies of work organization have tended to concentrate on the actual labour process which is divided between the machine and one or more workers, i.e. on what might be regarded as the objective, concrete flow of work. It is this concept of work organization which lies behind the argument that the introduction of new technology is Taylorizing women's work into smaller and smaller fragments, and that control of the work process is becoming more and more detailed. There is a tendency for women's work to become progressively deskilled, which further undermines their importance in the work organization. These research results which suggest that women's position is becoming increasingly marginal as a direct consequence of new technology are more or less explicitly based on Braverman's thesis (1974) of the degradation of work.

On the other hand, there is a school which argues the exact opposite and says that the introduction of new technology has a very positive influence on the qualifications of clerical labour: women now master certain aspects of the work process that their superiors do not. With their new qualifications, women gain a better position in working life and also better prospects of advancement in their career. (Karl and Ohm 1982).
There is also an approach which analyses changes in the content of work in relation to the dimension of a "good job" in more explicit terms. It is argued that a high level of autonomy on the job, social contacts, successful cooperation with colleagues, an optimal level of work load, and the opportunity to develop one's skills are factors that promote the worker's well-being (e.g. Frese 1978, Gardell 1971, Gardell 1976). This approach is therefore concerned with what work is like from the point of view of the individual's personality. Does it allow the individual to expand the sphere of her own activity, or does it impose restrictions? To what extent does it enable her to master and control the immediate environment?

A major deficiency of this approach is that it ignores the social world and the social structures within which people act and do their work. The labour process and the

various objective features of work content pose at least two important problems in studying the work of white- collar employees and women.
First, people do not work in social vacuum. Even when the organization of work is conceptualized as a formal division of labour between the worker and the machine, it concerns people as a group, as a chain, as some sort of a social construction. However, the different dimensions of the content of work - in the formation of which the organization of work plays a decisive mediating role - are individualized, concern individual workers; the content aspects are most clearly visible when we are observing one individual worker and one specific work task.
Second, the concept of work organization and the dimensions of work content have primarily been elaborated on the basis of empirical studies concerned with male industrial labour. However in many cases it is obvious that the dimensions of this sort of labour are not directly applicable to or adequate for purposes of analysing the content of office work, in which various social skills and organizing abilities tend to be more important: perhaps the most essential, albeit tacit, skill and requirement of office work is social interaction in which workers create continuously flowing procedure of office work (Wynn 1979, Lie and Rasmussen 1985, Goodman and Perby 1985, Rantalaiho 1985, Rantalaiho 1986, Suchman 1983). In other words, in office work it is what people do that matters most, how they orientate themselves. The content of office work is a kind of totality, an integrated understanding of how one is supposed to act and behave in a given situation (Olerup 1985, Rantalaiho 1986).

On this basis it has been argued that the consequences of the introduction of new technology for the position and work of female clerks are not particularly dramatic. Instead, there seems to be an extremely deep and broad gap between different visions, predictions and reality. As a rule new technology is introduced in such work organizations which in formal terms remain completely unchanged. At the informal levels of the work organizations, clerical workers have to manage on the strength of their own individual motivation. This is the conclusion proposed by many studies carried out in the Nordic countries which have focussed on the concrete work activities of clerical employees (Hedberg et al. 1987, Lie and Rasmussen 1983, Rantalaiho and Korvajärvi 1985). The personal goals and ambitions of female clerical workers and their structural positions intersect in the work organization.

3. THE INTRODUCTION OF NEW TECHNOLOGY AND THE ORGANIZATION OF WORK

The following analysis describes the kind of activities and strategies that office automation provokes in female clerical workers within their work organizations. More specifically, the question we attempt to answer is how are their goals and their structural positions affected by the introduction of new technology?
The research material was collected in 1985-1986 at four public-sector workplaces (hospital, social welfare office, employment office, and tax office) and five private-sector workplaces (two large industrial companies, department store, a bank, and the head office of an insurance company). The emphasis in data collection was on the so-called lower white-collar employees, but other white-collar groups are also represented in the sample. Various methods were used in data collection: questionnaires (424 respondents), observation of work, work organization and the workplace culture (37 observation days), interviews of clerical workers (31 interviews), interviews of their superiors (20 interviews), and discussions with clerical workers, both individually and in groups. (Korvajärvi et al. 1987).

Office automation seems to be advancing at an accelerating rate in Finland: at all the workplaces included in the study, new equipment and innovations were being introduced all the time. The introduction of new technology was not a once-for-all event with a beginning, a transitional stage and an end: after the first of the new equipment arrived, there were various changes which seemed to start a continuous process. Nevertheless as far as the daily work and work orientation of female clerical workers were concerned, there were no dramatic changes in the organization of work.

In a closer study of the process of office automation and the different stages in this process, we can see the hierarchic structures involved and women's position within these structures. In planning and decision-making, clerical women do not even have access to all the relevant information. They can discuss the matter amongst themselves and they are sometimes asked to state their opinion on certain important issues, but they have absolutely no say in the final decision. For example, at all workplaces we were informed by the management that they had plans to carry out changes at the workplace, but according to our questionnaires only half of the female clerical workers said they knew with certainty about the changes. Others had heard rumours or believed that no changes were going to take place.

Typically, then, the female clerical workers would arrive at work and find a new computer terminal on their desk, running a programme that had been designed for their tasks by someone who had no experience in them. The whole change was imposed upon them from the outside; in fact they were not even aware that such a change was coming. Naturally, they considered their training insufficient for work on the new machines. There was apparently no conscious policy on the part of the employer either in programming or in the introduction of computer technology to change the flow of work at the office, or to change the division of labour between people. It simply happened that part of the work tasks were taken over by the computer's memory, which meant that there developed a new division of labour between the machine and the human worker. Otherwise the policy was to retain the old formal organization of work, meaning that there was no change in the position of clerical employees.

Nonetheless the white-collar employees themselves felt that their work had become more demanding, and believed it would become increasingly so in the near future. We do not know exactly the source, from which the feelings of increasing demands rose up. One reason is certainly the fact that in the neighbourhood or in one's own environment the things were changing and one had to learn new ways to manage. But another reason are purely quantitative demands: the amount of work load, especially increasing pace at work. There were only very few of those who said that their job was now easier, and also few who suspected that their work would become more monotonous. However, most of the clerical women regarded the content of their work, its variability, independence, and social contacts especially customer service as the best sides of their job, and the work load, the stressful pace of work, and the poor wages as

But in spite of the demands imposed on women, there seemed to be no signs of real changes in their position at the workplace. They continued to occupy the lowest ranks of the organizational hierarchy, with no career prospects and no chances of exercising power. The female clerical workers also felt that the appreciation and recognition they received for their work had not increased, and they assumed there would be no changes in this respect in the near future.

4. WOMEN'S POTENTIALS AND RESOURCES FOR CHANGE

Nevertheless the clerical women we studied were willing to take an active part in the development of their work. There are at least two possible explanations for this. Firstly, they have a very profound knowledge of their own work process, of the integrated

understanding that is required in their work: they realize the importance of their own active contribution in the effort to introduce computer technology in the most efficient and flexible way.
Secondly, the content of work is for very many female clerks the most important source of work satisfaction; customer service and the social contacts involved in this kind of work were regarded as particularly rewarding. On the whole they considered office automation a positive development, and they were keen to learn how to use the new equipment more efficiently. Although they complained about their relatively poor wages, the work orientation of the women we studied was by no means purely instrumental: they regarded their job as a professional occupation, took pride in it, they wanted to develop their skills and also to advance in their career. In this sense their work orientation was one of identification with the contents of work.

But somehow, female clerical workers have to manage. And today they are trying to manage by being as flexible as they can within outdated work organizations, by integrating the new technology and the formal work organization into their everyday activities on the basis of their own work orientation - often with mixed feelings (cf. Wagner 1985) and on contradictory grounds (see Korvajärvi and Rantalaiho 1985). What is more, they are doing this at the expense of increasing stress and an increasingly hectic pace of work, and at the same time undermining their own chances of social interaction, that vital element of office work.

The rigid organizational hierarchies and the ambitions and resources of female clerks are at sharp variance with each other. Clerical women would quite obviously need channels through which they could further their own interests. One such channel is offered by the trade unions of salaried employees. In Finland unionization levels are comparatively high among lower-level female clerical employees (e.g., at eight of the workplaces concerned they were between 75 and 100 %), but the unions have not dealt with such questions as the content of work, the organization of work and the changes caused by the introduction of computer technology, as their main preoccupation is with various issues related to collective agreements, which do not include technology agreements. Beside, trade unions have not offered real alternative strategies in the effort to break the rigid hierarchies at the workplace level, nor do they have any clear strategy of how to use women's potential resources for the development of working life.

In Finland the development of working life and workers' opportunities to take part in decision-making at their own work places has always been a politically very sensitive issue. At present there is a lively public debate on what is described as a "controlled structural change", which of course is largely a euphemism for large-scale rationalization in industry. Recent studies have indeed shown that the number of industrial jobs is declining rapidly with the introduction of new technology (Kortteinen et al. 1987). Within the trade union movement too, the attention is focussed on the modernization of industrial work. The problems of white-collar work and women's work in particular have received only marginal attention because in these cases the loss of jobs would seem to be a less serious problem than in industry.
It should be obvious that the improvement of the position of female white-collar workers and the eradication of the present gender division of labour are problems that cannot be solved merely by new technology; the issue is much more complex than that and involves various social and cultural structures, such as industrial relations. My main argument is that the core of the problem lies in breaking the hierarchic and extremely rigid gendered division of labour. The key strategic question is, how to convert the responsibility shown by women for the development of their work into an energy that would change the the hierarchic structures prevailing at the workplace. If women really wish to pursue their own goals at work, this is what they should be striving at.

As far as scientific research is concerned, this means we should focus on the wage labour and the structures of wage labour from the point of view of the gender system and patriarchy. At the workplace level, it means we must attempt to answer the question of how gender identity and the gender relations are reproduced in wage labour. However, a great deal of this cultural content of the gender system in the work places is invisible (Hearn & Parkin 1987). The meanings, which express the relations between men and women are historically build on previous patterns and the gendered division of labour (see Walby 1986). Concerning the activities of women and men in wage labour such problems as what kind of work is seen suitable on the one hand for women and on the other hand for men and what kind of qualifications are valued in the organizations, are of great importance. What kind of patterns of feminity and masculinity both women and men in their every day practice in different hierarchical positions of the organizations reproduce and maintain during the introduction processes of the new technology?

ACKNOWLEDGEMENTS

I wish to thank Liisa Rantalaiho, Katalyn Nagy and Kea Tijdens for useful comments.

REFERENCES

Braverman, H. Labor and Monopoly Capital. Monthly Review Press, New York 1974.

Frese, M. Partialisierte Handlung und Kontrolle: Zwei Themen der industriellen Psychopathologie. In: Frese, M., Greif, S., Semmer, N., (Hrsg.) Industrielle Psychopathologie. (Fragmented activity and control: two themes of industrial psychopathology. In: Frese, M., Greif, S., Semmer, N., (ed.) Industrial psychopathology). Bern-Stuttgart-Wien 1978.

Gardell, B. Produktionsteknik och arbetsglädje (Production technology and enjoyment of work). PA-rådet, Stockholm 1971.

Gardell, B. Arbetsinnehåll och livskvalitet (Job content and quality of life). Prisma. Stockholm 1976.

Gunnarson, E. & Ressner, U. Från hierarki till kvinnokollektiv (From hierarchy to women's collective). Arbetslivcentrum. Forskningsrapport nr. 40. 1983.

Hearn, J. & Parkin, W. Sex at Work. Wheatsheaf Books. Brighton 1987.

Hedberg, B. & Elling, M. & Jönsson,S. (et al.) Kejsarens nya kontor (The emperor's new office). Liber, Stockholm 1987.

Heiskanen, T. Automation,Work Content and Work Requirements. Acta Universitatis Tamperensis. Ser. A, Vol 168. University of Tampere, Tampere 1984.

Karl, H. & Ohm, C. Textautomation: Männersache? Frauensache? (Women's issue? Men's issue?). Das Argument 135/1982, pp. 653-663.

Kortteinen, M. & Lehto, A.M. & Ylöstalo, P. Information Technology and Work in Finland. Central Statistical Office of Finland. Studies No. 125. Helsinki 1987.

Korvajärvi, P. & Järvinen, R. & Kinnunen, M. Työ, toimihenkilöt ja muutokset (The changing trends of white- collar work). TVK ry. Jyväskylä 1987.

Lie, M. & Rasmussen, B. Kan kontordamene automatiseres? (Can you automate office women?). IFIM Rapport. Trondheim 1983.

Lie, M. & Rasmussen, B. Office work and Skills. In: Olerup,A., Schneider,L., Monod,E., Women and Computerization. North-Holland, Amsterdam 1985, pp. 43-52.

Olerup, A. Office automation. Work and Skill. In: Olerup,A., Schneider,L., Monod,E., Women, Work and Computerization. North-Holland, Amsterdam 1985, pp. 15-21.

Position of Women, Central Statistical Office of Finland. Statistical Surveys, no. 72, Helsinki 1984.

Rantalaiho, L. Work Organization and its Impact on the Worker's Life in Work. Department of Sociology and Social Psychology. Research Reports 22/1977, University of Tampere. Tampere 1977.

Rantalaiho, L. Kvinnlig arbetsorientering och kontorsautomation (Women's work orientation and office automation). Sosiologia 1/1985.

Rantalaiho, L. Office Work as Women's Work. Paper presented in the ISA RC 30, Sociology of Work, Workshop on Theoretical and Methodological Trends in Sociology, Jablona, Poland. January 1986.

Rantalaiho,L. & Korvajärvi, P. Women's Office Work and Coping with Technological Change. Department of Sociology and Social Psychology. Working Papers B 16, 1985, University of Tampere, Tampere 1985.

Siim, B. The Scandinavian Welfare States - Towards Sexual Equality or a New Kind of Male Domination? Acta Sociologica 1987 (30), 3/4: 255-270.

Suchman, L. Office Procedure as Practical Action: Models of Work and System Design. ACM Transactions on Office Information Systems, Vol. 1, No. 4, October 1983, pp. 320-328.

Wagner, I. Women in the Automated Office. In: Olerup, A., Schneider, L., Monod, E., Women, Work and Computerization. North-Holland, Amsterdam 1985, pp. 53-64.

Walby, S. Patriarchy at Work. Polity Press. Cambridge 1986.

Wynn, E. Office Conversation as an Information Medium. Department of Anthropology. UCLA, Berkeley 1979.

BIOGRAPHICAL NOTE

Päivi Korvajärvi, Researcher, MA in social sciences, has sstudied several years the changing trends in clerical and white-collar work. Adress: Work Research Cente, University of Tampere, P.O. Box 607, SF-33101 Tampere, Finland

Women, Work and Computerization: Forming New Alliances
K. Tijdens, M. Jennings, I. Wagner, and M. Weggelaar (Editors)
Elsevier Science Publishers B.V. (North-Holland)
IFIP, 1989

WOMEN, WORK AND TECHNOLOGY: THE LABOUR MARKET

Mary Ann Madison and Vary T. Coates*
Office of Technology Assessment
USA

Employment levels and occupational choices for women have been heavily dependent in the past on national economic performance, demographics, and social/cultural environments. Technologies are now adding another dimension to these impacts on jobs. Women have increased their participation in the labour market at the same time that total population has exploded, economic performance in the U.S. has deteriorated and automation has restricted the number of workers needed to perform a given amount of work in many industries. Clerical work which employs large numbers of women workers has been particularly affected. Continuing automation of tasks in all industries and many more occupations will exacerbate the unemployment effects unless sufficient new jobs and industries develop to take up the slack. At this time, the largest future job growth points to service-type jobs such as janitor and waitress. Technical and information-oriented jobs will show large increases in percent of new jobs, but much less in actual numbers of jobs

1. INTRODUCTION

There have been tremendous changes in the past decade in the structure and content of tasks, jobs, occupations, organizations and industries throughout the world because of automation of offices. The changes have occurred in the United States concomitant with extensive changes in the economy, political and social values, and in the composition of the population. Because of these interrelationships, it is difficult to judge if the changes have been caused by technology and whether they are "good" or "bad" for particular industries, occupations and workers. However, it is clear that the way work is performed in offices is very different from what it was a decade ago and this is due, in large part, to automation.

This paper will outline the impacts of office automation on work in the United States, focusing particularly on women and the labour market. It is important to look at the history of technology, the economy, political and cultural values and employment to better understand the significance of the changes occurring now.

---

* The views expressed are those of the authors and not necessarily those of the OTA, Technology Assessment Board or Congress.

2. DEMOGRAPHICS

2.1 Population

The work experiences of each group of cohorts in the population has been strongly influenced by the size of the group because of the direct effect of supply and demand on the labour market. The rate of increase of the population of the United States has varied considerably over the past century with high growth in the mid 19th Century, falling to a historical low by 1940 at the end of the Great Depression and the beginning of World War II. Growth rose dramatically during the 1950s and early 1960s resulting in the famous "population explosion." A considerable decrease in fertility rates in the 1970s and a mini-explosion of births in the 1980s again reshaped the population profile.

2.2 Labour Force Participation of Women

Women's labour force participation rates have been affected by many factors-fertility rates, demand for labour in response to economic conditions, the need for income in addition to that of the male head-of- family, and shortage of men in the labour force during wartime. Women were allowed to work in the office for the first time during the Civil War when a shortage of male labour necessitated the "relaxation of standards" as to what women were suited for.[1] Office work, which proliferated with the expansion of industry in the late 1800s, brought to the young and unmarried women the opportunity to work in more pleasant surroundings and at higher wages than they had been able to obtain in the factory.

FEMALE LABOR FORCE PARTICIPATION RATES
(Female Labour as Percent of Total Labour Force)

1910	1920	1930	1940	1945	1950	1960	1970	1980	1986
20.0	22.7	23.6	27.9	35.8	33.9	37.8	43.4	51.5	55.4

* U.S. Statistical Abstracts, U.S. Department of Commerce, Bureau of the Census (Washington, D.C., Government Printing Office, 1987) and Historical Abstracts, U.S. Department of Commerce, Bureau of the Census (Washington, D.C., Government Printing Office, 1972).

A major change occurred in work force demographics when married women were accepted into the labour force during World Wars I and II. Declines in fertility rates and family size and increasing childlessness have contributed to the higher participation of married women in the labour force in this century. And higher divorce rates have contributed to women's greater participation in the labour market in the 1980s.
The dramatic future changes in the number of young workers available will most likely have far reaching effects on the labour market also. Because of their fewer numbers, younger women may have easier access to new occupations at beginning levels than had previous generations, but may lack promotional opportunities because of the large number of older employees still in the job market for the next 20 years or so.

2.3 Education

Education played an important role in the work experiences of women in the late 1800s and early 1900s when, because of their higher education levels, they were preferred over men for the new office work. This educational advantage for women helped them enter the work force but, until this decade, has probably not helped them much

1 Davies, Margery Woman's Place is at the Typewriter, (Temple University Press, Philadelphia PA, 1982)

otherwise. Even women's rapid increase in college degrees from 14 percent of bachelor's degrees in 1870 to 41 percent in 1940,[1] did not improve their career opportunities so much as it improved their selection of husbands. This was followed by an era of blatant discrimination towards women in higher education sharply diminishing their access to degrees after World War II. That handicap has been overcome as more women than men are enrolled in colleges in the 1980s.

3. CULTURAL/POLITICAL FACTORS

What is often thought of as the traditional culture throughout the history of the modern world, with men earning the income and performing the important political work outside the home and women performing the supportive, nurturing, family maintenance work inside the home, has seldom been true. But these beliefs have shaped society's and women's attitudes toward technology and work and may be responsible for the continued segregation in occupations. Technological change that relieved workers of heavy physical labour did not help women into a wider range of careers until very recently. Other political and cultural changes, such as equal pay and equal opportunity legislation, were necessary before women were able to take advantage of new technologies to move into some nontraditional jobs.

4. ECONOMICS AND TECHNOLOGY

The history of technology, of economics, and of work in the United States are closely related. This makes it difficult to isolate the effects on women of the use of new technologies in the office. A few simple principles are at work here--economic growth usually results in more jobs, although some jobs are lost because of technological change. Jobs and organizations may be restructured and relocated because of new technologies or new industries. And one type of labour is substituted for another--women for men or vice versa, higher level for lower level or vice versa and machine for human--during technological changes.[2]

Development of transportation and communications systems fostered significant changes in the economy, work and lifestyles in the past century. The telephone, for example, allowed companies to expand their businesses while centralizing their operations in urban areas apart from suppliers, customers and factories.[3] It facilitated more efficient and economical communication for businesses within cities. Employment patterns in the telephone industry shifted over time as the industry matured; boys were the first operators but were replaced by women. This opened up a new opportunity for women to obtain dignified, clean work, even though at low pay. The number of employees increased for three decades as the population of telephone users increased, but automation drastically decreased the employment of telephone operators in the 1960s and 1970s. The number of supervisors has also decreased as automation has expanded the span of control. With electronic monitoring, a supervisor can monitor a large number of employees continuously. This monitoring and speeding up of the work pace has increased the stress in telephone operator and other clerical jobs. The felt need by firms for more security against mischief and theft, especially of trade secrets, indicates that levels of electronic monitoring of employees and workplaces will continue to increase.

Cost-cutting measures by businesses in response to the severe economic pressures of the 1980s contributed to the unemployment of large numbers of workers. Other workers

1 Digest of Education Statistics, U.S. Department of Education, National Center for Education Statistics, (Government Printing Office, Washington, D.C., 1983)

2 Hartmann, Heidi I., et. al., Computer Chips and Paper Clips, (National Academy Press, Washington, D.C., 1986), p. 24-29

3 Hartmann, op. cit., p. 27

could only find lower paying part-time or temporary work. New technologies, particularly automation of work, have been partly responsible for the employment problems of the 1980s and have also contributed to the high demand for products and services. Because of the high rate of growth in the tertiary sector, some businesses would have been unable to service their customers without automation. This is especially true of the insurance and banking industries where increased productivity because of automation permitted greatly expanded service, and also contributed to the displacement of many workers in these two industries.

5. AUTOMATION IN OFFICES

The majority of U.S. workers are now white-collar workers and about 45 percent of all U.S. workers work in offices. In 1980, 63.5 percent of employed women were white-collar workers and 82.1 percent of office clerks were women.[1] Office work is rapidly being "automated", or computerized. What does this mean for the productivity of office workers and the number of office jobs that will be available in the future? What skills will office workers need? What job ladders will be open to them? Will the quality of their working life be better, or not as good? What new opportunities and new problems can we expect, as a result of sweeping technological change in offices? The dominant trends in office automation are likely to be:

* A continuing strong movement toward microcomputers and toward distributed data access and data handling, usually superimposed on rather than superceding centralized automatic data processing;
* more powerful, easier to use, software;
* a strong trend toward linking and networking of microcomputers, minicomputers, mainframes, and peripheral and supporting systems;
* increasing choice among technological options for accomplishing information handling objectives;
* more and more capture of data at the point of origin, decreasing the need for repeated keyboarding and centralized data entry; and
* growing capability for communication, between devices, between organizations, and between locations.[2]

There are important differences between this technological revolution and previous technological advances. It is generally true, now, that the time from scientific breakthrough to development and implementation of a new technology has been greatly shortened and new technologies tend to diffuse more rapidly. Office automation, particularly, has spread quickly and is used in all sectors and all industries which multiplies its impact.

PURCHASES OF OFFICE, COMPUTING AND ACCOUNTING MACHINERY
(billions of constant 1982 dollars)

	1960	1970	1980	1984	1986
Office, computing & acctg. mach	1.8	2.3	15.3	49.1	71.9

* Source: Survey of Current Business, U.S. Department of Commerce, Bureau of Economic Analysis, (July 1987). Note: Computing machinery includes all computers, not just office computers, but is indicative of the level of growth of the industry.

1 Hartmann, op. cit., p. 18

2 Automation of America's Offices, (U.S. Congress, Office of Technology Assessment, Washington, D.C., 1985), p. 11

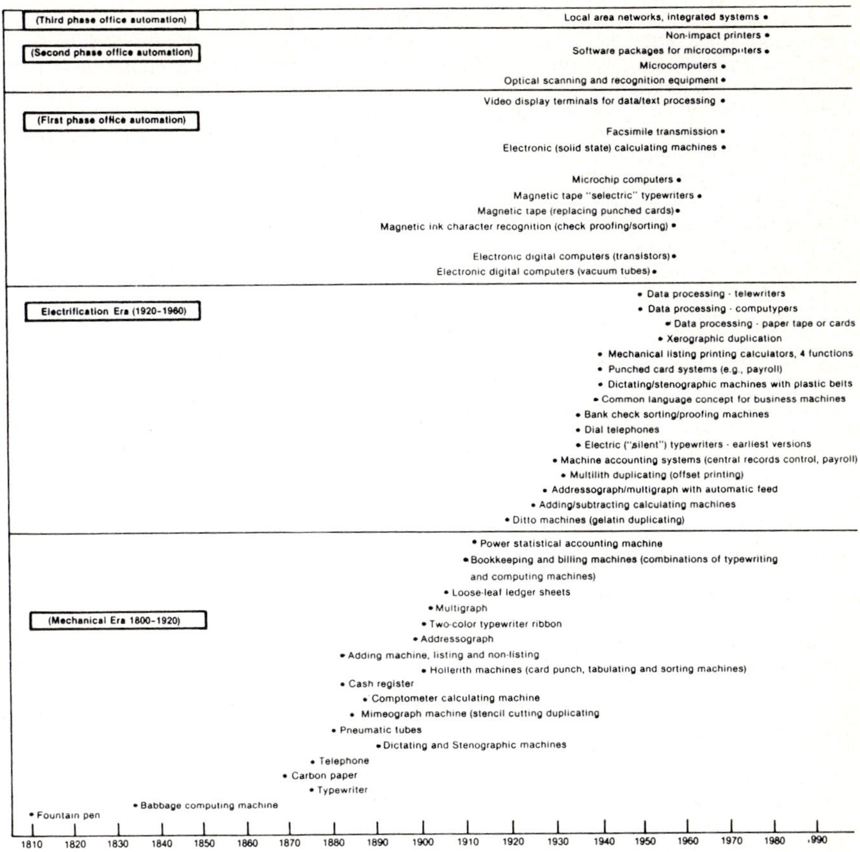

HISTORY OF TECHNOLOGIES USED IN THE OFFICE

5.1 Employment Levels

Technology has had direct and indirect effects on employment. The direct effect has been the saving of labour. This can be accomplished by substituting machine for human labour and/or by aiding human labour, making it more efficient. The goal of automating in most organizations is to reduce the labour required to perform a given amount of work. Lower level clerical skills were the earliest to be transferred to electronic equipment. This was most evident in the industries which were using large numbers of clerical employees--insurance and banking--and in the tasks which were most easily transferred--customer records and billing. Some jobs, such as customer service representative, could not be directly transferred to automated systems, but were enhanced by the addition of automated records systems.

Other direct effects of the automation of offices are the ability to offer more services, the integration of processes within the firm and the development of new products. The banking and insurance industries, for example, could process their high level of new accounts and policies and offer new financial services in the past twenty-five years only because of automation in their offices. Travel agents have experienced greatly increased productivity by means of computerized reservation systems. Membership information for associations can be processed in a fraction of the time required before automation and firms can produce more professional looking proposals, reports and other documents. Numerous new businesses are offering new products and services that have been developed from the computer.

The indirect effects of automation of offices are the changing structure of organizations and new locations for performance of work. The structure of organizations was changed first by the division of jobs into automatable tasks and then the reintegration of some tasks into new processes. Managerial tasks were often transferred to upper level clerical employees and some clerical tasks were transferred to managerial employees. Job ladders were often changed because of these new structures. New locations, such as the home, regional and local offices and foreign countries were made possible for performance of work.

5.2 Occupations

The content of jobs, the structure of organizations and the employment levels of industries will all be affected by office automation. However, the changing participation of women in industries and occupations is not, for the most part, changing because of automation. Despite great gains made by women into professional and nontraditional occupations, much of the work force is still segregated, with women in the lower level clerical jobs in the office, in the teaching rather than administrative jobs in education, and in the service-oriented industries rather than the goods-producing industries. More than 36 percent of employed women work in ten occupations, nine of which are categories dominated by women.[1] Women account for 25 percent of machine operators, assemblers and labourers but their participation in other occupational categories of the goods-producing industries is negligible.

Office occupations which have been most affected by computerization are the lower level clerical positions, followed by those who supervise and manage the lower level employees. The Office of Technology Assessment (OTA) said in its 1985 report that:

> "The reduction in labour will be most significant in the clerical/support occupations, especially those that predominantly involve data entry. Fewer lower level workers would in itself indicate a need for fewer first line supervisors and managers. But the span of management control can also be

---

1 Hartmann, op. cit., pg. 19

broadened by automation of the work process. Thus, fewer managers may be needed (again, for a given volume of information-handling). Some of the tasks of lower level managers can be automated, or be taken over by clerical workers who are lower paid, so this tier of jobs is again likely to shrink. Some of the tasks of paraprofessional or technical workers may follow the same route. Professional occupations are less vulnerable, but not immune to the substitution and adaptation effects of automation."

The OTA report predicted decline in the growth of clerical jobs, which was beginning to be noticeable in 1984, and an absolute decline in numbers of jobs by 1995. This prediction was based on progress made in collecting more information electronically at its source. This is the result in banking, where, for example, customers use electronic teller machines, leaving a record for the bank to process electronically. Future declines are attributed to the increase in use of electronic equipment at the source, the increasing efficiency of communications between computers (within and among organizations), the likelihood of significant breakthroughs in optical character recognition (OCR) technology and the completion of the transfer of information from large libraries of paper copies to electronic form. This is occurring with legal libraries, the Library of Congress, etc. During the transition period, larger numbers of data entry operators may be employed; however this is a temporary phenomena.

CLERICAL JOBS, 1950-87

Year	Number of clerical jobs (millions)	Growth in the number of jobs	Percent of total employment
1950	6.6		11.3%
1960	8.8	33.0%	13.4
1970	12.9	46.0	16.4
1980	16.9	31.0	16.9
1982	16.8	-0.6	16.8
1984	16.7	-0.6	15.9
1986	17.7	0.6	14.8
1987	18.3	0.3	15.0

* Sources: U.S. Department of Commerce, Bureau of the Census, Census of Population, 1950, 1960, 1970, 1980; and U.S. Department of Labor, Bureau of Labor Statistics, Current Population Survey Annual Averages, 1982 and 1984; Employment & Earnings, U.S. Department of Labor, Bureau of Labor Statistics, December 1987.

Managerial and supervisory work will be affected by office automation to the extent that it is found in industries which are information intensive. Manufacturing firms have relatively low levels of managerial staffing, while insurance carriers, securities, computer and data processing services have relatively high levels. In organizations with high levels of clerical employees, the managerial and supervisory work will decline as the numbers of clerical workers declines. Women are found in the greatest numbers in these types of supervisory jobs, so will feel the greatest loss of jobs with expanding automation. The number of management jobs will also decline as computers facilitate the aggregation and analysis of computer data and the development of reports. Corporate hierarchies will tend to be flattened.

Women have doubled their participation in architecture, computer systems analysis, economics, and legal work in the past 10 years, however their growth in these occupations has, unfortunately, recently slowed. These professional workers are using electronic databases, expert systems and computer models to aid them in performing their work more efficiently. Professionals, like managers and supervisors, may find themselves doing more of their own clerical tasks, such as data entry and typing of text as personal computers proliferate.

Skills required for new jobs in automated offices vary considerably. Low level clerical jobs which have been broken down into minute automated tasks require low levels of skill. Many secretarial jobs now require high-level skills to operate the complex software and the managerial tasks which may be incorporated into the new positions. The standardized skills acquired in these automated positions make moving laterally to other firms easier; the knowledge of the operations rather than knowledge of the firm or industry has become the important skill. At the same time, this subdivision of tasks has deprived some clerical workers of the industry and company knowledge that used to help them move up the job ladder. The trend towards hiring at the managerial level from the large pool of young college graduates rather than promoting from within has also eliminated promotional opportunities for clerical workers.

5.3 Hours and Location of Work

The reorganization of office work and the need to lower the costs of doing business has resulted in a noticeable movement toward part-time, temporary and contract labour, sometimes called a contingent work force. This movement was aided, in part, by the new office technologies that permitted decentralization of work. One large U.S. data processing firm had a five-year goal of transferring one-third of its work force to these alternate statuses. These groups of employees traditionally have fewer company-paid benefits and can be laid off easily when workloads decrease. This practice also allows the firm to maintain a minimum of office space.

Women are particularly affected by these policies promoting alternative work strategies. Some firms hire women to work at home, often on computer equipment rented from the company and at lower rates of pay than those in the office. Mothers of young children, have found the flexibility of part-time and temporary work to be more suited to their needs, but the lack of health, vacation/sick-leave and pension benefits in many of these jobs is often a severe handicap to these workers. During times of high unemployment, many women clerical workers were forced to accept part-time and temporary employment because full-time work was unavailable to them.

PART-TIME WORK FORCE
(Percent of Total Employment)

	Involuntary	Voluntary	
Year	Total	Workers	Workers
1970	16.4	3.1	13.3
1972	16.8	3.3	13.5
1974	17.1	3.5	13.6
1976	18.0	4.2	13.8
1978	17.7	3.8	13.9
1980	18.4	4.5	13.9
1982	20.2	6.4	13.8
1984	19.4	5.5	13.9
1987 (Nov.)	18.9	4.7	14.2

* Sources: 1970--Labor Force Statistics Derived From the Current Population Survey: A Data Book, Volume 1, (Washington, D.C., U.S. Department of Labor, Bureau of Labor Statistics, September 1982, Bulletin 2096), p. 682. 1982-Employment & Earnings, Vol. 30, No. 1, (Washington, D.C., U.S. Department of Labor, Bureau of Labor Statistics, January, 1983), p. 194. 1984-Employment & Earnings, Vol. 32, No. 1, January 1985, p. 43. 1987-Employment & Earnings, Vol. 34, No. 12, December 1987, p. 35.

While professionals such as consultants, artists and authors have often worked at home, and computer systems certainly facilitate this option, most professionals employed by firms still tend to work in the office. The face-to-face encounters are important to many of these employees and, some women especially feel, promotional opportunities are limited by lack of presence in the office.

Off-shore performance of data processing work has been going on for two decades or so. Because of labour costs as little as 15 percent of U.S. wages in Haiti, Barbados and Mexico, among others, it has become an inexpensive way for firms to accomplish large amounts of data processing at low cost. It is particularly used to input data from discount coupons, airline reservations and whole libraries of legal and other documents. It has deprived U.S. workers of some jobs, although the total number is relatively small. There may be inherent limitations to off-shore data entry as more information is collected electronically at the source and as new technologies such as OCRs, facsimile and other machine reading of documents make it feasible to translate printed material into electronic form.

6. EMPLOYMENT PROJECTIONS

The U.S. Department of Labor, in its 1986 projections for the year 2000[1], predicts that all new net employment growth will be in the service-producing industries. These are the industries where women are most heavily employed. Occupations requiring higher education levels, such as executives, professional workers and managers, are projected to grow most rapidly. Clerical positions in offices are expected to have declining growth rates to about 11 percent, largely due to automation. Clerks in services such as hotel registration, real estate, brokerage clerks, information clerks, customer service clerks are expected to increase by considerable amounts. Many people believe, however, that Department of Labor forecasts underestimate the impacts of technological change because this factor is not well accommodated in the agency's models.

The fastest growing occupations will be paralegal personnel, medical assistants, physical therapists and data processing equipment repairers. However the fastest growing occupations may not be the same as those with the greatest number of new jobs, because of their smaller initial size. Occupations with the largest expected job growth will be retail salespersons, waiters and waitresses, nurses, janitors and general managers.

It is clear from these projections that positions that cannot be readily automated will provide the most new jobs and positions that can be automated will provide fewer new jobs. Women's fortunes in the workplace will follow these patterns insofar as they are in the affected occupations. Since women will be heavily affected in the clerical occupations, they may be motivated to seek and more strongly demand, employment in more nontraditional fields.

BIOGRAPHICAL NOTE

Mary Ann Madison is a research analyst at the U.S. Congress' Office of Technology Assessment where she has worked on studies on office automation, particularly concentrating on the impacts on small businiess, women and minorities, and on technology and constitutional gouvernance. She has a master's degree in consumer economics.

Vary Coates is a Senior Analyst and Project Director in the U.S. Congress' Office of Technology Assessment, U.S. Congress. She has directed projects on office automation, information systems modernization, and the constitutional impacts of techology. Coates is now directing a study of information technology and financial markets.

Mary Ann Madison and Vary T. Coates produced a report on Automation in America's Offices, for the Congress in 1985.

---

1 Monthly Labor Review, (September, 1987), whole issue.

Women, Work and Computerization: Forming New Alliances
K. Tijdens, M. Jennings, I. Wagner, and M. Weggelaar (Editors)
Elsevier Science Publishers B.V. (North-Holland)

WOMEN'S LABOUR MARKET POSITION FROM THE PERSPECTIVE OF TECHNOLOGICAL CHANGE

Anna-Maija Lehto
Central Statistical Office
Helsinki, Finland

An extensive interview survey carried out in Finland has studied the spread of new information technology and its effects on employment and on job content. On the basis of this study combined with structural data on labour, this paper considers the position of women active in the different sectors in the economy in the context of technological change.

1. INTRODUCTION

It has long been apparent from forecasts concerning the effects of information technology that the application of information technology will mainly centre on the so-called female occupations, as represented by the different kinds of clerical work, for instance. In addition to the view that information technology will focus on female occupations, these forecasts also contain the assumption that the demand for labour in these occupations will decline rapidly. To give an example from the Federal Republic of Germany: according to forecasts from the early 1980s, no less than million secretaries and typists are expected to lose their jobs by the year 1990, which would mean a reduction of about 40% in these occupations. Subsequent forecasts for many other countries also show employment reductions of 30-40% in the number of clerical workers. In Finland, according to a report of the government committee on information technology, office automation is expected to result in increased structural unemployment, especially among women.[1] Fore casts like these also imply that there will be a turn-round in the development of working life: what used to be growth areas will start declining rapidly.

Sweeping forecasts also characterized the first few sociological studies concerning the effects of information technology. Apart from the effects on employment, these studies also discussed changes in job content. Sharply divergent views were presented: some researchers took a very favourable view of the information society, while the attitude of some others was negative (examples of these including Toffler, Braverman, Masude, Gorz, Naisbitt, Kern-Schuman and others).
As the views presented amounted to speculation based on estimates, there later arose in social research a need for in-depth analysis of the effects of information technology that would be based on accurate scientific observations. In Finland, the effects of computerization have been studied with reference to the paper and paperboard, textile, clothing

1 Korvajärvi, P., Rantalaiho, L., Toimistoautomaatio ja toimistotyö (Office automation and office work.), Tampereen yliopisto. Yhteis kuntatieteiden tutkimuslaitos, sarja B 40/1984, p. 18.

and electronics industries; clerical work related to pay-roll computation, social service, the justice department, banking and insurance; word processing; and so on.[1 2 3] Only a few of these surveys consider the effects of information technology separately from the perspective of men and women or especially from the perspective of women.
Although these surveys have produced plenty of important information, they have not provided an overview of the effects of information technology because they have only been concerned with selected jobs or with selected systems (such as the pay-roll computation system). Of the effects on employment in particular, generalizable information has been difficult to obtain.
For this reason in part, a comprehensive survey was carried out in Finland at the end of 1984. In the survey, a total of 4,502 persons were interviewed representing all employees. The survey was carried out by the Central Statistical Office of Finland, but the survey-report was prepared in collaboration with a group of researchers from the University of Helsinki.[4]

One reason for undertaking this survey was that Finland, like most other countries, lacked numeric estimates of the spread of information technology and of the distribution of its use across different population groups. Another reason was the desire to find out whether the large-scale application of information technology at work marks a clear change in the division of labour in society. In other words, answers were sought to the question: what kind of changes are taking place in the demand for labour and in the content of jobs?
To enable to assess any possible change in the development of the division of labour caused by information technology, the survey first set to describe the changes that had so far taken place in the division of labour.

2. STRUCTURAL CHANGES IN WOMEN'S LABOUR MARKET POSITION

Working life in Finland is characterized by a high rate and intensity of women's labour force participation. Women's labour force participation reached a high rate quite early in Finland, much earlier than, for instance, in the United States and the Federal Republic of Germany. This appears from the following table[5] documenting the proportions of women in the labour force of these countries:

	1950	1970	1982
USA	28.9	36.7	42.4
FRG	35.1	35.9	38.1
Finland	40.6	43.8	46.9

A comparison of the Nordic countries shows that, at the present time, Sweden has a slightly higher labour force participation rate for women than Finland and Denmark, and

1 Koistinen, P., Teknologiset uudistukset ja työvoiman käyttö : tutkimus Suomen paperi- ja kartonkiteollisuudesta. (Technological Innovations and Use of Labour Force: A Study of Finnish Paper and Paperboard Industry), Tyovoimaministeriö, Työvoimapoliittisia tutkimuksia 47/1984.

2 Kasvio, A., Piispa, L., Tekstiilitehdas ja sen naistyöntekijät. Vaatetustehdas ja sen naistyontekijät. Elektroniikkatehdas ja sen naistyöntekijät. (Textile Plant and its Female Workers. Clothing Factory and its Female Workers. Electronic Factory and its Female Workers) Tampereen yliopisto, Yhteiskuntatieteiden tutkimuslaitos, Sarja D 80, 82 and 83, 1985, 1986.

3 Järvinen, P., Tyllilä, P., Erään atk-systeemin käyttöönotto. (The Incorporation of an ADP System), Tampereen yliopisto, Matemaattis ten tieteiden laitos, Raportti A 42, 1980.

4 Kortteinen, M., Lehto, A-M., Ylöstalo, P., Information Technology and Work in Finland, Central Statistical Office of Finland, Studies 125, 1987.

5 OECD EF(80) 3, Statistical Annex, table 28, OECD Observer No. 127 (March 1984).

much higher than Norway. The intensity of women's labour force participation, however, is higher in Finland than in the other Nordic countries because Finnish women's working days are longer, i.e. part-time employment is less common than in Sweden, Den mark or Norway.
Although is has long been common for Finnish women to participate in the labour force, the most characteristic feature of the change over the last few decades has been the steady growth in the female labour force. To give an example from 1970 to 1984, two-thirds were accounted for by women (see figure 1).
The growth in the number of female employees has been especially rapid in administrative, clerical, health care and social work (the first two major groups). These sectors already account for one half of female employees, while the proportion of women engaged in the manufacturing industries has shrunk to 15%.
The occupational distribution of men shows the same trend. For men, too, growth has been fastest especially in technical, scientific and administrative work (the first two major groups). The aggregate proportion of these groups has grown by nearly 10 percentage points from the year 1970. Corresponding, the proportion of the manufacturing industries has shrunk from 52% to 44%.

The division of labour in society can be described not only by occupational distribution but also by socio-economic status (see figure 2). As a general observation on the trend over the last few decades, it can be noted that the proportion of white-collar workers among employees has been growing rapidly. This is, however, partly a question of definition: the number of white-collar workers depends on what occupations are included in the category. Nevertheless, the figures reflect at least the general trend.
At present, two-thirds of female employees are white-collar workers, while as late as 1960 more than half of them were blue-collar workers. Another notable feature of the trend in women's socio-economic structure is that the proportion of upper white-collar workers has grown substantially since 1970.
The decline in the proportion of blue-collar workers has been slower among men than among women. Still today, 62% of male employees are blue-collar workers. This is not surprising in view of the fact that in 1984 manufacturing accounted for 44% of male employees. For white- collar workers, the same trend can be observed as for women: the pro portion of upper white-collar workers has been growing especially since 1970.
There has also been a marked structural change in the use of the labour force as regards the distribution of employees across the different employer sectors (see figure 3). In the case of female employees in particular, almost all of the growth since 1970 has been absorbed by the public sector. Municipalities in particular have been hiring women. In 1984 nearly one-third of female employees were in the employ of municipalities.

Of the structural changes described above, at least the change in socio-economic status prompts the question to what extent it is linked to changed in employees' educational level. Figure 4 illustrates the changes in the educational level of men and women according to the 1977 and 1984 studies of working conditions. The educational level of both male and female employees seems to have risen considerably between 1977 and 1984. For both sexes, the proportion of persons with more than a basic education grew by 7 percentage points. It is notable that the educational level of men and women is about the same, though the pro portion of persons who have completed only an education at the lower level of upper secondary education is somewhat higher among men. This group includes persons with, for instance, a vocational education.
In recent years, the socio-economic status of women seems to have improved to better reflect their educational attainment, for in 1984 upper white-collar workers already accounted for 12% of female employees and women with a graduate degree for 14%. Among male employees, the percentage of upper white-collar workers continues to be higher (17%) than the percentage of men with a graduate degree (13%).

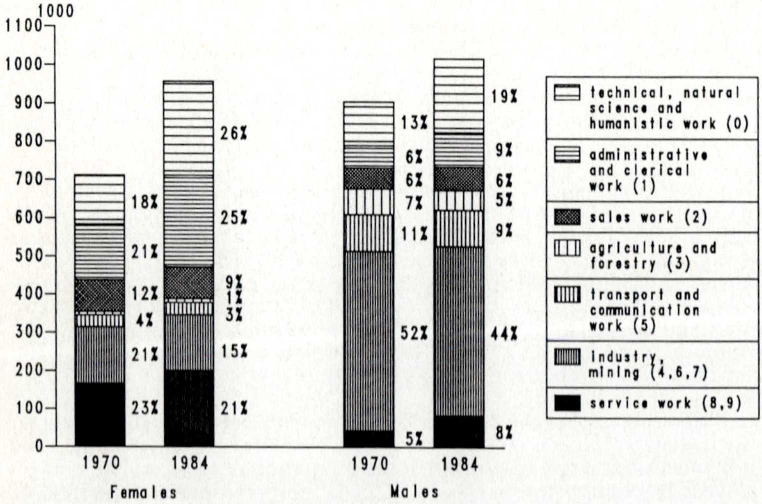

Figure 1. Employees by occupation in 1970 and 1984

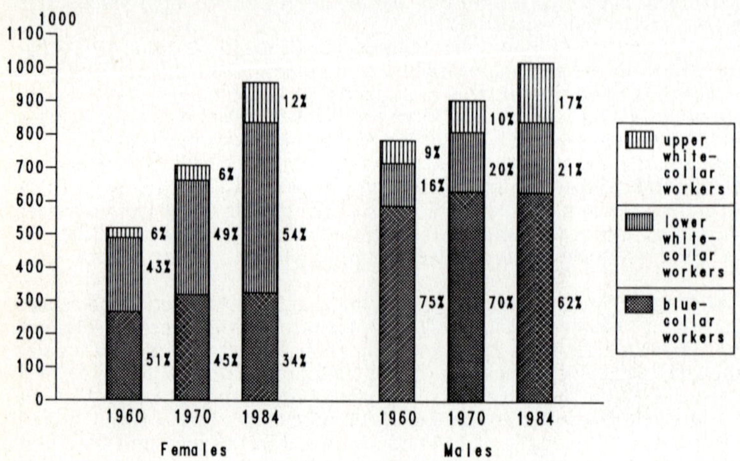

Figure 2. Employees by socioeconomic status in 1960, 1970 and 1984

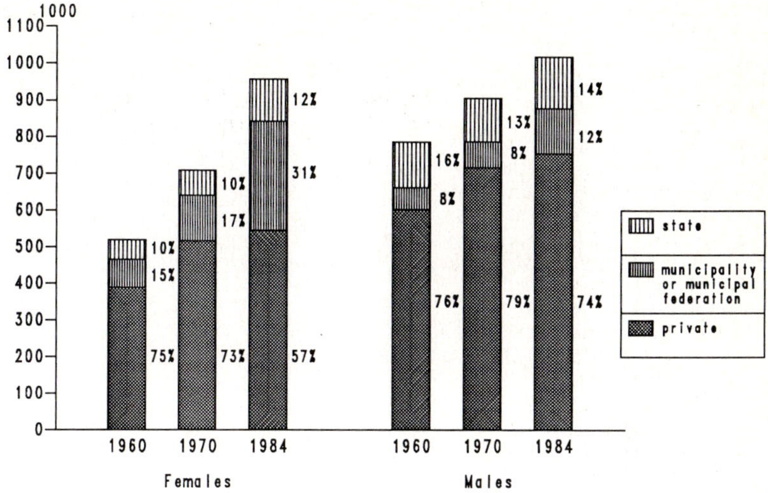

Figure 3. Employees by type of employer in 1960, 1970 and 1984

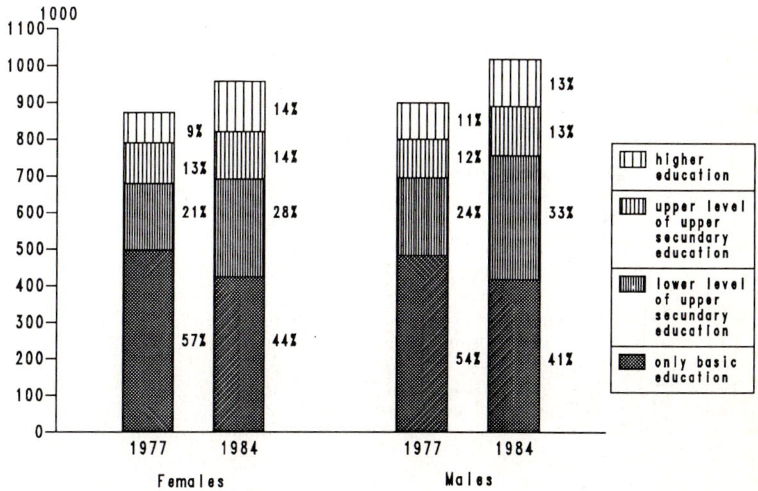

Figure 4. Employees by educational level in 1977 and 1984

3. SPREAD OF THE NEW INFORMATION TECHNOLOGY IN FINLAND

The spread of information technology has been very rapid in Finland. In 1984, when the survey was carried out, 332,000 persons, or 17% of all employees, used information technology in their work. Of them, more than half had become users only during the last three years. During the past few years, growth has averaged 50,000-60,000 persons a year. The number of Finns using information technology at work today can be estimated at more than half a million, or about one-fourth of all employees. However, the finding that 17% of employees used computerized equipment at work only indicates the average extent of use. The figure conceals the great differences between occupations. At the time of the survey, the list of the most computerized occupational groups looked as fol lows:

- Banking, proportion of users, 83%
- Chemical processing and pulp and paper making, 48%
- Bookkeeping, secretarial work and typing, 45%
- Other clerical work, 44%
- Miscellaneous office work, 37%
- Sales representation and marketing, 34%
- Technical and scientific work, 32%
- Work in the social sciences and the arts, 30%
- Administrative work, 28%

The list demonstrates that the rate of diffusion of information technology is highest in clerical work, just as had been forecast. In manufacturing, the longest traditions in the use of computer technology are those of the process industries.
Though the number of persons using information technology was nearly the same for men and women, the distribution of use among men and women showed certain differences according to socio-economic status.

WOMEN, total	18%	78,000
Upper white-collar workers	17%	20,000
Lower white-collar workers	28%	143,000
Blue-collar workers	5%	15,000
MEN, total	15%	154,000
Upper white-collar workers	28%	51,000
Lower white-collar workers	27%	58,000
Blue-collar workers	7%	45,000

Among upper white-collar workers, the use of the new equipment is far more common among men than among women. On the other hand, the majority of women using information technology are lower white-collar workers.
In addition, it appears on the basis of the educational background of users that men use the new equipment for educationally more demanding tasks than women. Of men, 26% hold graduate degrees compared with 14% of women. Among women, the proportion of persons with only a basic education is twice as high (41%) as among men (20%). Yet, among employees on average, the educational structure of men and women is much the same.
Regarding the purpose of use of the equipment by sex, the responses showed that men use it for considerably more demanding tasks. Of them, 31% write programmes on the machine, compared with only 9% of women. Data entry was far more common for women than for men (76% as against 52%).

4. THE IMPACT OF INFORMATION TECHNOLOGY ON THE DEMAND FOR LABOUR

Regarding the structure of the employee population, two questions arise. Does the large-scale application of information technology mark a clear change in this structure? What is the trend in employment like for each sex?

In the study of working conditions, the effects on employment were studied by asking the respondents questions about the trend in employment at their own place of work. The results show that in the past few years the demand for white-collar workers has been rising rather than falling and that, on the other hand, reductions in employment have been common among blue-collar workers. These trends have been accentuated by the introduction of information technology.
To put it in another way, the trend in clerical work and, hence, in much of the work done by women has not been so alarming, at least not yet, as has been assumed in the forecasts. This also appears from statistics. On the other hand, it is apparent from the responses that even in the office sector reductions in employment have occurred, but they have been carried out by "natural wastage" or attrition, i.e. new personnel has not been hired to fill vacancies.[1]

One reason for the surprising information that the number of whit- collar workers has been growing during the time information technology has been applied may be that the nature of information work is different from manufacturing work. As new office equipment is acquired, every effort is made to put it to as efficient use as possible. Hence, the amount of work has increased considerably and the personnel cuts promised by the vendors have not materialized.
In manufacturing, on the other hand, employment prospects are very bleak because the application of new information technology has barely begun in the sector: about 9% of persons engaged in the manufacturing industries were users of the new equipment.
The textile and clothing industries are typical female industries where the application of information technology has so far been quite modest. Yet the existing clothing industry applications relating to the patterning, grading and cutting functions have already resulted in some measure of job loss. On the other hand, it is usually impossible in manufacturing to separate the effects of different factors on employment; so, the reduced demand for labour in the textile and clothing industries is due to the aggregate effect of such concurrent factors as international competition, cheap imports, reduced trade with the Soviet Union and the application of new technology.[2]

In such typical male industries as chemical processing and pulp and paper making, the application of information technology has already advanced quite far. However, according to the census data of 1975-1985, even in these industries labour cuts have affected women more than men. The results of the study of working conditions would also seem to indicate that labour cuts in the manufacturing industries have affected women more than men.

---

1 Rantalaiho, L., Korvajärvi, P., Women's Office Work and Coping with Technical Change, University of Tampere, Departement of Sociology and Social Psychology, Working Papers B 16, 1985, p. 8.

2 Kasvio, A., Piispa, L., Tekstiilitehdas ja sen naistyöntekijät. Vaatetustehdas ja sen naistyontekijät. Elektroniikkatehdas ja sen naistyöntekijät. (Textile Plant and its Female Workers. Clothing Factory and its Female Workers. Electronic Factory and its Female Workers) Tampereen yliopisto, Yhteiskuntatieteiden tutkimuslaitos, Sarja D 80, 82 and 83, 1985, 1986.

5. CHANGES IN JOB CONTENT

Information technology has generated a seemingly endless process of speculation concerning the content of work: is work becoming more diversified and demanding or is it becoming less demanding, a mere routine? All sorts of generalizations have been presented in answer to this question, such as the view of a one-way development in either direction or the view of a polarization of skill requirements. On the other hand, based on the latest studies on information technology in particular, the view has been presented that it is impossible, if not senseless, to generalize about the effects because they are a function of the particular terms of application and that the whole issue is a product of technological determinism.

There is no avoiding the tendency to generalize, however, if any assessment of the trends is desired. In the study of working conditions, an effort was made to solve the problem of excessive generalizations by the expedient that correlations between the use of information technology and job content were considered, for instance, according to different socio-economic groups.

With respect to several factors bearing on job content, such as the repetitiveness, diversity and independence of the work, lower white- collar workers using information technology turned out to be worse off than their colleagues who did not use information technology. The same finding was made in a separate investigation on female lower white- collar workers engaged in administrative and clerical work. Their work had become more repetitive; there was less opportunity for influence; their work demanded more speed and dexterity than such traditional clerical skills as organizational skill, the capacity to attend simultaneously to several things, and so on.

With the exception of independence, information technology would seem to have had the reverse effect on the job content of blue-collar workers, especially in the manufacturing industries. Through the new equipment, jobs have become more demanding and such typical characteristics of manufacturing jobs as dexterity and speed have lost some of their importance. On the other hand, comprehensive control of the object of work has become more important.

The effects of information technology on the jobs of upper white-collar workers are more difficult to assess. In any event, no changes were revealed in measurements of the kind that were used in considering the job content of lower white-collar workers and blue-collar workers. The reason for this is perhaps that, when using information technology, upper white-collar workers use it less than other employees and mainly as a supplement to other work. Thus, there are only minor changes in job content.

On the other hand, part of the upper white-collar workers using information technology are computer professionals. In this group, the position of women differs a great deal from the position of lower white-collar workers. Women do not experience their position as being inferior to the position of men: though lower in salary level, their work is just as demanding in content as men's. This may be due to the fact that, because of the industry's short history, there is no established male/female role division in industry.

6. SUMMARY

Table 1 shows the main findings of the survey. For female employees, the most important effects center on lower white-collar workers, who constitute the majority of all lower white-collar workers and who account for 80% of the use of new technology among women.

	Blue-collar workers	Lower white-collar workers	Upper white-collar workers
Labour demand	Falling at increasing rates	Growth continues	Growth continues
Content of work	More demanding, greater diversity	More fragmented and repetitive	No fundamental changes
Independence and control, incentives	Better opportunities for promotion and training for training	Less independent, better opportunities	Better opportunities for promotion and training
Job satisfaction and problems	Satisfaction grows, mental stress and social conflicts more common	Satisfaction drops, mental stress and social conflicts more common	No differences in satisfaction, criticism becomes more marked

Table 1.The significance of the introduction of information technology in various socio-economic groups -biggest differences.

According to the results, the main problem associated with the use of information technology by women is that women have insufficient opportunity for influencing the planning of the introduction of information technology and that they receive insufficient training in data processing techniques. Women using information technology had received seven days of training in data processing on average, compared with 13 days for men.

Although the new technology is obviously labour saving in character, the results would seem to indicate that the jobs of white-collar workers are not in danger. In recent years, similar surprising (from the point of view of forecasts) findings have been made in studies carried out elsewhere, such as those concerning the OECD countries[1] and the office sector in the United States.[2] However, in service industries, such as wholesale and retail trade, banking and insurance, the new technology has been found to be implicated in the increase in part-time employment. On the other hand, when "natural wastage" is applied, the focus of employment problems shifts to persons wishing to enter or re- enter the labour force.[3] The latter group contains fewer women in Finland than elsewhere, for the practice of leaving the labour force to re-enter it at some later date is less common in Finland.

Among blue-collar workers, especially in the manufacturing industries, the application of information technology has been, with respect to employment, much more problematic. In numeric terms, this applies more to men than to women, for 44% of men were engaged in the manufacturing industries, compared with only 15% of women. However, the job reductions to date have in large part affected women.
In conclusion, women's position on the labour market has improved in that better education also means a better position in working life. An increasingly small portion of women are engaged in heavy industrial work with low pay and inconvenient working hours. On the other hand, there have been conflicting developments in the white-collar worker sector. The distance between lower white-collar workers and upper white-collar workers. Among women, too, the well-educated are in a position to utilize the advances

---

1 Werneke, D., Microelectronics and Office Jobs: the impact of the chip on women's employment, ILO, Geneva, 1983.

2 Automation of America's offices, 1985-2000. Congress of the United States, Office of Technology Assessment. (Washington D.C., 1985).

3 Paukert, L., The Employment and Unemployment of Women in OECD Countries (Paris, OECD, 1984), p. 55.

in technology, while the less educated are faced with the prospect of an increasingly subordinate position.

BIOGRAPHICAL NOTE

The author is a sociologist with an M.A. in the social sciences from the University of Tampere. She serves as a pPlanning Officer in the Central Statistical Office, where her duties cover ad hoc studies on the labour force. Correspondence address: Central Statistical Office,P.O.B. 504, SF-00101 Helsinki, Finland

Women, Work and Computerization: Forming New Alliances
K. Tijdens, M. Jennings, I. Wagner, and M. Weggelaar (Editors)
Elsevier Science Publishers B.V. (North-Holland)

THE TIMES, ARE THEY A-CHANGIN'?

Frida de Jong
Delft University of Technology
Loes Steensma
Netherlands Postal and Telecommunications Services (PTT)
The Netherlands

New technologies change working procedures and hence also the way in which work is organized. Yet, technology plays a secondary role in theories about occupational segregation. The authors investigate how technology and stereotypes of women and technology have again and again confirmed the division of labour between the sexes. They describe the feminization of clerical work in the first half of this century in relation to the introduction of technical equipment in the offices. Using the theories of Sullerot as a point of reference, they compare this historical process with the current automation in The Netherlands and especially at the Dutch PTT. They find many similarities. However, changing patterns in society are indicated that women can handle today to improve their position on the labour market, to influence the organization of work and to introduce technology in an acceptable way.

1. INTRODUCTION

New technologies change working procedures and hence also the way in which work is organized. On the one hand, new equipment creates employment opportunities for individuals, including women, who learn to use it. On the other hand the introduction of such equipment renders superfluous much work - mainly of a routine nature - which in many cases used to be performed mainly by women.

When duties are re-allocated because of the introduction of new technologies, this is mostly done in accordance with Tayloristic principles such as rationalization and efficiency. The principal concerns of profit and non-profit organizations are to keep investment to a minimum, maximize returns and avoid resistance and demotivation on the part of employees. The result is that specialization increases and work becomes more fragmented. Re-allocation of duties related to the introduction of new technology also involves a segregation of the sexes: work is allocated in accordance with the particular characteristics which are attributed to all the members of one of the two sexes.

Technology has played a decidedly secondary role in theories about sexual segregation. It is our intention to investigate how technology and thinking about it have again and again confirmed the division of labour between the sexes.

At present we see segregation in the field of automation. At a previous stage in history, industrial and office mechanization had a similar effect. A comparative study of both processes may improve our understanding of the position of women in an employment market which is changing rapidly because of technical innovation.

Our ultimate aim is to use this understanding to develop strategies on how and at what points in time these mechanisms can be influenced to be beneficial for women.
For this purpose we shall compare current developments as regards automation at the Dutch PTT with the mechanization of clerical and administrative work at the turn of the century. We believe that historical knowledge can be useful as a guidance for developing new strategies.

In our study, we observe new technologies being introduced and at the same time new areas of "women's work" came into being. But at the present, society is changing rapidly and new patterns at the labour-market and in the relationship between the sexes are emerging. We will conclude our article by pointing out the perspectives for women to make use of the new opportunities in automation.

2. THE FEMINIZATION OF THE OFFICE ENVIRONMENT

Office work changed radically in the first half of the 20th century. Such new equipment as adding machines, addressing machines and typewriters came into use. The use of the telegraph and telephones improved communication. Efficient card index and storage systems made it possible to store and process information more quickly and efficiently. The growth of government, the intensification of international trade and the rise of several large multinationals led to the establishment of large office departments and even of separate office blocks. In less than half a century, office work was almost completely feminized, a process most clearly exemplified in the growing numbers of secretaries employed.
In 1900, offices in The Netherlands were still largely a male preserve. Offices were small; many were still attached to the homes of those who owned and ran the business. They were mainly associated with commercial activities such as trade, banking, insurance and stock-brooking. The activities involved consisted of writing letters, drawing up invoices, ensuring that they were paid, and keeping accounts. Records were handwriting, using ledgers. Legible and pleasing handwriting was an important criterion for appointment, as was appropriate linguistic ability. Clerks worked their way up within a business from teaboy to book-keeper. They had a 'clean' occupation, involving contact with the upper classes, and they formed part of a culture based on writing. Lady office workers were a rare sight at the turn of the century. The National Association of Commercial and Office Employees (1897) considered that it "would not be beneficial to encourage more ladies to take up office work". Conservative unions such as the Rotterdam Mercurius (1882) and the association based on denominational principles did not permit women to become members.[1]

With the mobilization which occurred during the First World War, women were recruited as a reserve work-force for offices, for example to cope with the large quantities of extra work generated by ordinances concerning the rationing of essential supplies. Under the pressure of the unusual circumstances, traditionally-minded bosses abandoned their patriarchal attitudes and, for the time being adopted a more modern and more rational point of view. After the war, international competition - especially during the crisis of 1921/1922 - asked for continuing this rationalization. This so called 'efficiency movement' called for young, cheap employees. Single girls with suitable education met the demand well. Accordingly, more and more women were recruited for office work during the 1920s. The unions had already started to accept women as mem-

---

1 Reinalde, B., Bedienden georganiseerd. Onstaan en ontwikkeling van de vakbeweging van handels- en kantoorbedienden in Nederland van het eerste begin tot in de tweede wereldoorlog. (The organization of junior employees: the origins and development of trade unionism along shop assistants and office staff in the Netherlands from the beginning up to the Second World War), SUN, Nijmegen, 1981, p. 86.

bers during the war. During the 1920s they campaigned women to join them so as to ensure that they did not exert too much downward the slogan, "parents and guardians! Don't let your child end up as an office worker!" By "child" it would appear that they implicitly meant "son". In any case, women were running in at the same time. Between 1920 and 1940 the image of office work was in decline. During the 1920s, office wages fell remarkably rapidly in relation to industrial wages. This being so, they could hardly fall much further during the 1930s - at least, not male office workers. It was felt that women could still put up with some further deterioration. [1]

Nonetheless, many girls aspired to work in an office. It was clean work, unlike factory work. In comparison with shop work it had the advantage that you could sit down. Many coming to it for the first time thought of it as an occupation in which they would be ministering to others' needs and working in a civilized environment. This conception was frequently based on an outdated and highly idealized picture (not unlike those which exist of other occupations mainly taken up by women such as nursing or being an air hostess). In view of their socialization in the home, the expectations of their parents - often working class people who wanted their children to do something better - and the schooling they had received, the most that many girls could hope for was to find an office job, ideally just for a view years. (According to a special law from 1923 unto; 1955, all female civil servants had to resign the day they got married).
By 1950, local government, industry and public utilities (the giro bank, gas, electricity and water boards, the postal and telecommunications service, pensions funds etc.) had separate rooms for use as offices, typing pools and telephone switchboards. Work was highly specialized and in many cases consisted purely of standardized form-filling, reference work and filing, rather than the more creative tasks of writing letters and keeping the books. Typing was increasingly taking the place of writing. At this period the typical office worker was a young woman who had no intention of staying in the occupation for long. Employees received little introductory training and above all could readily be replaced. Accordingly, there was very little vertical mobility.
In the first half of the 20th century office work not only increased dramatically in volume: it also changed in character on account of changes in technology and communications. In addition, it came to be regarded as women's work.

3. WOMEN'S WORK

There are various labour market theories which take account of the differences between groups of employees. Doeringer and Poire (1971)[2] sought to explain these differences by formulating their theory of the double labour market. They assumed that two separate markets existed: a primary (strong) and a secondary (weak) one. These markets or segments were to be found in sectors of the economy, individual enterprises and jobs.

Various feminist researchers have observed that women and men occupy different positions within the same segment. They have different careers, even if they start out from the same point.[3]
Sullerot [4] has studied women's work through the centuries. She concludes that there are always characteristic differences between women's work and that done by men in the

1 Idem, pp. 358-359.

2 Doeringer, P.B., and Piore, M.J., International labor markets and manpower analysis (Lexington, 1971), in: Tijdens, K., Automatisering, interne arbeidsmarkt en segmentering, Social Science Informatics Department, Amsterdam University, June 1986, pp. 14 et seq.

3 Tijdens, pp. 21 and 22.

4 Sullerot, E., Geschiedenis en sociologie van de vrouwenarbeid (History and sociology of women's work), SUN REPRINT, Nijmegen, 1979.

various periods and cultures considered. Women often do simple manual work and housework, their work is virtually always low in status and in financial rewards and without much adventure, risks and satisfaction. Sullerot identifies three fundamental arguments which have been used to justify the sexual segregation of occupations over the centuries:

- the "natural" differences between men and women, notably as regards physical strength. Women are also deemed to be better suited to active occupations which do not require a high degree of thought;
- the idea that people need a clear sexual identify and that an established order which allocates separate spheres to men and women provides such clarity;
- views on what is decent and respectable for women. Women are often expected to stay at home on account of their sexuality. Too much contact with the outside world is dangerous and liable to prostitution.

In the 20th century the situation and the ideology remain largely unchanged. Although technology has made physical strength less and less important in the work place, women continue to do different jobs from men. Women's wages are also structurally lower than those of men doing similar work. Only in times of severe labour scarcity (such as during the two world wars), women have the opportunity to do "men's work". Conversely, during the slump of the 1930s women as a group were the first to be compelled to withdraw from the labour market or at least to accept low wages.
Sullerot's theses are as follows:

- the labour reserve: as and when the economic cycle requires, women are employed at the lowest possible wage rates;
- the law of prestige and returns determines how jobs are divided between men and women: women are given low-paid, low-status jobs, while men get jobs which are new, enjoy greater prestige and are better paid;
- a job which becomes bureaucratic is left to women; men are always on the look-out for new (risky, honourable) fields of employment;
- men (fathers, employers, bosses) do not let women take expensive or interesting courses or enter the professions to which they give access.

If we consider developments in office work in the light of the theories outlined above, we see a number of mechanisms being confirmed. Women take jobs at offices because of political and economic circumstances; they form the reserve labour pool and mainly work in the secondary segment. They have little power because they are poorly organized. They are employed at low wages. The law of prestige and returns operates in the type of work which women do in offices and in the associated ideology. Office work has become bureaucratic and poorly unionized. Unions have even advised anybody who is ambitious to avoid it. It is on account of patriarchalism that women are neither immediately admitted to senior positions (or the necessary courses leading to them) nor allowed to work their way up to them.

4. WOMEN'S WORK AND TECHNOLOGY

Sullerot's theory and the labour market theories do not explicitly deal with the role of technology. Nonetheless, technology has always influenced the way iwork is organized.
Classical employment structures grew up around the turn of the century under the influence of industrialization. They are marked by rigid hierarchies, far- reaching division of labour (thinking, decision-making and executive work are separated, and specialization prevails) and a direct link between performance and reward (piece rates). The name Taylor is irrevocably associated with these structures. A major premise of Taylorism is

that there is only one scientifically ascertained way of doing a job. Tools are important conditions and as such worth studying.
In later technological determinism, the role of technology is seen as more dominant: in this view, technology is believed to develop autonomously and to determine the mode of production, the form and extent of the division of labour and cooperation and hence the structure of the work-place. However, this view has been critized by various commentators (e.g. Van Hoof [1]). Of course, technology undoubtedly influences the organization of work: where machines are used they soon come to determine the character of the work, the way in which it is performed and its speed. But this does not necessarily mean that the development of technology itself is autonomous. choices can be made according to the aims of an organization. The outcome of these choices is ultimately determined by the balance of power between employers and employees.

5. THE MECHANIZATION OF OFFICE WORK

5.1 Women and new equipment

In early 20th century office work too, processes were rationalized as new machines were introduced. At the same time, office work was increasingly feminized. In the narrower sense, technological development in the office can be described as the succession of new machines and the consequent process innovations. Once new machines are available on the market and shown at fairs, they catch the eye and often become status symbols for those who own or use them. Initially, the new machines tend to be relatively expensive. There is often a stage during which companies play and experiment with them. This makes work new and exiting. Women are conspicuous by their absence during this stage and at this level. Only once the novelty has worn off are women set to work on these machines. By this time it is well-known how the new machine works, what its capacity is and where it can best be used. Superfluous human labour around the machine is eliminated; so is any inclination to experiment. Work is organized around the machine, it is predictable and the production norm is soon established. In this stage of introduction, mainly directed by men.

5.2 The typewriter

The first people to start using typewriters in the late 19th century were writers and journalists - in other words, men who understood the advantages of producing clearly legible copy. The military too were quick to make the most of them, given the importance of unambiguous orders in military communications. In 1984, the Dutch trading office of Van Eeghen took delivery of its first typewriter, a Harmonia; "the boss was the only person who could use it at all well".[2] By about 1910, every office in The Netherlands had a typewriter, although at this stage it was still a machine for general use; the telephone had previously passed through a similar stage, and the photocopier did so later. As with every other machine, increasing productivity soon became the aim. This was achieved on the one hand by perfecting the machine technically and on the other hand by increasing the speed at which the operator worked. In the early 1920s, typewriters were operated by both men and women, and the masculine form of the world "typist" was used to describe the occupation.

1 Van Hoof, J.J.: Op weg naar humanere arbeid? (Towards more human work?). In: De Galan, C., Van Gils, M.R., and Van Strien, P.J., Humanisering van de arbeid, Van Gorcum, Assen, 1980.
2 Reinalda, p. 99.

In the 1930s the development of office work increasingly paralleled that of industrial production work. It was said that an office revolution had taken place comparable to that which had taken place in factories during the industrial revolution. It was estimated that between 1/4 and 1/3 of all office staff would be made redundant as a result of mechanization.[1] At this time, specialist literature in Germany was now using the feminine form of the word for typist instead of the masculine form previously in use. Tests and courses for typists from this period indicate that speed and accuracy were the most important requirements of typists, together with "a little imagination and inclination to daydream" because the work was so soul-destroying. In the 1920s, the emphasis had been far more on good manners and linguistic skills.[2]

5.3 The telephone

The invention of the telephone in 1876 and the rapid spread of its use since then can certainly be described as revolutionary. Before telephone was automated, connections had to be made manually by a human operator, in practice always a young girl. Although it was a modern, technical occupation, the work was routine, monotonous and, particularly in the early days, irritating because of the large amount of background noise on the line, as well as other faults in connections. The prospects of promotion for a telephone operator were virtually zero. The work was nonetheless regarded as attractive to women because it could be done sitting down. The main criteria for employment were being well-spoken, speed and dexterity.

5.4. Image

As noted earlier, the feminization and mechanization of work have often been explained on the basis of theories about the division of labour and the balance of power on the labour market. Sullerot's contribution was to show to what extent the balance of power between the sexes likewise affected it. Social science research in the 1960s and 1970s pursued this line of enquiry further, developing such concepts as role patterns and the information of underlying images. The feminization of office work and the position of women in relation to the introduction of new technology need to be viewed in the context of stereotypes of women and technology. In the upbringing of children and in advertising, particular equipment is associated with males or females. Research has shown that radios, music centres and the like are relatively sex-neutral in this respect. Other equipment for use in the domestic environment is quite otherwise, however. Equipment for carrying out impressive household improvements which need to be made only once and are then finished with (drills, saws) or for relatively infrequent tasks (lawn-mowers) is mostly operated by men. Equipment for less spectacular routine work (washing clothes, ironing, vacuum cleaning) is mostly operated by women. In this way new technology is incorporated into highly traditional patterns. In the office, mechanization and feminization went hand in hand. Illustrations from for example the Dutch periodical "Administratieve Arbeid" (Office Work) demonstrate this more than the content of this article does. The question of women's role is disregarded, and women are never quoted while the illustrations show mainly women operating office machines. They clearly tell their own story.

---

1 Administratieve Arbeid, Maandblad voor rationele werkmethoden (Office Work, Monthly magazine for rational working methods), 1930, p. 267.

2 Inter alia C. Bunt, 1923: J.M. Lahy, 1924: D. Bieneman,1923.v

6. AUTOMATION AT THE PTT AND OPPORTUNITIES FOR WOMEN

6.1 Women and automation

In the present day, the pace of technical advance is accelerating. It is proceeding so fast that employers are hard pushed to keep up with it, both as regards updating their equipment and as regards training staff to develop, maintain and operate it. New jobs are created, and there are many who suggest that the shortage of skilled staff presents an ideal opportunity for women to catch up with men in this sector of the labour market. In the following, we shall examine this suggestion with special reference to the situation at the PTT.

In general, in The Netherlands automation is a field where far more men than women are employed. This tendency is actually increasing: whereas in 1976 9.4% of the work force in the sector were women, by 1987 the figure had fallen to 6.1% [1](study of the Automation Market.[2] The survey by Berenschot in 1986 [3] quotes a figure of 6.6% (and 7.7% in 1987), nothing that women are mainly employed in the more lowly strata of the field: in 1982, 12% of programmers and 16.2% of junior programmers (and 17.2% in 1986) were women. In 1982, women occupied 3.5% of the more senior posts, and in 1986 a mere 1.9%. This is in spite of the fact that, according to the Study of the Automation Market, relatively as much men as women have graduated from a higher education level. Yet, the percentage of men in higher and managerial jobs is 3,5 times as high as the percentage of women in that kind of jobs. On the other hand, automation is a field where there is a severe shortage of staff at present.
The Berenschot survey (1986) showed that 14% of the vacancies in The Netherlands are jobs related to automation. Although, as this indicates, there is considerable demand for experts on automation and related subjects, women are evidently not successful at penetrating into the field.
The PTT is no exception in this respect: most of its automation staff are men. Twenty-two percent of the PTT's total work force are women; the majority of these are employed in clerical and service posts, such as post office counter work and directory enquiries. Two thirds of the women work part-time, as against one sixth of the men.

In 1984 the Board of the PTT explicitly announced that it was adopting a policy of appointing more women to jobs where they were underrepresented. The decision was based on two considerations. One was social trends: the PTT likes to consider itself a modern employer and attaches importance to its image and to pursuing an enlighten staff policy. Tho other was the realization that for economic and competitive reasons it was essential to make maximum use of the potential available, both inside and outside the PTT, and that this included the female section of the labour force. Staff shortages are becoming increasingly pressing, particularly in managerial and automation posts. The PTT is encouraging women to take up automation work as part of its more general policy on creating equal opportunities. In an experiment designed to counteract role patterns, 16 women were trained as operators; a number of specific recruitment drives have been organized, and information is provided both to management and to women themselves on the opportunities which exist for women in the automation sector.
Over the years, various jobs have been created in automation and related fields at the PTT. They may be classified as follows:

---

1 These figures exclude data entry staff.

2 Branche-onderzoek Automatiseringsmarkt (Study of the Automation Market). In: Computable 20, (1987), 27, 3 July, p. 17, 18.

3 Idem

- Technical posts: e.g. operators and system programmers. These are mostly at computer centres.
- System development: e.g. programmers, systems analysts, systems developers. These are mostly in automation departments.
- Policy and management: e.g. systems managers, help-desk/microcomputer advisers, information analysts, information advisers, information managers. These posts are sometimes at the automation department, sometimes at the organizational department and sometimes in the user organization itself.

There are very few women in any of these jobs Apart from the 16 women operators from the experiment, there are a few female information analysts and help-desk staff.[1] The following arguments are adduced to explain sexual bias in this sphere:

- certain characteristics peculiar to women suit them particularly for certain jobs. People who use this argument mean that women could be microcomputer advisers because this entails serving others, and that they could work as information analysts because they are capable of working with precision;
- women do not have the right previous education (too little mathematics at school, for example);
- women themselves are not interested in genuinely technical work.

From the above "arguments" it seems that the blame for the limited participation of women in the automation sector is largely laid at their own door. However, there are strong counter-arguments.
As regards the educational argument, for example, a study of a Dutch retraining project for unemployed school-leavers (the ISI project, which was specially concerned with training school-leavers to work in informatics) has shown that it made no significant difference whether trainees has studied mathematics or not: the results they achieved in training were unaffected. There is also a certain amount to be said about women's interest in the field. It is striking for example that more than 300 women applied for a job as systems manager in the Groningen Telecommunications District when it was advertised specifically as a job for a woman. Evidently they were not afraid of the job as such - or of the company. The advertisement stated that no special training was required, as this would be provided; that the job could be done part-time, that nursery facilities would be available. Furthermore, the advertisement was specially addressed to women, which may have inspired confidence in the seriousness of the PTT's intentions.[2] There was also great interest in the experiment to counteract role patterns by training women operators. The 16 women who take part all completed their training successfully and are now working at the PTT's computer centres. However, considerable difficulties were experienced in arranging practical experience during training and in finding posts for the women once they had been trained, on account of resistance by managers of the computer centres.
Thus the limited number of women working in automation cannot be attributed purely to women's own attitudes. Many structural and ideological obstacles exist within the organization which must be overcome by women.

6.2 New opportunities for women?

The indications are that the familiar segregation of the sexes, with women taking jobs which are low-paid, carry little status and are of a routine and/or service nature, is

---

1 "Ik wist niet dat jij technisch was?" (I didn't know you were technically minded?). Brochure issued by PTT Headquarters, The Hague, 1987.

2 Werft........en gij zult vinden. Intern rapport Centrale Directie PTT. (Recruit...... and you shall find. Internal report, PTT Headquarters), The hague, 1987.

emerging in new areas of automation. Apart from similarities, developments in society are leading to new patterns at the labour-market and in the relationship between the sexes. Analysing the similarities and the differences between the actual development in the field of automation and the pattern in the past, can deepen our understanding of the underlying mechanism. Also, it can help us break away from the historical pattern of sex segregation and to strengthen the position of women on the labour-market.

Sullerot's "laws"or conclusions seem to be borne out in the development of automation. The l*abour reserve* is drawn upon in times of scarcity. The shortage of automation experts is one of the principal arguments (within the PTT as well as elsewhere) adduced in favour of encouraging women to work in automation. In the case of certain types of post, the company pays for training (this would be inconceivable if there were no labour shortage).

The *law of prestige and returns* operates, determining the time at which women enter certain occupations. At present, for example, most of the programmers are men. However, the status of this job is on the decline and it has even been predicted that no programmers will be employed in ten years' time. In fact, women have entered theses jobs lately; in The Netherlands, programming is the job in automation that counts the highest amount of women (14%).

Bureaucratism is occurring in the increasing division of labour in automation. First, programming was separated from systems analysis; next, programming was increasingly standardized. The work of operators has also gradually changed: whereas they started out as mechanics, they now increasingly manage and supervise systems which themselves provide information on what should be done next.

As regards *male resistance*, the "arguments" used for allowing women to enter particular occupations indicate that it is indeed a matter of men allowing women to enter them; clearly it is men who decide what women are good at. In automation we likewise recognize the principles of division of labour and labour-market theories which have been discussed above. We see that the main principles of Taylorism are still at work: as the field of work expands, job specialization, standardization and the reduction of tasks to the elements of which they consist occur. In consequence, the number of employees who require more specialized knowledge and training is reduced to a minimum and the majority of the work force can be employed at the lowest possible wage levels. By confining work to a limited number of routine operations, work is intensified. (In extreme cases it may only be possible to do the job part-time as a result - as has happened with directory enquiries, for example). Most jobs which have undergone such a process are occupied by women. In this way, wage costs are kept to a minimum and employees are readily interchangeable and replaceable. Moreover, labour time and operating time can be separated from each other without difficulty (shift work) so that optimum use can be made of capital goods.

However, in jobs where higher qualifications are required it is not so easy to draw on a labour reserve which is interchangeable and will accept any type of work and terms of employment.

Specialist knowledge is in short supply at present; this strengthens the position of those staff with the right knowledge and skills. Companies are therefore greatly concerned to tie specialists to the organization they are working for by improving the quality of jobs, providing further career opportunities etc. At the same time, companies seek to eliminate the shortage of specialists as quickly as possible, and one way of doing so is to recruit women as well as men.

When it comes down to it, it seems that the views on women and technology have not changed much either. Women are entering particular jobs within automation, namely those which are not regarded as genuinely technical. Developing and administration of software and especially the so called 'intermediate jobs' are regarded as pre-eminently suitable for women. Women are supposed to have the serving attitude and the social skills that are needed to support the users.

6.3. Cultural changes

The above similarities do not sound very optimistic about the possibility that sexual segregation in the automation world may be overcome. Yet there are also differences, which make it possible that events may take a turn which would be unexpected in the light of past history.

The main differences lie in the socio-cultural sphere; also we can observe developments that influence the balance of power between men and women and between employers and employees.

Feminism: Women are increasingly aware of the importance of economic independence. It is more common for women not to leave the labour-market as soon as they get married or have children. Women enter higher education and aspire a career. Also, women are increasingly aware of patriarchy and of the necessity to organize themselves to counterbalance it. This is illustrated by the emerging Women's Networks with general and/or jobrelated purposes.

Demographic trends: Families are becoming smaller, and it is easier to plan families. People are becoming more mobile and opting for alternative living arrangements. AS a result of all this, more and more women are continuing to work even after they have children. Small families and the ageing of the population is causing shortages on the labour-market, particularly of young people, and this strengthens employees' position on the labour-market.

Labour-market trends: as a result of the on-going process of automation, the amount of unskilled work which needs to be done, particularly in industry and at offices, is decreasing. Technological advances are creating a growing need for experts with higher education. Employers will have to recruit and, if necessary, train women in order to overcome shortages. The number of occupations in which physical differences between the sexes can be used as an excuse for preferring men will continue to decline.

Changing organizational theories: Companies and other organizations are increasingly confronted with a rapidly changing environment. To react to this adequately, the traditional Tayloristic principles are not suitable anymore. The actual Human Resource Management emphasizes the highly qualified, flexible employee, who must be able to be responsible for his or her job. Especially in jobs in the field of automation, concerned with the development and introduction of new technologies, Tayloristic principles will be detrimental for the creativity and flexibility that is needed for this kind of work.

7. STRATEGIES FOR WOMEN

Difficult though it is to affect the course of history, we believe that there are ways in which the above-mentioned factors which favour change can be exploited. Changing patterns in society open new perspectives to break away with the division of labour between man and women and to fight patriarchy. Women can follow several strategies to realize these perspectives.

* Responding to contemporary opportunities and trends.
  Advantage must be taken of shortages on the labour-market at the right moment, notably as an opportunity for women to obtain technical and managerial jobs instead of acquiesing in selective employment in particular jobs regarded as suitable.
  Shortages on the labour-market and the ageing of the population must be exploited to eliminate the structural obstacles to the employment of women. This means e.g. claiming (temporary) part-time work, flexible working hours and day-nursery facilities; offering training and education to women who are entering jobs in the field of automation.
* Girls must be encouraged to opt deliberately for courses which present prospects. Where necessary, courses leading to employment in areas of automation where

women have not traditionally been accepted should be organized by and for women.

* Combating stereotypes.
  The stereotypes which exist in the minds of both men and women should be combated by improving information material at schools and labour organizations.
* Qualified jobs.
  Women should fight for broad, highly qualified jobs and for the re-valuation of social skills; these skills are up till now mostly underestimated in women's work. Coalition with the unions can be made at this issue.
  Also, personnel advisers and system could be influenced.
* Increasing women's power.
  The organization of work is not determined purely by technology but also by the decisions which are taken; these decisions are the resultant of the balance of power on the labour-market and between women and men. In order to improve the position of women on the labour-market it is necessary to adjust the balance of power in women's favour. Women must organize via networks and representative organizations in order for example to ensure that women are not sent home again once the shortage has passed, in order to secure better terms of employment, in order to create career prospects and so on. For many women, networking will also be a possibility to break away from their isolation and to increase their self-confidence.

ACKNOWLEDGEMENTS

We owe special thanks to the following people who provided critical insights as referees during the conference: Cynthia Cockburn (City University London, United Kingdom); Martha Kirezcyk (Free University Amsterdam, The Netherlands); and Ina Wagner (Research Centre Labour and Technology, Vienna, Austria).

BIOGRAPHICAL NOTE

Drs. F. de Jong is a historical,working as a researcher at the Faculty of Philosophy and Social Sciences (WTM) of the Delft University of Technology, P.O. Box 5060, 2600 GB Delft, The Netherlands.
Drs. C.L. Steensma is information-consultant at the Information Management Department of the Dutch Telecommunication Services, P.O. Box 30000, 2500 GA The Hague, The Netherlands. Until recently she has been a researcher at the Social Research Department of the Netherlands Postal and Telecommunications Services (PTT).

Women, Work and Computerization: Forming New Alliances
K. Tijdens, M. Jennings, I. Wagner, and M. Weggelaar (Editors)
Elsevier Science Publishers B.V. (North-Holland)

PART 2

STRATEGIES FOR SYSTEMS DEVELOPMENT

Mary Jennings
United Kingdom

INTRODUCTION

The theme of women, work and computerization not only refers to the work that women do and how it is being affected by computerization but also to women who work in the field of computerization itself. Women are traditionally seen to be outsiders in terms of technology, we are not supposed to like or enjoy working in technology. However as with many myths about women this is only partially true. There are many women working in the field of computing as computer programmers, systems analysts and developers, trainers and computer scientists. For those of us who work within the field of computing the workshops on strategies for system development were a very exciting component of the conference. I have included comments from referees and reports from the workshops in this introduction.

METHODOLOGIES

Kari Thoresen in her opening speech to the conference raised major issues about the state of system development today when she posed her question "Why are computer systems impractical to use?" The reasons were not that systems developers were not interested, it was mainly that there was a lack of suitable design approaches for developers to use. The methods available are biased towards what the computer can do, 'the machine approach' which Thoresen compared with a 'tool approach'. A tool is an instrument that is refined over time by the user and the user remains the source of professional expertise. Thoresen demonstrated the methodology of the two approaches to a project to convert maps and aerial photos into computerized digital maps. From a system development viewpoint there was a problem with the tool approach however. The refinement of systems, 'maintenance' had low status in the profession. She concluded by saying the conference of a design approach had different implications for the user to develop autonomy and competence.

The concern with developing methodologies for system designers was continued in several of the papers. Fiorella de Cindio and Carla Simone introduced their 'five universes of discourse'. Participants in the design process, from designers to users, use different languages, or 'lexicons' to describe their interactions with computer systems. Each language is separate and does not convey useful meanings to the other participants.

The aim of the 'universe' framework proposed by de Cindio and Simone is to create a reference system in which the concerns of the different 'lexicons' can be compared. The universes are as follows: the 'physical' is concerned with the environment eg. screens and keyboards; the 'operational' refers to such things as accounts systems and fourth generation languages; the 'procedural' refers to the products of the system and the work performed to achieve them; the 'commitment' refers to the group structure and interactions in which work takes place; finally there is the 'individual universe' which is concerned with people's subjective perceptions, attitudes and expectations. The authors have used their 'universes' framework with women workers affected by computer-based technologies. They suggest that the 'five universes' approach could be used to analyse the techniques in system development, 'a new tool in the tool kit owned by designers and users to enhance their mutual understanding'.

Margrit Falck argues that system design 'by the user' does not yet have a developed methodology. Her proposal is for an 'actional strategy' which takes the job-activity and its motivation as the starting points for system design. The user, she maintains, who is willing to carry out the task and is part of the social organization. This approach, as it is not focused on the technicalities of the system, puts women who are mostly in non-technical fields, more into the centre of system design.

However, one of the problems with putting women's work at the centre of the systems development process is that the expertise that women have is consistently ignored, undervalued and not given status. Lucy Suchman and Brigitte Jordan ask the question "How can we make women's ways of knowing visible as the basis of design?" They compare the introduction of medical technology into the childbirth practices of the Yucatan Indians and the introduction of computer technology into office work. In both instances they find that the introduction of technologies fails because of the assumptions embodied in the new technological innovation. They argue particularly against the 'fallacy of the empty vessel', which assumes there is no pre-existing technological and social patterns with which 'new technology' has to contend. The 'empty vessel' approach can lead to the new system being ignored by its intended targets. Suchman and Jordan suggest two methods to illucidate women's expertise and knowledge: ethnography, borrowed from social anthropology, to study activities in a social setting and interaction analysis to investigate the relationships among human beings and with technology.

Whereas some attention has been given to the 'users' of computer systems, hardly any research has been done on the system developers themselves. Susanne Bødker and Joan Greenbaum, in their ROSA project, investigate the informal relationships among system developers. They use the 'gender perspective' that is an awareness of gender as a dynamic in the workplace and not a focus on difference. As researchers and system developers they are interested in constructing useful computer systems in supportive ways, in high-lighting the social construction of computer systems and the informal rather than the formal working practices of systems developers. The questions they focus on are: what is the role of intuition, 'know-how', in the system development; what learning takes place?; what types of language are used by system developers; what roles are played by system developers?. Their methods are multi-disciplinary, from action-research from the social sciences to conversational analysis from linguistics.

Finally two papers investigated whether trade unions can interact successfully in the system design process to the benefit of women. Leikny Øgrim studied the process of computerization in a book publishing company. From her research she argues that some of the informal work organization in the system development may be advantageous to women particularly with user participation schemes. As women have a weak position in trade unions anyway, trade union negotiations may not be relevant to women workers. Whereas management can use the introduction of decentralized computer systems as a means to weaken trade union power, these decentralized systems because of their local nature may be used by women to gain some control. The research of Anne P. O'Brien and Monique Ramioul, focuses more on the problems of women in trade unions than on the role of trade union in system development. They suggest that new technology agreements put women in a more vulnerable position than their male co-workers and that women need to organize their own 'channels for opposition'.

WORKSHOP DISCUSSIONS.

The workshop attracted computer scientists, social scientists and users, a good mixture of groups who are affected by system development.
The new technique of ' the framework of the five universes of discourse' was discussed in terms of its practical usefulness to system developers. The workshop agreed that using the framework was one way of getting into a user organization. It could be used to order the thoughts of system developers so that they did not make too many easy assumptions about solutions to problems in user organizations. The 'framework' could also be seen as, and used as, a complete prescriptive method but that was not its intention. There is a conflict between seeing the 'framework' as a generally effective tool and between using it as a perspective for understanding one specific workplace. There was discussion about whether a all-inclusive method was required or whether what was wanted was a way of working together so that many people can contribute to the understanding of a situation.
Some comments were made about the 'universes' themselves. Could there be more than five universes? Using the 'framework' in new situations may make the original choice of universes seem arbitrary. The 'commitment universe' is identified with women's concerns and this seems to correspond with responsibility for the consequences of our own actions upon others, the environment and society. The comment was made that perhaps a discussion on the 'individual universe' could be expanded as properties belonging to that universe are often used against women. The focus on problems in each universe may mean that the 'framework' is most suited to analyzing hat happened after a new system is installed and in many cases this is too late. The openness of the paper proposing the 'framework' was appreciated especially its tone of optimism.

The culture in women's work environments and in user organizations was the focus of another discussion. Current technology in the user organization is the basis for competence. 'Old technology' is often suited to the concrete situations in which it is being used. There are reasons for the old way of doing things even if system developers cannot see these reasons as they come from the outside. There is 'no empty vessel'. The question was asked 'How could computer scientists get to know what's in the vessel?' Some points that emerged from this discussion were: each user organization is special in some way, there is no overall generalization that applies to all user organizations. There will always be a history and a culture in the organization and this should be the basis for change. This existing technology is a part of the basis of the organizational culture. The question of who should decide how the user organization is to be changed was discussed. The workshop agreed that computer scientists must learn to relate to other people's settings and culture. They should relate to both the ideology and the practice within the organization and to the formal and informal organization structures. Computer scientists are therefore much of the same situation as the social anthropologist - both types of work are related to analyzing other people's organizations. Hence computer scientists may use the techniques of anthropology in systems development.

More techniques were discussed for 'getting into situations'. Using video technology, making a video of one specific process, discussing it with other people and with the people in the video can provide several explanations of the same phenomena and can elicit some of the hidden 'tacit knowledge'. Deliberately making a 'bad' system demonstration in order to get users to explain what is and what is not useful can be helpful to system developers. The sharing of working space may also improve communication in the work process. There was a discussion about how to deal with the time and resource limitations of system development. If it is possible to reach the goal within the limits, it is better to say that the goal cannot be reached rather than rushing into creating a 'bad' end-product. The question was raised 'Why discuss system development?' It was agreed that influencing and changing system development is a strategy for improving women's working conditions.

'Do women computer scientists design better computer systems for women?' There was discussion about differences between men and women that may influence computer science and that perhaps female computer scientists may bring other ways of working into the field of system development. However some reservations were expressed about such questions that apply to many other similar issues. Women need to identify and describe properties particular to our situation as women. We can then know our positive strengths and good qualities and incorporate them into our working methods, computer systems, and society in general. However the moment we present a list of 'women's properties and preferences' this may be used against us. We can then be vulnerable to sexist arguments such as those from sociobiology, like 'It's in our genes and therefore our inferior position in society is natural'. A focus on 'women's properties' or 'women's values' may make us not question whether they are idealizations formed by oppressed people. In addition such a 'list' may function to unify women and set up norms for women when in fact many women may not identify the 'list'. It is important to investigate motivations of researchers and writers on issues of 'female properties'.

Can trade unions influence the design process? Do trade unions represent women's interests sufficiently? These were the central questions arising from the workshop on trade unions.

The first general question led to a distinction between two levels of influence in the design process: the conditions under which the new system is introduced and the content of the design process. Before the system is introduced, trade unions can influence conditions like salaries and affirmative action procedures particularly in countries where new technology agreements are required by law. Even after the implementation of a new system, trade unions still have some influence, for example over the use of monitoring systems. Trade unions are traditionally strong in bargaining for these types of conditions. Influencing the content of the design process, both system design and job design, is more difficult. This can only happen by participation at the departmental level. But managers ad system designers are becoming aware that the people who do the work have the necessary knowledge to help design a successful system. Participation os the trade unions is sought but often without giving them any real influence. It was suggested that the trade unions could bargain over the conditions under which they would participate in the design process saying that no information would be given to designers unless some real power of influence was given to the unions in return.

In terms of the second question, the weak position of women within all trade unions was generally recognized. The organization and overall membership of unions varied considerably from one country to another. The highest membership rates are in the northern European countries, about 70%, whereas in France union membership is quite low, about 14%. It was not considered desirable that one type of union organization should be used as the model for all countries but that trade unions should be specific for local conditions as with system design and research.

The stronger participation of women in trade unions was seen as a hopeful development, as with recent important strikes in Norway and in Britain within traditional women's fields like nursing and teaching. The question is not how to mobilize women to take part in the unions but how to mobilize the unions to take a stand on women's issues. Affirmative action within the unions is necessary to strengthen women's' position. As long as women are in a minority at the highest representative levels, trade unions tend to 'eat' their active women. As far as automation is concerned, women and men have different interests. Before and after automation different tasks are always assigned to men and women. Participation in systems design was seen to be attractive to women as women's values and concerns are traditionally more concrete and specific.

STRATEGIES

Several strategies were mentionned in the workshops:

* System development from a women's perspective is a strategy for improving women's working conditions.

* We need to understand women's work in order to design 'better' systems, but there are certain dilemmas. The first concerns the general versus the specific: computer scientists are trained to look for similarities and routine in order to generate computer solutions but on the other hand we always have to adjust the computer system to concrete local settings in order to make a useful system. The second dilemma is the process versus the product: being product-oriented means looking for technical solutions but we also want to be more process-oriented in doing system development by focussing on designing a 'good' system development process that will result in a computer system.

* Strategies for going into situations: the 'framework of discourses, video technology, share workspace, using both observation and interviews, elaborating alternatives, investigating the use of current technology, encouraging social scientists to transform some of their knowledge and techniques for the use of system developers.

* Mutual learning as we cannot separate tools used in system development work from the ways we organize systems development work, mutual learning between all participating users, computer scientists and social scientists. Learning and communication is part of system development and is a long term strategy. This has implications for the methods and tools we use as well as the way we organize projects.

* A short term strategy is the conscious re-definition of the tasks to be solved in system development. This includes conscious participation and multi-disciplinary teams.

* A recognition that there is not one gender perspective but several, combining looking at the whole and not the parts, looking at problems and not solutions, looking at all users of the system and recognizing that system developers too are part of the system. System development is an exploratory process.

* Discussion between system developers and sociologists. There is now a lot more disagreement between system developers and sociologists. We disagree more than ever but this is good because it means that both know more about each others work. It used to be sociologists and users on the one side of the fence and system developers on the other. But now there are any more links between us. We are together as part of a process working for e feminist political strategy towards a better technology for women.

* Trade union action at central levels to set conditions should be combined with local level involvement in the design process. Local levels can provide more room for women's involvement. Carry bargaining and negotiation into local conditions of participation in the design process. Develop strategies for trade unions to influence job design.

* Develop strategies to get systems developers with humanist perspectives and with local knowledge, understanding and local involvement.

* Use the strategies of education and training about new technologies to expand traditional trade union training into the system design process and to involve more women in trade unions.

* Develop support networks of women within trade unions, between unions and outside unions in the community.

* To address the structural position of women in the labour market and in the unions: to mobilize women in unions we must mobilize unions towards women.

Women, Work and Computerization: Forming New Alliances
K. Tijdens, M. Jennings, I. Wagner, and M. Weggelaar (Editors)
Elsevier Science Publishers B.V. (North-Holland)
IFIP, 1989

SYSTEMS DEVELOPMENT ALTERNATIVE DESIGN STRATEGIES

Kari Thoresen
Norwegian Computing Center
Box 114, Blindern
0314 Oslo 3

Computer systems may be designed according to different principles. The paper outlines two approaches to system design: the machine approach and the tool approach. By means of an example from map production, the two approaches are illustrated. It is argued that the machine approach may lead not only to unfortunate working conditions but also bad product quality.
To change current practice of systems development, laws and agreements allowing participation and negotiations are necessary. In addition systems developers should have increased opportunities for direct feedback from the users/operators, both during the design activities and during the use of the system. Also, new models for evolutionary development are needed.

1. INTRODUCTION

This paper takes a rather different outlook compared to the other papers. The paper will try to illustrate some of the choices that systems designers may face in their work, consciously or not. My reasons for choosing this position are several, but I will concentrate on one particular reason.
Norway has a rather well developed set of laws and agreements for participation and negotiation concerning computerization in the work place. By and large these rights are accepted by the employers and to a large extent used by the unions. However, there is a growing realisation, also on the part of the unions, that negotiation and participation is only one side of the question. A number of systems development projects with well functioning cooperation between systems designers and union representatives have not given the anticipated results with respect to system quality, in particular when it comes to the characteristics of the system in use. We may end up with systems which are impractical and interfere with the users' work in ways which nobody was able to foresee. In other words: why do things go wrong when we do things right ?
As concerned systems developers, it seems natural to ask ourselves what is wrong with the way we design systems, and how can we improve it.

2. WHY ARE COMPUTER SYSTEMS IMPRACTICAL TO USE ?

There are of course a number of reasons. The discussions of technological or social determinism point to some of the reasons. However, neither technological nor social determinism is sufficient as a general explanation for what happens when computer systems are used in daily work. I would like to point to another position, not yet properly developed, which concentrates on the use of computer systems: how they are used by humans, and how they are made to fit into the primary work of the users. Understanding the use of computer systems demands insight in the design of the system, in the actual work in which the system is used, and in the context in which the work is performed.

Methods and techniques are inadequate.

Seen from inside the systems development profession, it seems obvious that some of the reasons for systems which are impractical and cumbersome to use, are the methods and techniques available to the systems designers, and the values and understanding inherent in these methods. Their origin in technical-mathematical disciplines tend to focus on the technically elegant solutions, seen from within the computer. The "outside" becomes less interesting, and the use of the systems is of marginal concern.

The reason for this is not only lack of interest on the part of the designer. It is also due to a lack of methods and techniques to design systems with particular characteristics of use (which must not be understood as "user-friendliness"). Even those who are interested and realize the importance of this field, are in practice quite restricted because of lack of adequate design approaches. The computer profession has an inadequate understanding of the concept of "work", which makes it difficult to design systems with high quality characteristics of use.

When I maintain that low quality systems design is to a large extent a result of the systems designer choices, it does not imply that a systems designer wants to design low quality systems. It is my conviction that most systems designers want to make systems of high quality with regard to the users' working conditions. However, methods and techniques are biased towards what the computer can do, not what the people using the computer can do. Based on such a construction logic it is natural to neglect the use of the system, it is the construction which is fascinating. In addition, the designers are often severely restricted by economical, organizational and technological limitations.

The machine approach.

One point of departure for systems design is "the machine approach", which underlies many of the visions and nightmares of automation. It looks upon a computer system as a machine intended to automate a work process, and the designer's task is to find out how far it is technically possible to automate. The user/operator comes into the process only when the computer is not able to deal with some particular task. The operator thus becomes an extension of the computer. The operator's job will be a combination of tasks which the computer cannot handle: a seemingly random mixture of activities, determined by the (present) weaknesses of the system. This view manifests itself by the attitudes expressed by some systems designers when they complain about having to rely on the users for controlling parts of the system. The designers feel that they have not done the job properly, they did not succeed. In this view, humans are "not yet redundant" parts of a machine.

One system - two design approaches.

To illustrate how the machine approach will manifest itself in system design, I will use an example from a project in which I took part a few years ago. Based on project discussions, I will give an example on how a part of a computer system may be designed according to a machine approach and a tool approach, and how these approaches may influence the work situation for the users.

The example is from the area of map production. The project was a systems development project, where the final computer system should be used to convert existing maps and aerial photos into "digital maps". When the existing maps or aerial photos are digitalized into databases, it is easy to change the maps, or to produce different versions of the map, e.g. seen from different angles, or different thematic variations.
One part of the work process would be to digitize ordinary topographic maps. This is done in several steps, here I will concentrate only on the two first steps.

1. The map is scanned by a laser scanner. The laser beam sweeps the map systematically and produces a raster image, a map copy in raster format stored in the computer. Instead of lines we now have long, narrow chains of dots (fig.1a and b).

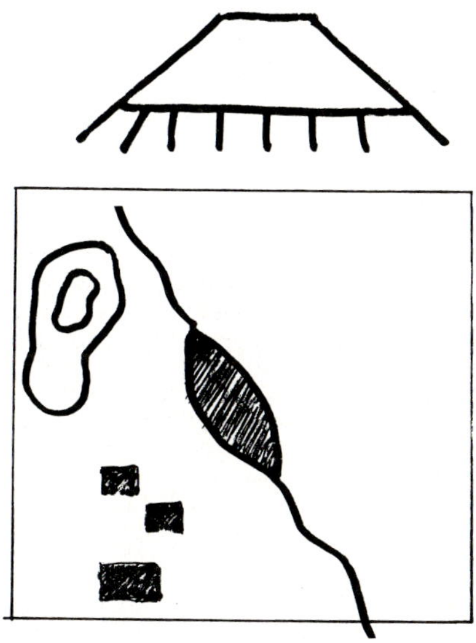

Fig.1a Scanning the map

Fig.1b.Producing a raster image

2. The raster image is transformed to a line image. The long areas of dots are transformed to lines which are mathematically defined and are stored in the database The lines may be retrieved for updating, or displayed in different ways (fig.2, next page).

Fig.2. Displaying a version of digital map

The system may be designed in several ways. The machine designer would ask herself how far it is possible to automate the process. The first step is straightforward: start the scanner which scans until some pre-defined limit. The second step is somewhat more complicated, since the question of judgment arises: What are lines and what are areas ? An example is a river running into a narrow lake (Fig.3a). A person, and in particular a cartographer, will have no problems in distinguishing even very narrow lakes from rivers. The computer system however, has no possibilities to distinguish between them, it registers different shades of grey, in the form of dots only. It cannot decide whether a small expansion is a small lake or just a heavier line (fig.3b).

Fig.3a. A river running into a narrow lake.

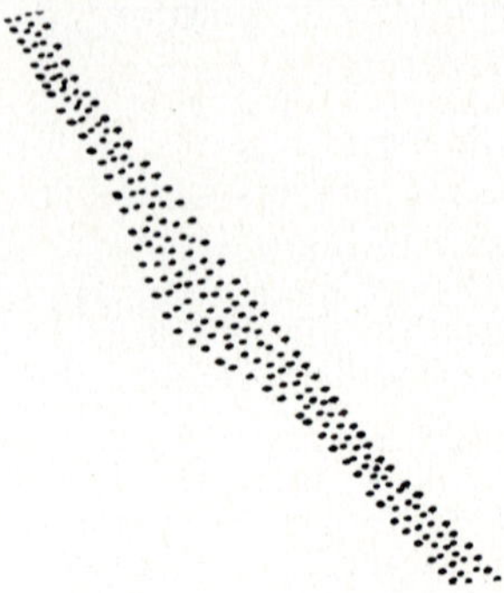

Fig.3b. The computers view of a river running into a narrow lake.

One way of solving this problem for a machine designer, is to use a "standard limit". If the distance between the outer dot on each side of the dot chain is greater than, say, 2 mm, the chain is transformed to a lake. If the distance is less than 2 mm, the chain is transformed to a river.
The designer has built into the system a decision built on judgment, and so far she has been able to keep the operator out. The operator does nothing but place the map in the scanner and wait till the scanning is finished. Her job is that of surveillance.
This works smoothly for a while until a new series of maps is being fed into the scanner. They are from a mountain area, with high and steep mountains and very narrow valleys. Most lakes are narrower than 2 mm (the standard limit), and 80 % of the lakes are (erroneously) transformed into rivers. Now we have not only a monotonous surveillance job, but also low product quality.
The systems designer understands that this is a general and recurring problem. It is no use just changing the standard limit to a smaller value, because new situations where judgment is needed will always occur. It is impossible to foresee all types of situations. There is no way around it: the operator must take over this particular part.

A dedicated machine designer may do it as illustrated below:

> She leaves it to the system to transform all "safe" cases, i.e. where the chain width is well below 2 mm (they are transformed to rivers) and considerably more than 2 mm (they are transformed into lakes). When there is doubt, the operator is called upon. The systems halts, displays the case on the screen and asks for a decision: river or lake. The operator decides and enters an appropriate command, and the system resumes the processing.

It is no longer an automatic process, the operator is a part of it. The control lies with the system, however. The system decides what she should do and when. The operator cannot decide when something is displayed, it is the construction logic which decides. Her work is determined by deficiencies in the system
The machine approach is based on Tayloristic ideas from the early part of this century, but it is still alive as a part of our design basis. Not as explicit methods, more as an underlying drive, disguised as a professional challenge to see how far you can push the limits of your system.

The tool approach.

But there are alternatives. An popular metaphor is the "tool". At present it is mainly a hard-selling word among many other words in the computer business, but it is interesting to explore it a little further.
The tool metaphor is used for different purposes. On the one hand, it may be used as a pacifier, to diminish the importance of the impacts of the computer system for worried users ("Relax, it is only a tool"), and to restrict systems designers insisting on understanding the contexts in which the system will be used ("You should design a tool for a specific purpose, no need for analyzing everything else"). Such a use of the tool concept makes a critical position necessary. It hardly gives an adequate picture of the importance and impacts of computer systems in organizations.

On the other hand, it may also be used as a metaphor to guide systems design aimed at improving a work situation. This is the aspect of the tool metaphor I would like to explore here. The designer will try to extract the characteristics of a tool considered crucial for work performance and use these characteristics as guide-lines for design.
The point of departure for a tool designer is not what the system can do, but what the operator can do, and how her professional competence could be augmented by use of a system (Thoresen, 1983; Ehn, 1988).

A tool designer will take for granted the professional skills and knowledge of the operator, both regarding her work and her tools. Furthermore, the control over the tool should lie with the operator. It is her choice when and how she will use her tools. A tailor knows when she must use cutters or embroidery scissors, and she knows how to use them. Professional use of tools are based on knowledge about both the work process and the tools.
The tool designer will start with the work of the operator: what does she do, and what kind of knowledge does she have to support what she does ? And here is the first stumbling stone for systems designers. Our methods are insufficient to analyze work and professional knowledge. They are based on a mechanistic understanding of work, what is knowledge, and how can we transform our analyses to systems design. Therefore, it should not be left to computer professionals only.
It goes without saying that the potential operators are the ones who possess most of the knowledge we look for. Participation is a must. But participation does not solve everything. The field of work analysis and design is one of the many areas of systems development where we need interdisciplinary cooperation. Work sociologists and psychologists have worked for many years analyzing work and knowledge, but their results are not always usable for systems designers. The basic difference between "analyzing/understanding" and "design/construction" is a problem, in systems development as in many other areas.

Assuming that the same system was to be designed as a tool, the approach might be as follows.

> The tool designer in cooperation with the operator goes through the operator's work and gets an impression of the operator's knowledge about maps and map production. The designer soon realizes that the operator has a vast knowledge of her work which the designer can never even hope to survey. However, one important thing is certain: The operator knows that maps come in different scales, quality, types, series etc. Scanning them all through the same standardized system procedure would in all probability give bad product quality for some of the maps. The operator needs opportunities to evaluate the maps and adjust the scanner according to the map type and quality.

A heavily contoured map, say from a ragged mountain area, contains a lot more information than a map from a flat moor area with few contours and roads. These maps need different adjustment of the scanner. The mountain map needs high resolution on the scanner to catch the density of contours, while the moor map needs lower resolution. The tool designer will build adjustment functions for the scanner into the system. The choice between these possibilities is the operator's, based on her professional judgment of the maps she intends to process.
One criterium for a tool is that it should be under the operator's control. Part of this control is being able to use the system either as a tool or a machine. If the operator has 20 maps she wants to process, she will perhaps use 4-5 maps before she is content with the result. She uses these first maps to experiment to find the optimal combination of parameters and values necessary to produce a quality map.
When she is satisfied, she should have the opportunity to give a command for automatic processing of the rest of the maps. It may be successful or not, but the choice to use the system as a machine is hers. With time and autonomy she will learn to control the new tool to proficiency.

Tools are changing.

Another criteria for a tool, seen historically, is that it is under development. Its present appearance is a result of a long social process where users design, use, change and redesign the tool. The number of different hammers used by different trades have taken a long time to develop, and the development is a process of continual interplay between a number of technological and societal factors. This ability to change points to an important condition: it is not enough that a tool is designed according to the tool criteria. It does not deserve the tool label until it has been used in practice, changed, and used again.

For computer systems this means that the demands for gradual development becomes crucial. However, there are conflicts between the demands for change and the way many systems development departments are organized. The development of a system beyond the first version is classified as maintenance. It is low status work, commonly done by other persons than those responsible for the original design. This means that the important part of development taking care of the gradual changes of the tool is transferred to maintenance, which is not considered important by the computer profession.

These two approaches are meant to illustrate not so much a question of designing user-friendly dialogues. A machine approach may provide an ergonomic perfect dialogue part, according to some of the principles for user-friendly design. Rather, they are meant to illustrate the question of how different design approaches may open up for different work organization, different types of autonomy and different opportunities for developing professional competence.

Whether the computer system will be used the way it is intended by the designer, is a different question, related to the organizational conditions at the work place in question. Systems design will, however, define some of the limitations and some of the opportunities.

3. SOME REMARKS ON STRATEGY.

How to change the practice of systems development ?

First of all, laws and agreements providing a basis for participation and negotiation, are a necessary fundament. Changing practice by law is a slow process, however. Organizations which want to move systems development in the desired direction, may use different strategies.

One strategy should be what I would term "Make them (i.e. systems developers) see for themselves!" Systems developers are often systematically shielded from contact with the users/operators. They rarely do field work for analysis, and they seldom see the results of their own work. Once a development project is finished, they move to a new one. Increasing the opportunities for systems developers to get in contact with the users/operators would give them the feedback that everyone need in order to learn.

Another strategy has to do with the way systems are tested out. Systems should, to a much larger extent than is the case to-day, be tried out in practice. It is not easy to see the consequences of a system, particularly the organizational problems, just by evaluating a prototype, or by an acceptance test. Again, this is a challenge to the systems development profession: to try out models for different types of evolutionary development, e.g. how to design and develop prototypes which can be used in production, not only as a means of demonstrating screen lay-out.

Changing systems development is, however, a battle over paradigms. It has to be fought on many arenas:

* research, to improve our ways of analyzing organizations and work;
* education ,to ensure that new understandings find their way to the students; t
* he unions, to take steps to update their current agreements.

It is not a battle for systems developers only.

LITERATURE

Bermann, T.: Data Technology and Women's Employment, Office: Technology and People, 3, 1987.

Ehn, P.: Work-Oriented Design of Computer Artifacts, Arbetslivscentrum, Stockholm 1988, Almqvist og Wiksell International.

Fossum, E. (ed.): Computerization of Working Life, Ellis Horwood Ltd., Halsted Press, 1983.

Gasser, L.: The Integration of Computing and Routine Work, ACM Trans. on Office Information Systems, vol. 4, no. 6, 1986.

Pape, T. Chr. and K. Thoresen: Development of Common Systems By Prototyping.In G. Bjerknes, P. Ehn and M. Kyng (eds.): Computers and Democracy, A Scandinavian Challenger. 1987. Avebury, Aldershot.

Rosenbrock, H.: Technology Policy and Options. ECC Fast Conference on "The Information Society: The Distribution of Benefits and Risks Associated with Micro-electronic Applications, 1982, London.

Suchman, L.: The Role of Common Sense in The Process of Interface Design. In D. Marschall and J. Gregory (eds.): Office Automation: Jekyll or Hyde? Working Women Education Fund, 1983.

Thoresen, K.: Design of Jobs in Future Cartography. In H. Opheim (ed.): Science and Fiction in Automated Carthography, 1982. Norwegian Computing Centre, Report no. 715.

BIOGRAPHICAL NOTE

Kari Thoresen was educated as a geologist at the University of Oslo. Worked as a research scientist since 1972 at the Norwegian Computing Center in Oslo in the fields of systems development in an organizational perspective, working conditions with VDUs, and participation questions in relation to information technology. Norwegian representative to TC9.

Women, Work and Computerization: Forming New Alliances
K. Tijdens, M. Jennings, I. Wagner, and M. Weggelaar (Editors)
Elsevier Science Publishers B.V. (North-Holland)

A FRAMEWORK FOR UNDERSTANDING (WOMEN'S) WORK AND ITS COMPUTERIZATION

F. De Cindio and C. Simone
Department of Informatics
University of Milano, Italy

Designers and users, involved in the design of socio-technical systems very often realize that people describe the same phenomenon by using different 'lexicons', different linguistic dimensions. This can lead them to look for the solutions along the wrong dimension or to consider only one side of the problem disregarding other possible dimensions. What is missing is a reference system in which the different aspects of problems and solutions can be compared on a sound basis, i.e., by confronting commensurable values, in the aim of finding a good balance among all the dimensions. What we propose in this paper are five 'lexicons' which allow one to recognize five linguistic dimensions (physical, operational, procedural, commitment and individual) for analysing (office) work, skills, technology and some specific characters of women's work attitude and practices.

1. A CONFUSING TERMINOLOGY

People, both designers and users, involved in the design of socio-technical systems very often realize that others, and themselves too, describe the same phenomenon by using different 'lexicons', different linguistic dimensions.
For example, let us consider the answers we can get if we ask someone using some kind of VTU device: "What are you doing?". S/he can reply:

- "I am tapping on a keyboard and looking at the screen: my eyes ache";
- "I am using a wordprocessor which is not provided with the 'undo' command: some mistakes become a disaster!"
- "I am drawing the monthly report on the budget: I am late because some data are wrong and I have to finish it before the Committee meeting tomorrow !"
- "Mr. Jones asked me to prepare the budget report. I probably need the help of Mrs. Kleene to be able to describe the marketing aspects too."
- "I'm angry since I agreed to do the report for a colleague of mine: it will never happen again!"

Each one of the previous answers is 'per se' meaningful, but contains quite different information about the phenomenon under consideration. The first one mentions physical objects and physical movements; the second one a facility and its use; the third one refers both to specific steps (the report preparation, the meeting) of a more complex process (budget handling) and to their mutual relationships (time constraints); the fourth emphasizes the motivation (a request by Mr. Jones) and future cooperation with other people (namely with Mrs. Kleene); the last one points out distress over having been involved in an unpleasant job.

The words, the 'lexicon', used in each answer stress different dimensions of the work, i.e. different perspectives from which the phenomenon "using a VTU devise" can be considered. In reality, people are more likely to mix the various linguistic dimensions since they leave in all them, they hold the five perspectives at the same time. More probably we had get as answer: "I'm typing on the keyboard but I'm very late in finishing the work I promised for tomorrow". Then, if on the one hand this mix is natural, on the other hand it could lead to some difficulties when the language is used not in the aim of chatting, but to express needs, problems and their solutions in the system development process. For instance, often users describe their problems and uneasiness in terms of eye fatigue and of general stress after a working day. This can lead the designers *to look for the solutions along the wrong dimension:* e.g., to assume tiredness and stress problems would be solved by changing the VTU device and/or by reducing the time spent using it. Empirical research [1] shows that in a number of cases only the redesign of a set of procedures can actually solve them; *or to consider only one side of the problem* disregarding other possible dimensions, as in the following example. A group of researchers claim: "We would like to have a copier on our office floor, to avoid having to spend hours trying to use the one on the first floor!" The natural solution of buying a new copier and locating it on the researchers' office floor considers the technical or logistic aspects and disregards the organizational ones, as: who is responsible for copier maintenance if it is located outside the Secretariat room?

What is missing then is a reference system in which the different aspects of problems and solutions can be compared on a sound basis, i.e., by confronting commensurable values, in the aim of finding a good balance among all the dimensions. What we propose is that the five 'lexicons' can allow one to recognize five linguistic dimensions for analysing (office) work, skills and technology. This idea was first developed in [2] by the authors together with G. De Michelis, giving rise to a framework which, in this paper, we intend to use for identifying some specific characters of women's work attitude and practices.

The paper is organized as follows. In Section 2. we introduce five different *Universes of Discourse,* and discuss for each one of them some significant factors. We will then use this framework, in Section 3., for discussing some specific attitudes relative to women's work. The paper ends with a few final comments which try to relate the women's perspective with some issues concerning system development.

2. THE FIVE UNIVERSES OF DISCOURSE

The above five answers are given by using different words, different "lexicons", which characterize different *Universes of Discourse*. Each one of them describes the considered phenomenon by using a different set of terms and of relationships binding them.

2.1 The five Universes definition

The first is the *Physical Universe,* where the answer "I am tapping on a keyboard and looking at the screen: my eyes ache" makes sense; the lexicon describes the phenomena in terms both of the *physical environment* (e.g., the kind of furniture, the facilities layout, etc.) in which the work is done, of the *movements* the people perform in doing their job and of their *senses.*

The second one is the Operational Universe, where the answer "I am using a wordprocessor which is not provided with the 'undo' command: some mistakes become a disaster!" makes sense; the lexicon describes the phenomena in terms of the kind of *operations* and *facilities* that can be used to perform these operations: e.g., to write a text using a wordprocessor. The kind and granularity of the operations depend strictly on what is considered a significant unit of resource/facility.
The third one is the *Procedural Universe,* where the answer "I am drawing the monthly report on the budget: I am late because some data are wrong and I have to finish it before the Committee meeting tomorrow !" makes sense; the lexicon characterizes the phenomena in terms of the *products* (reports, letters, invoices, orders, budgets, etc.) and of the *input/output transformations* performed to produce them.
The fourth one is the *Commitment Universe,* where the answer "Mr. Jones asked me to prepare the budget report. I probably need the help of Mrs. Kleene to be able to describe the marketing aspects too." makes sense; the lexicon characterizes the phenomena in terms both of the *conversations* necessary to (re)define people's mutual *commitments* and of the *roles,* defined in terms of the conversations they are involved in, which characterize the *group structure.*
The last one is the *Individual Universe,* where the answer "I'm angry since I agreed to do the report for a colleague of mine:. it will never happen again!" makes sense; the lexicon characterizes the phenomena in terms of people's *subjective perceptions, attitudes and expectations.*

The five universes can be used to describe any of kind of activity inside an organization (an office, a family, a team, etc.). Our emphasis is however focused on the office work environment in which we are going to discuss the woman perspective.
Let us note that the Procedural and Commitment universes provide the linguistic framework for characterizing a *group* work, and then for characterizing *each specific office,* in terms of the rules governing both tasks/actions execution and commitments definition and fulfilment.
The phenomena occurring in the office can be described and evaluated in each universe separately, and there is no dependency or hierarchy between any two universes. At a first level of abstraction, using a metaphor derived from geometry, the five Universes define a five-dimensional space in which each co-ordinate autonomously contributes, in a complementary way, to the qualification of any phenomenon in the space. We will use, in the following, "dimension" as a synonym of "Universe".
The five Universes can therefore be used as a grid, a structured framework in which:
- the *support* : resources, facilities and *technology* ;
- people's *skill* ;
- the *problems* and *disciplines* necessary for studying them, and the *parameters* for comparing problems and solutions in both a qualitative and quantitative way;

which make sense in each universe can be identified.In regard to technology, we will refer mainly to computer-based (CB) technologies, and in particular to CB technologies supporting clerical work.

2.2 The grid based on the Universes

In the *Physical Universe* the *support* to take into consideration are the office equipment (desks, cabinets, partition panels, lamps and whatever kind of furniture), and their physical interfaces: in particular, in regard to CB technologies, this physical interface consists of mouses, touch and high resolution screens, vocal analyzers and so on.
Problems concern the office layout, physical environment, illumination: they can therefore consist of the consequences deriving from the use of the facilities like physical fatigue, distress due to bad positions, eye aches, etc.

The workers' *skills* significant in this dimension are resistance to physical fatigue, manual precision and dexterity.
The *disciplines* are work environment architecture and ergonomics; the assessment in this universe is based on *parameters* like the mutual dislocation of objects, materials, machines, their ready-at-handness for workers, environment illumination, ventilation, sound-proofing.
In the Operational Universe in regard to *technology,* we can mention the applications typical to personal computing (wordprocessors, spreadsheets, databases, graphics, personal agenda, project management packages), applications devoted to specific professions (accounting, budgeting, CAD/CAM, scientific applications; Decision Support Systems) and also query languages and fourth generation languages; the interfaces with the applications typical to the Procedural and Commitment Universes.
Parameters significant in the operational dimension concern both the available facilities and the operations whose accomplishment these facilities support. Parameters concerning the operations are: their relevance, frequency, duration. Parameters concerning the facilities are: how they are easy-to-learn and easy-to-use; their access/answer times; the reachness, integration and flexibility of their functionalities; availability of built-in functionalities supporting error/exception handling.
In the line proposed by [1], [3] we call stress the generic *problem* which can arise in this dimension: it can be, from time to time, specialized in terms of the above listed parameters. So we can observe stress due: to out-of-standard duration of some operation; to the use of a wordprocessor without the undo command; to the use of an application with very low answer times; and so on.
The *discipline* useful in dealing with these problems is the cognitive ergonomics, and examples of the workers' *skills* significant in this dimension are the following: competence, experience and level of autonomy in the use of the various facilities; the number of applications of the same kind and different complexity one is familiar with; the resistance to stress.
The support *technology* significant in the Procedural Universe includes all the classical EDP applications (for handling warehouses, personnel, invoices, customers and suppliers), possibly integrated in the Management Information Systems and with some data banks (e.g., medical, statistic, territorial data banks); the forms/document handling applications; the integrated CAD/CAM/CIM applications; etc.
In this universe the assessment *parameters* refer to qualitative and quantitative aspects which have to be balanced to achieve efficiency and effectiveness. They concern: procedures reachness/complexity (e.g., type and number of used facilities; types and volumes of products; type and number of different roles involved); their control structure (e.g., degree of sharing/competition on resources; activities synchronization by mutual exclusion or blocking or concurrence; resources access times); their flexibility (availability of features supporting error/exception handling and the effects of unforeseen events); the level of autonomy in the solution of conflicts (e.g., if the criteria are predefined or partially discretionary; if the conflicts solution is a local or a cooperative process); the nature of the involved communication (synchronous/asynchronous, active or passive, structured or unstructured); their duration and costs; the quality of the received/sent products from/to other procedures; etc.
In this universe the *problems* are then due to the difficulties in finding the best balance between the values of different assessment parameters which can be mutually alternative. So, for instance, the highest concurrence does not give the best results, since it implies high complexity and then high probability of mistakes.
The *discipline* for analyzing work in the procedural dimension is the scientific work analysis as historically developed and applied both to productive [4] and to clerical [5] work.
The most precious *skill* in the procedural dimension concerns the ability to control the flow of activities and the variances, e.g.: the ability to isolate the current problem so that

the overall procedure can restart; the ability to find the best scheduling and resource allocation; etc.
The most frequently used *technology* for working in the Commitment Universe has been, up to now, the telephone with all its extensions (telex, telefax, teletex, etc.). In some environments (some highly developed countries, large multinational companies) we observe also broad use of Electronic Mail Systems based on Geographic Telecommunication Networks and on Local Area Networks. Furthermore, in recent years, people have been giving increasing attention and effort to develop software for supporting the communication, conversation and commitment networks of which all non-routine work consists, i.e. for supporting the work group dimension (that is why they are often referred to as "groupware"), or, in other words, for supporting people's co-operation. For a quite comprehensive review of these innovative proposals, see[6],[7], [8].
In this universe *problems* concern the efficiency and effectiveness of the negotiation which are related to the complexity of the office conversation network and to the definition of the office roles. The complexity of the office conversation network depends on the difficulty of identifying and reaching the partner, on the adequacy of the used support technology, on the number and frequency of conversations together with their sub-conversations. Role definition is crucial since it determines the level of flexibility of the overall office and the autonomy of the role itself in terms of the domain of possibilities each one has open; of degree of visibility of the overall process one contributes to govern and of the nature of such govern (co-ordination, control, support, consultancy); the degree of ambiguity and conflictuality between roles and the like. All the above mentioned technologies aim at supporting their users in dealing with these problems, while the existing technologies (telephone and EMS) leave all the co-ordination charge to the user.
Some of the *parameters* significant in this dimension have been used in the above description of the problems, while a more complete list can be found in [9].
What characterizes workers' *skills* in this dimension is their autonomy, as the ability to live in complex human relationships networks and in interpreting with competence and flexibility their role.
In this universe the main *discipline* is the phenomenological work analysis, and we must mention [10] as the most complete effort to develop this discipline for the analysis of (office) work.
In the *Individual Universe* it is meaningless to speak about support technologies. The *problems* here are related to people's personal reactions to the organizational, technological, social setting constituting the work environment, as it results from the composition of the factors occurring in the four previous dimensions. The related *disciplines* are work sociology and psychology, and therefore out of the authors' professional competence.
We have used this framework in a number of different contexts: for educational purposes, in order to make people - managers as well as employees, professionals as well as trade unionists, and women too (see § 3.1) - aware of the combination of activities the(ir) work consists of; for evaluating the different dimensions of (office) support technology usability [2]; for characterizing a set of quantitative and qualitative parameters to be used in office analysis [9]; for defining new ways of acquiring linguistic competence by young children and therefore as a different characterization of linguistic knowledge [11]. Others [12] have used this framework for a characterization of professions.
After all these essays, it results to be a powerful tool. In particular, for the analysis of office work and its computerization it avoids the misleading mixing of different dimensions and then helps to prevent falling into the problematic situations described in Section 1.

3. UNIVERSES 'BY/FOR WOMEN'

The above-sketched framework has also been used by the authors for a number of courses addressed to women for the purpose of introducing and discussing the work computerization and the major consequences of the related changes for their work. In the following we summarize the reasons which make this a positive experience.

3.1 A framework for understanding both technological and organizational innovation

In Italy working women, especially but not only the youngest ones, show a generally positive attitude forward information technology [13]. This positive bent is often constrained by the kind of education usually women are provided with when new information technologies are introduced changing their work environment. In fact in most cases it is a low-profile education which gives only hints about how to use the new devices (operational universe). This is the standard education for office and production workers where women are in the great majority, while more conceptual education (Procedural/Commitments universes) is provided to the higher levels of the hierarchy where women are the large minority.

One could question if this choice is profitable and far-sighted even from the management point of view: actually a low-level education makes the operator less autonomous w.r.t. the CB system and able to perform only a limited number of tasks, often a subset of the learned ones. A minor change in the CB system can then require relevant further training. Nevertheless, this is what happens. One direct consequence on courses addressed to women is that we often must use a few additional hours w.r.t. the planned ones to provide women with some 'tools' for a better understanding of their work computerization.

A second consequence is that the lower level office and production workers, and then women, have a low visibility of the organizational structure they live in, and their understanding often concerns their office/shop only.: e.g., they do not know the mechanisms governing the company and the major trends characterizing its evolution which can change the women's position in the internal labour market. But, on the other hand, perhaps more than the other workers, women use informal relationships with people outside their workplace to know what is happening there. What they lack is a conceptual framework for making their information effective knowledge.

In our experience the Universes of Discourse are effective in this respect since they give a framework for understanding and discussing not only the technological aspect of work computerization, but also the organizational context in which such change occurs, and the interrelations between technological and organizational innovations.

The Universes are language-based, then they are intuitive, easily understood and facilitate people to recognize the major dimensions of their current work, and in comparing it to the past experience and the possible innovation.

For example, the pattern companies go through, from EDP to Personal Computing and Individual Productivity Tools, toward a computerization considering also the communication aspects can now be read as an evolution from a computerization mainly set in the Procedural Universe to a computerization which sets side by side with this latter the applications typical to the Operative Universe and, sometimes, the first applications significant in the Commitment Universe. Through the grid presented in the previous sections, the corresponding organizational changes and the consequences for workers' skills become much more evident.

For example, such visibility gives the women a key for reasoning better about their career evolution. In one course addressed to keypunchers for becoming programmers in a large software house, this way of interpreting its future trends allowed the women to point out some company's present weakness on the market place and to foresee possible future risks for their work

3.2 A framework for identifying some specific characters of women's work attitudes

In the last two decades an increasing number of women have tried to overcome problems and limitations due to a long history of inequality/exclusion from work and to take advantage from the opportunities offered by the computerization process. During this process they have recognized that there are differences which should be preserved and turned to better account, if they want to refuse both to occupy a subordinate position and to accept a complete homologation to the existent (male-oriented) organization of work and system of values.

To achieve this goal women need investigation and guide-lines both for identifying what are these differences, these specific characters; and for discussing which of them women want to be preserved, turned to better account.

From the very beginning of our reasoning [14], [15] on the Conference themes we have tried, in connection with the Italian women's movement, to identify the specific positive and negative characters of women work and to distinguish women's and men's work disposition and practice . But we have had difficulties in communicating our ideas and then in verifying them with working women by taking into account their work experience. We had also difficulties in motivating our "optimistic" view which focus from the very beginning on opportunities rather than on problems.

The use of the Universes of Discourse has appeared as an effective tool for overcoming these difficulties, and allowed the women we met to enrich with original suggestions the grid we are going to present.

Before doing this, it is worthwhile to note some points.

1) We are aware that there are men showing feminine attitudes and vicecersa. Furthermore, woman/man is not an abstract category, as each woman/man has her/his own identity and specificity, her/his own "culture of work". Nevertheless, we choose to take the risk since, in our opinion, this is the only way for checking whether "difference versus homologation" is just a slogan or an effective strategy.
2) The following grid is more oriented toward identifying characteristics "in positive" rather than "in negative". This follows from the above-mentioned goal: if diversity has to be checked as background for an effective strategy one must first find some advantage, and only afterward cope with the possible disadvantages. If the first step fails, then the second is not useful. The women we met naturally shared this approach: they first tried to recognize their points of strength, and then were more open to recognizing their weaknesses too.
3) In presenting the grid a * will put in evidence the abilities which seem to be directly imported into the work environment from the house/family environment. In fact it is strategic to recognize when a factor derives from the "double work" women perform.

In the <u>*Physical Universe*</u> two characteristics specific to women's work are *manual precision and dexterity*. They are well-recognized and often used against women who are asked to execute highly repetitive jobs, since they have great *endurance for physical stress* and a constant *power of concentration* when doing jobs where physical strength is not involved, but where other senses - sight, feeling, sometimes hearing - are heavily stressed, e.g., as in the clothing industry, in some branches of alimentary industry, in electronic productions, where women are heavily exploited (for as long as they are hired!) until a more sophisticated technology arrives.

On the contrary, the attention women pay to the *physical setting of the work environment** is usually not considered as a relevant factor, and open irony often is addressed to women who put plants and other ornamental objects on the machines they use, or who spend time in rearranging things on their desks. Nevertheless, this attention can become a productivity factor (in the next universes) because a pleasant and tidy workplace enhances the comfort and reduces the time spent for finding lost objects or papers. Finally the *attention* women pay to *health problems* could be turned out for focusing the physical ergonomics problems related to the CB technology. This topic should not exhaust the health problems concerning the technology, as sometimes happens, since it covers only one dimension, but on the other hand it should not be neglected, as too often happens.

We would not suggest to make a big effort to turn to better account manual precision and dexterity since these stressing characteristics will be embedded in very reliable automatic high-precision machines. Nevertheless, we would not exclude taking advantage of the women's work experience and competence for designing the introduction of these new machines. They could act as consultants for organizing the logistics of the new work environment, for optimizing supply and for designing the kind of control to be applied to the new system (a weak and spread control instead of a dedicated attention?). This strategy would allow saving not only a number of women's jobs, but building the automatic production on the basis of a year's concrete experience of work.

In the Operational Universe the main specific character we have identified refers to a difference in the *relationships with objects and machines*. For men they are mainly a status-symbol to possess: as per the car, this holds also for the computer to exhibit on the desk top. On the contrary women evaluate objects and tools they use for the support they give. Women do not make a myth of computers as well as they do not make a myth of their washing-machine and the like.

This more pragmatic and concrete attitude often gives rise to a greater *capability of exploiing the various tools,* by possibly enriching a low-level training (see § 3.1), to an availability in accepting their features, in finding tricks for dealing even with the worst ones, and, more in general, to a greater *endurance to the psychological stress,* which can be observed in workers forced, e.g., to use software tools with a low-level of usability [3].

Sometimes women arrive at exhibiting a love for the machines they use for long time: so, for instance the women working in a clothing industry were angry when the management decided substitute the old machines. This reaction can lead to a rejection of the new technology, but can also be viewed as an antidote *against the waste** so common with CB technologies. The stress here and in what follows on this topic strongly depends on the Italian situation where, as in all historically non-rich or recently enriched countries, people and working women in particular are used to face the problem of few resources, tools and technology.

Starting from the a 'laical' manner by which women use all the available facilities for supporting the various office work operations, we would urge women to plunge into the CB technologies, so that the previous skill could give rise to a new profession: from typewriters to experts of wordprocessors applications, from archivists to experts of the textual spread-sheet packages, from graphic workers to computer graphic experts and so on.

In the Procedural Universe the characteristics mentioned in the Operational one result in a women greater *reliability* and attention to avoiding waste during the production process. Reliability plays, in general, a major role for enhancing the quality of the product, while a reduced amount of waste enhances overall productivity. We would attribute to women a big responsibility in this Universe. For a long time people were inclined to consider routine work as the major part of office work, each invoice, each order equal to the previous and to the next ones, but for minor changes (the variable fields). Now people are discovering that each instance of object is unique. Women, who are the large majority of the workers who have had to cope with procedure rigidity and prescriptiveness, have already personally gained this experience and then can say which parts of the implemented procedures are effective, and which ones have to be changed and redesigned. In the process of reducing the rigidity and prescriptiveness of the CB technology designed in the procedural dimension, the possibility of modifying and reusing part of the amount of existing software is a crucial task.

Other characters that can be viewed both as a means to overcome the rigidity imposed by the procedures and as ingredients necessary for working in the *Commitment Universe.* Since it is difficult to present them separately, we are going to describe some of them by discussing how each characteristic belongs to the two dimensions.

Let us begin with *the ability in organizing and performing a number of activities concurrent by applying a kind of weak control over them and optimizing the time spent* * which is typically imported in the work environment from the house/family environment. Let us consider a woman who comes back home in the evening: she has a number of different jobs to perform in a short range of time: cooking, looking after the children, starting the washing-machine, tiding up the house. Everything should be accomplished around dinner time. She starts more or less all the various activities concurrent and applies to them a weak control by intervening when necessary, i.e. when some break-down occurs, for instance, by resceduling her activities. We agree with [Men] when she claims: "Women execute the various household jobs in a non-hierarchical and apparently unordered way". Isn't this apparent un-order an important way for introducing concurrence and therefore for optimizing job execution? Furthermore, in order to deal with the possible breakdowns (children's illness, two husband's friends at dinner and the like) women always have available different kinds of supply, the minimum possible in order to avoid waste.
A number of women have recognized that, even unconsciously, they actually use the same scheme in their office time. In the Procedural universe, if a woman is involved in more than one procedure her first concern is to identify the 'most effective' scheduling of them, e.g., by giving the highest priority to the tasks which if not completed hold up others. In the commitment dimension this ability is crucial for living effectively inside the complex network of interrelated conversations opened to deal with break-downs [10]. Women who often have to handle 'double work', but who do not want to reduce all their life to them, have developed a culture against time waste since time is a precious resource. The large majority of women consider as positive this ability which they mention as a major point of difference with respect to men's work attitude (they often claim: 'Men are monoprocessors!'). They are also aware that often it allows them to be more productive than men without this greater productivity receiving recognition.
Another characteristic which is related to the previous one can be called the *nurturing attitude* *. Women derive from the role played in the house an availability to consider their duty the solution of problems of others. This characteristic is highly used , e.g., by managers' secretaries who become their nurses and alter ego; however, it is often present in the way all women work, since it is often difficult for women to lose the 'bad' habit of nurturing. Without denying solidarity as a positive value, the large majority of women we have met is highly skeptic about the value of this attitude since even when formally recognized (as per the secretaries), it leaves the women in a subordinate position. Anyhow, they recognize that this skill is sought-after and then it could bring some economic advantages.
Finally let us come to two characteristics, typical to the Commitment Universe, whose discussion could be really broad since it would involve on the one hand some of the present trends of company structure evolution, on the other hand the relationships between women, company structure and power. We characterize the first one through the dichotomy *intuition versus rationality;* the second one through the dichotomy *cooperation versus competitivity.* In regard to the first dichotomy, let us quote Herbert and Stuart Dreyfuss from their book [17]: "...intuition is the product of deep situational

involvement and recognition of similarity... Intuition or know-how, as we understand it, is neither wild guessing nor supernatural inspiration, but the sort of ability we all use all the time as we go about our every day tasks, an ability that our tradition has acknowledged only in women, usually in interpersonal situations, and has adjudged inferior to masculine rationality.". Before them, Simon had said that intuition is the main way the manager has for overcoming the bounded rationality he has to deal with. These two claims do not mean that women can become managers without reckoning with 'rationality tools' like mathematics, computer science or economy foundations [14]. Nevertheless, they say that intuition is an important 'a tout' women can play.

The dichotomy *cooperation versus competitivity,* or better still, a good balance between the two is a crucial point for the companies in the '90 which will choose the network structure. In fact, they have to be both highly competitive on the marketplace, and highly cooperative and synergic in regard to the mutual relationships between the network companies. The dimension of these latter will possibly decrease and big hierarchical structures will be broken in smaller and more effective units. The increasing attention devoted to the development of the mentioned CB technology supporting cooperative work confirms this trend. Such organizational and technological contexts could make more likely the basing of the individual career on the acknowledgment of an increasing competence, greater autonomy and relational abilities (cooperation), without surpassing other colleagues (competitivity). On the one hand the described evolution could reduce the discomfort a number of women feel with respect to the male-oriented organizations and work culture of today [14], but on the other it would oblige them to deal with a less protective organizational structure (non-hierarchical) and to be personally involved in the company power structure.
All these considerations allow women to look at their characteristics as points of strength. In particular they explain well why women do not make career: in the present organizations career is bound to responsibility, and responsibility in its turn is defined in terms of the number of people one has under his (her) control rather than in terms of her/his competence on fields of activity.
They say also why the organizational and technological change we can foresee for the next future gives women an opportunity to change this unpleasant situation. Professionals become key figures more than middle managers and therefore companies are forced to shift the formal system of skill evaluation toward a greater acknowledgement of professional competence. Furthermore, the trend is toward the automation of the repetitive (elementary) tasks, while people will have assigned the richer tasks of handling break-downs in the context of the several socio-technical systems people belong to.
Furthermore, a number of recent papers [18], [19] claim that companies prosperity and survival depend on an effective balance between their organizational structure, technology support and ability to turn to better account human resources. Since an increasing percentage of the labour force consists of women, the ability to turn to better account their specific characters can be viewed also as a part of the strategy of company management for better employing the "woman as a resource".
It is extremely challenging that the most significant even less recognized women's characters are identified at the commitment level. With a drop of optimism we can say that the future opens to women a big opportunity for becoming the designers and managers of the new organizations and new technology supporting them. It is mainly their choice to enter or not in this game. A positive signal comes from the wide presence, so unusual in the computer science world, of a relevant number of women (also in leadership positions) in the '86 and '88 CSCW Conferences [6].
This remark can be push to a more strong claim, which we propose for discussion. We believe that the *Commitment Universe is the "women universe"*. What does it exactly mean? We have seen that it is in this dimension that women can exploit at the best their more significant characteristics to take advantage of some challenging opportunity at the

organizational and technological level. But there is also something more that we feel but cannot well explain yet. It concerns the real nature of women attitude in the life, in their relationships with other people, in their ability to catch situations in the whole, without reducing them to 'algorithmic' terms.

4. CONCLUSIONS

In this paper we have proposed a framework for analyzing the various dimensions of work and its computerization and used it for identifying and discussing some specific work attitudes and practice of women. This work takes rise from the need of developing conceptual tools so that women can effectively face with work computerization by using their specific characters as a strength rather than a weakness.
As final remarks we want to stress two different points.

4.1 The Universes and the system development process

In the introduction we stressed the need for finding the appropriate lexicon in which to formulate problems and their effective solutions so that they can be confronted on a sound basis. The metaphor of the spatial co-ordinates, depicted in the figure, suggests how to use the universes in the system development process. Let us distinguish the case in which we are active actors in the system development, so that we can use the grid in order to design the system, from the case in which we are asked to understand, evaluate and question a posteriori the changes introduced in women work by an already developed or installed computer-based technology.

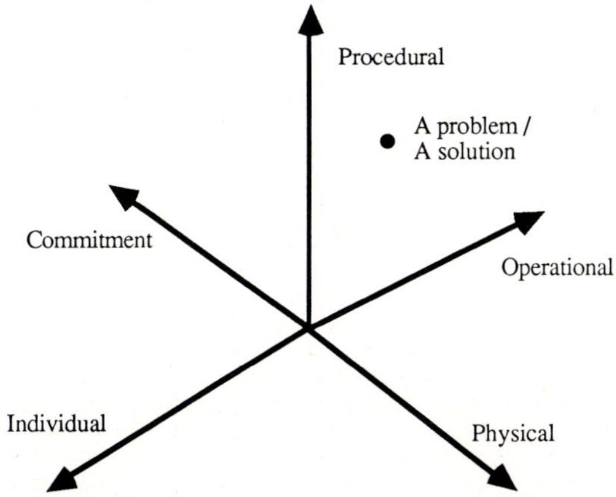

In the first case, *first of all* we can decompose problems along each dimension to enhance the tractability of their complexity. *Secondly,* in each dimension we can develop the proposals and evaluate the changes we are introducing in terms of work contents, efficiency and effectiveness, workers' skill and autonomy, support technology, and how these changes impact women work. *Thirdly,* we can recompose the proposals within each universe into a comprehensive solution whose final evaluation can be done on the basis of a weighted balance of the measures 'computed' within each universe. This allows us to avoid confusing improvements/innovations in one dimension with the solution of the whole problem. For example, a graceful and user-friendly interface, even

well accepted and useful, could distract the attention from the efficiency/effectiveness aspects of the whole application; the setting of a nice office well equipped with user-friendly personal computing devices might not necessarily solve the needs of social relationships which require a shared room and/or some co-ordination support.
In regard to women work, since it is often difficult to win on all fronts, and choices and compromises are unavoidable, the universes frame of reference allows system and work designers to balance positive/negative changes in one dimension with complementary changes in another one. For example, one would conciously accept to decrease the work agreeableness in the physical or operational dimensions, if one could enhance women autonomy. This can be the case of introducing a new technology in the commitment dimension even if it has a poor human/machine interface. Viceversa, if the innovation inside the group (both procedural and commitment) dimension cannot become the leading strategy, one could ask as counterbalance some innovation in the operational dimension, by choosing technologies which enhance the capabilities of self-organizing and of local processing owned by the users.

In the case of a posteriori assessment of the technology effects, our experience inside the Affirmative Action group which conducted a field test at Italtel [20], tells us that the grid allows the designers together with the involved women to reconstruct how the innovation process has influenced the work in the various dimensions. This experience says that the biggest and most significant changes have occurred up to now in the operative and procedural dimensions. Once again this could become an input for an individual or collective bargaining in order to recover valuable skills and competences or, more often, to set a program for developing women attitude in the new environment.

These quite general guide-lines could become more effective after a further work would be carried out. It consists in: a) using the five Universes to identify the scope of the several analysis and design languages and techniques proposed for the system development process, so that one can better understand which universe specific analysis/design languages and techniques are concerned with; b) recognizing the necessity of new approaches for the uncovered office work dimensions, e.g., the commitment one, and developing them; c) realizing the necessity of methods or guide-lines for recomposing the various dimensions (second step above) and providing them.
Let us note that the this work would avoid a double risk: to use too rich/heavy tools for dealing with 'simple' problems, or tools inadequate, not pertinent in relation to the problem. For example: the choice of some PCs to be used individually do not require a detailed analysis of the work flow; methods oriented to data processing are inadequate to handle problems in the commitment universe and viceversa; the redesign of a user interface requires disciplines other than computer science [21]; prototyping is surely very useful to design the application interface (physical/operational and individual universes) since it provides users reactions and wants, but it can show some weakness in determining the internal structure, the office model underlying the application itself, since it not easy to evaluate on its basis other more organization oriented parameters (procedural/commitment universes); etc.
In this way the Universes might become a new tool in the tool kit owned by designers and users to enhance their mutual understanding and the understanding of the problem in whose solution process they are called to cooperate with their different perspectives.

4.2 About a strategy for highlighting women's work

Before finishing, we must consider a final and major difference between men and women: that is, their different ability in turning to advantage their specificity [15]. There are, indeed, also specific characters of men's work attitude and practice. Even if we did not develop an analogous grid for men's work attitude, let us mention three examples:

the greater availability of men to physical toil, to work traveling and to night-work. In all these cases workers, with the support of Trade Unions, have made these characteristics matter of a negotiation on the basis of a sturdy and sure sense of possession of a skill which has to be recognized within the well-founded and accepted hierarchical system.

This is not the case for women. While men's endurance to physical toil is rewarded, women endurance in supporting monotonous jobs is recognized (managers use this ability and leave women for a long time in such work) but never rewarded. The same holds for manual precision and dexterity: the low level of salaries of women performing precision works speaks for itself. Let us now consider nurturing: it is to some extent recognized for secretaries which often have good salaries even if not always a corresponding good category. But when this attitude becomes the kernel of a profession, as for nurses and for nursery and elementary school teachers, the salary does not show a high evaluation of such professions.
This means that the formal recognition -in term of career, rise, social rewarding- of women specific characters would require profound changes: a change in the current way employers and Trade Unions classify work in terms of skill, and therefore in the organization's game rules; a more difficult change in the underlying culture; finally, a very difficult change, for women, in their relationships with the power dimension of the work.

ACKNOWLEDGEMENTS

The authors wish to thank all the women they met and the 'Positive Actions' Research Group of ITALTEL who provided us with several examples derived from the field; Gro Bjerknes, Tone Bratteteig and Guri Verne of the University of Oslo, and Giorgio De Michelis, our colleague at the University of Milan, for their stimulating comments on the draft of the paper.

REFERENCES

[1] F. Novara, R. R. Rozzi, G. Sarcinelli, Psicologia del lavoro (Il Mulino, Bologna, 1983)
[2] F. De Cindio, G. De Michelis, C. Simone, Dimensioni dell' Usabilita', Universi del Discorso e Software d'Ufficio, in S. Bagnara, A Stajano (eds.) Software Amichevole (CLUP, Milano,1987)
[3] S. Bagnara, L'Ergonomia del Software: una condizione per il successo nei processi di automazione, in S. Bagnara, A Stajano (eds.) Software Amichevole (CLUP, Milano,1987)
[4] F.W. Taylor, The Principles of Scientific Management (Harper & Brothers, New York, 1911)
[5] H.A. Simon, The New Science of Management Decision (Prentice-Hall, Englewood Cliffs)
[6] CSCW'86, Proceedings of the Conference on Computer-Supported Cooperative Work, Austin (1986)
[7] Office Technology & People, vol.3 no. 2 (1987)
[8] ACM Transactions on Information Systems, 5.2 (1987)
[9] C. Simone, A language to speak about parameters for office systems evaluation, in Proc. IFIP Conf. on Languages for Automation, Vienna (1987)
[10] T. Winograd, C. F. Flores, Understanding Computers and Cognition (Ablex, Norwood, 1986)
[11] G. De Michelis, private communication

[12] F. Butera, Dalle occupazioni industriali alle nuove professioni (Franco Angeli, Milano, 1987)
[13] F. De Cindio, E il computer parlera' al femminile, il Sole 24 ORE, 122.35 (1986)
[14] F.De Cindio, C. Simone, Women and work in te age of computers: (many problems), one opportunity and two challenges, in A. Olerup, L. Schneider and E. Monod (eds.), Women, Work and Computerization: Opportunities and Disadvantages (North-Holland, Amsterdam, 1985)
[15] A. Catasta, F. De Cindio, C. Simone, An educationl Experience toward Actions for Equal Opportunities, presented at the 2nd Conference Women, Work and Computerization, Dublin (1986)
[16] L. Menapace, Donne e computer: il nuovo alfabeto al femminile, Il Manifesto, XIV 172 (1984)
[17] H. Dreyfuss, S. Dreyfuss, Mind over Machine (The Free Press, NewYork,1986)
[18] K.A. Golden, V. Ramanujan, Between a dream, a nightmare: on the integration of the human resource management and strategic businness planning processes, Human Resource Management, winter (1985)
[19] H.C. Lucas, Utlizing the information technology:guidelines for managers, in Sloan Management Review, fall (1986)
[20] B. Osimo, An experience of Affirmative Actions at ITALTEL, in this volume
[21] L. A. Suchman, Plans and situated actions: the problem of human/machine communication (Cambridge University Press, Cambridge, 1987)

BIOGRAPHICAL NOTE

Fiorella De Cindio took her degree in Physics in 1976 and since then on is senior researcher at the University of Milan, Department of Computer Science. Her activity concerned languages and techniques for the development of computer-based systems of different nature, including office support systems and in particular co-ordination technologies. In the course of this activity she has given attention to the impacts of computerization on work, being involved in educational experiences for companies and Trade Unions. In the last years she has given particular attention to the relationships between women, work and computerization participating to all the three IFIP Conferences on Women, Work and Computerization and being involved in the Italtel Affirmative Action project.
Carla Simone took her degree in Mathematics in 1972. She is actually senior researcher at the University of Milan, Department of Computer Science. Among her areas of research there are the methodological aspects of office support system analysis and design and the development of technological supports for the cooperative work. She has been involved in several educational programs about the impacts of the technologies on the office organization, some of them specialized to the women work modification. She is actually involved in the Italtel Affirmative Action project.
Correspondence address: Dipartimento di Scienze dell'Informazione, Università di Milanovia Moretto da Brescia 9 - 20133 Milano (Italy)

Women, Work and Computerization: Forming New Alliances
K. Tijdens, M. Jennings, I. Wagner, and M. Weggelaar (Editors)
Elsevier Science Publishers B.V. (North-Holland)

SYSTEM DESIGN IN SOCIAL ORGANIZATION AND METHODOLOGICAL BACKGROUND - PROBLEMS AND CHANCES FOR WOMEN'S PARTICIPATION

Margrit Falck
Division of systems design and Automated Information Processing
Department of Theory of Science, Humboldt University of Berlin
German Democratic Republic

Various strategies of system design correspond to different degrees of participation. Beside the technical and the socio-technical approach of system design, the paper discusses the actional strategy of user-oriented system design in social organizations as an integrative concept of system design BY the user. It refers to job-activity and motivation as the starting points for system design and includes design of system, of work and of organization. The most important advantages of the concept of job-activity for women's participation will be shown.

System design for, with and by the user was the title of the first conference on social questions of system design and the situation of its users held in Riva del Sol/Italy in 1983.
FOR, WITH and BY the users are abbreviations for various strategies of system design whose features are "..... decisively determined by the organizational theories and by the view of the world and in particular of the human being which it explicitly and implicitly represents." [4].
In his organizational analysis FUCHS-KITTOWSKI differentiates between the following main strategies:

1. a technically task-oriented technological strategy;
2. a socio-functional (socio-technical) strategy;
3. and actional strategy (directed at personality development).[1]

Strategies are the philosophies of using methodological instruments. Grossly simplified, the following connection can be established:

* the purely technical, task-oriented technological strategy is based on methods of software-engineering and uses them - FOR the user -;
* the socio-technical strategy bases itself essentially on prototyping and software-ergonomy in combination with the existing methods of software-engineering and uses them - WITH the user - and
* the actional strategy - understanding system design as social action - does not yet have a further developed methodology, which takes social requirements into special consideration and is able to use them - BY the user -.

---

1 Actional, in contrast to functional, oriented to social activity; activity or action are understood in a wider sense as practical action and verbal behaviour, creating information, forming symbols and communicating values.

Our research aims to fill this gap with a methodology for a *user-oriented system design in social organizations.*
The necessity to widen the socio-technical strategy to a actional appraoch results from the increasing complexity of information systems. The technology extends over areas and hierarchies and penetrates social structures and processes in an organization with technical structures and processes. Compatibility between social forces and resources - like readiness, competence, qualification, interests, abilities, needs and motives - on the one hand and the goals of technical systems on the other is more than ever needed for the efficient use of technology.

Although many methods use attributes such as user-oriented, user-centered, user-friendly of also participative, a more exact analysis shows differences in the degree of participation.
Technical system development aims to automate the operations, which occur in the work procedures of many users. It is directed at the use-value of a technical product. Therefore, it restricts itself to using a picture of the user-groups as users and it assumes a technical ability to use the systems. Intentionally neglected in the development of systems is the organization of work procedures, which differ from user to user , from task to task and from firm to firm. Socio-technical system design is directed at support of whole work procedures of professional user-groups and at dealing with the instrument - computer and/or dp-system (instrumental actions). Thus socio-technical system design sees the user as a co-operative partner who articulates his/her need for action (system functions and knowledge) so that the system can be designed in line with its operational and cognitive features. The user participates as a bearer of a role, thus expressly abstracting from him/her as a person. The functional aspect of the organizational environment for the task-execution is emphasized. (15), (9), (3)

The user is expected to have technical and task oriented knowledge and instrumental skills as s/he would need for using any other tool. Computer and/or dp-system, however, are not the customary instruments or tools, which enlarge the organic forces. In contrast to those tools the computer can replace a person and can degrade a job.
Therefore, the user-oriented system design in a social organization is directed at work design and is based on job-activity which is psychically regulated and is influenced by the user's motivation.
Women and men are parts of social networks. In particular as users of systems they are a part of the social networks within an organization of work. Their individual motivations for job-activity are connected with personal needs, group-specific interests and with work-conditions, which result from structures and processes of technically supported *social organization*. The base for structures and processes of the organization are the division of labour and the assignment of functions to technology. System design - by the user - does not simply programming oneself or merely improving the utility of the primary system, but it, above all, involves organizing the division of labour and activities. It involves projecting and designing jobs. The users become active controllers and designers not only of the system, but also of the organization.

The difference between an actional strategy of user-oriented system design and the socio-technical approach becomes clearer if we consider the structure of human labour. One task results from the purpose and aims of an organization. Motivation for an job-activity is based on the individual's identification with the task given. Components of the job-activity are actions, the aims of which grow out of the motivation, that means that the user is willing and able to carry out on the background of his/her interests and abilities. Part-aims are fulfilled by part-actions or operations.

The outcome, simplified on the basis of Hacker (7), is the following hierarchy of pairs:

purpose	task
motivation	job-activity
aim	action[1]
part-aims	operations

Apart from the fact that tasks are the starting point for system analysis and design in general, the user-oriented system design begins within this hierarchy job-activity and its motivation, whereas the socio-technical system design refers to actions and their aims which no longer have a direct connection to the motivation. The technical system development refers only to operations without any real connection to aims (fig. 1)

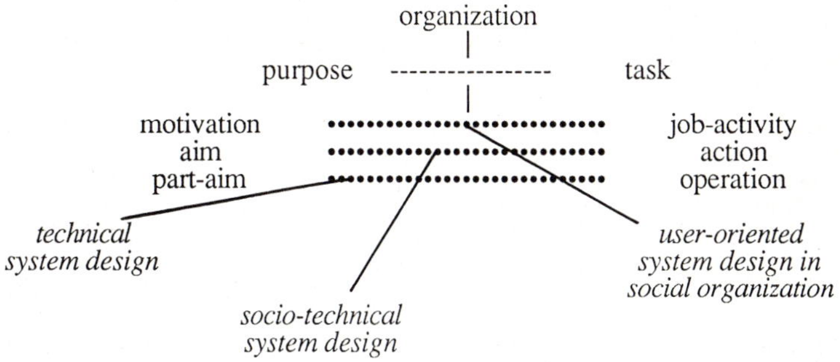

Figure 1 Structure of human labour and approaches of different methodologies in system design.

Leontjew writes:"....Job-activity can loose the motivation out of which it grew; then it becomes an action that may carry out completely different job-activity. (.....)Finally, an action can be transformed from a procedure to achieve an aim into an operation which is capable of carrying out various actions". [12]
As long as system design is based on functional roles and actions which abstract from the individual and his/her motivations, attitudes, morale or from group specific interests, women will have less chances to help shape their professional environment in accordance with their needs. Even if equal rights are stipulated in the laws, the historically conditioned role-sharing between men and women still determined, because of the traditional attitudes which remains as a subconscious factor.[2]
Only if the logic of human relations between the subjects of organization become transparent and also determine the logic of technology, can technology really be controlled.

The basic advantage of an actional strategy as the further development of the a socio-technical is the simultaneous societal and individual orientation of job-activity.
The most important advantages of a design concept based on job-activity are:

* it includes both the cognitive realization of the work and its motivation;
* the individual job-activity is part of a totality and makes it possible to see groups, organizational units as well as individuals;
* the concept includes activity conditions, its means and its environment;
* it is a psychological conception, that includes psychic processes, circumstances and properties;
* it is a conception of self-development of the psyche that includes development of the personality. [8]

---

1 Here is seen in a narrow sense as an element of work procedure in job-activity, not as an social action.

This makes it clear, that this concept is more comprehensive, that a connection to a surrounding organization becomes possible and that it is easier to grasp the focal content of a user-orientation.
I would make the following proposals for a phase-model of user-oriented system design:

1. Analysis of individual job-activities and motives with sociological and labour-scientific methods. (14), (16)
2. Design of individual job-activities and conditions which are necessary for it with regard to the system to be planned (work-design).
3. Drafting an organizational conception for a team or for an organizational unit connected with bargaining for the goals of the system based on individual drafts from 2. (organizational design).
4. Development of a prototype on the basis of the organizational conception. Cyclical installation of the prototype and its evaluation with reference to the organizational conception (socio-technical system design).
5. Realization of the system with full performance (technical system development).
6. Installation of the system, evaluation and further development (then begin with 1 again).

The totality of this process describes an integrative model, whose macro-structure shows the linear phase-model. This includes the development of both the prototype and of the final system in their significance as profounder processes of specification and implementation. The prerequisite for this is the requirement analysis by work-design and organizational design. (fig. 2)

The whole process of user-oriented system design from the first planning steps to the permanently installed system takes place on three levels:

* the level of social organization (division of labour);
* the technological level of task-execution (assignment of functions to the computer);
* the purely technical level.

Prototyping and the realization of the system are complex processes, which are, in turn, described through phase-models.
For prototyping, the process-model for integrative system design by Ch. Floyd and Keil (3) is used. They describe two parallel processes, which influence and support each other. One refers to the "(.....) description of the development of user activities and the surrounding organization", which means organization on the work-place. The other is "(....) the description of (technical) functions of dp-supported system to be strived for and interaction with the activities of users". Both processes go through cycles in which developers set up system versions and users set up evaluations, whereby the versions can be analyzed and revised.
In this way, the mutual dependence between job-activity on the technical labour conditions created through the system, becomes visible. The model is suitable to describe empirical research into an adequate division of functions between the human being and the machine as well as technical labour conditions which encourage productivity and individual task-execution.

Organizational analysis with the single steps of the analysis of job-activities, design of job-activities and the finding of the organizational concept including the aims of the technical development, provides a requirement-definition of a completely new quality. The new quality arises not only through participation, but also through the priorities given to the social and psychological aspects within the economic and technical requirements. The socio-technical approach to system design does not become separated from the system, because an evolutionary and cyclical process comes into being through a prototype or an existing system. This process results in technological design, which perform indirect organizational design. In the actional strategy, design of job-activity

Figure 2. Phase-model of user-oriented system design in social organization.

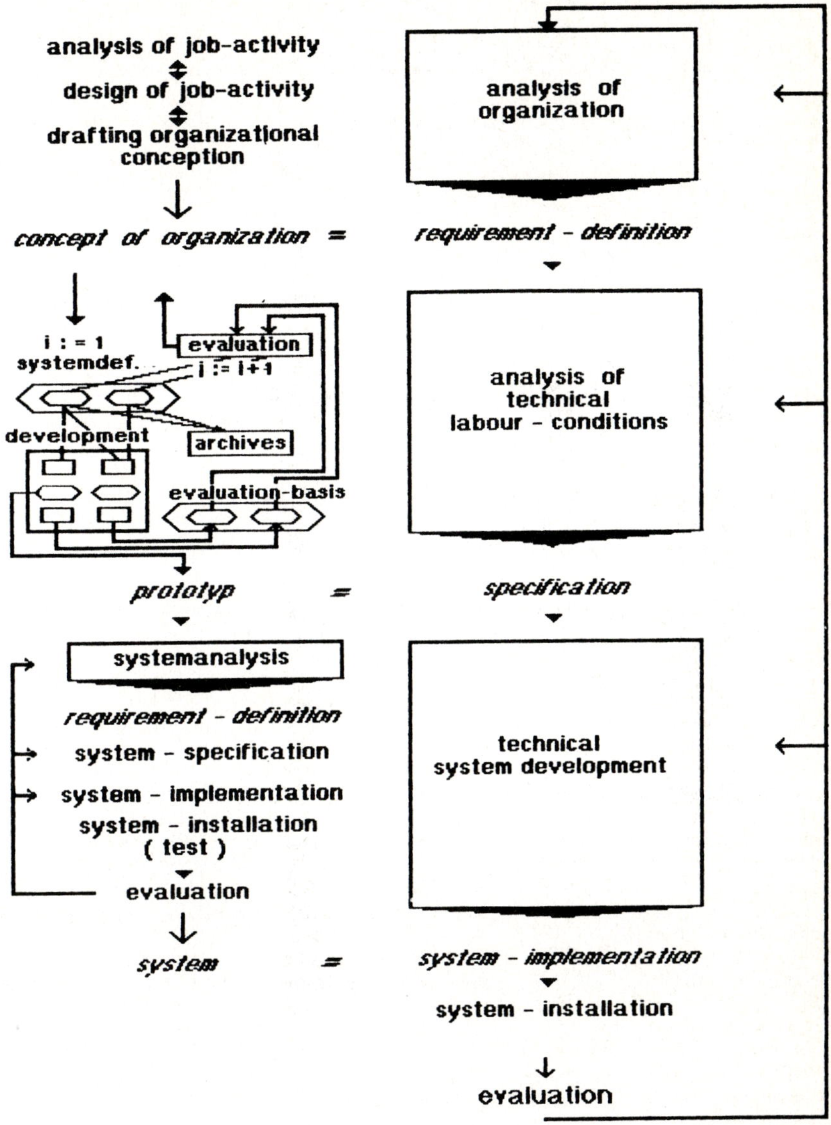

and organization are given priority over technological design. This corresponds to E. Mumford's demand

> "(....) The design task starts, not with the assumption that a computer-based solution will be introduced, but with the question 'do we need to change? Is our department optimally organized now to achieve its mission or could organizational and/or technical change make us more effective?'(....)".[13]

We agree with E. Mumford and try to give a better methodological basis for realizing this demand, than does the socio-technical approach.

For women, too, this brings clear advantages. The narrow technical view gradually broadens to one that includes the social sciences and society. The starting point of system analysis changes, in so far as it becomes based on both individual activity as well as the totality of the social organization. This puts women in non-technical fields, who up now have found themselves on the periphery, more into the centre of system design.

The over-emphasis on technical knowledge is reduced by a greater knowledge about social relations and inner-office organization. [1] The authority, that now determines the status of experts in development has to become a communicative relationship on a partner level between users and developers and on a collective level between colleagues with other motives and between groups with other interests. However, the greatest advantage is the motivational aspect, which is bound to the individual and makes it possible for women to contribute their specific feminine motives and attitudes. The problem, especially for women, is: they are not encouraged to express their individual needs, interests and motives. Therefore we need procedural methods and communicative conditions, which can help to grasp and to articulate these on the individual level and to offer them on the collective level. The new quality arises with the further development of democracy in work. (5)

In line with the various levels of design the whole process contains three different methodological instruments of which the cybernetic has up to now been most efficiently developed. The instruments of socio-technical system design are still being developed. These include, above all, the software-ergonomic interface design. For the user-oriented system design in social organization we need instruments that integrates the different levels of information and human communication on a background of cognitive psychology and other social sciences.[6]

For labour and organizational design in system design there exist only a few instruments which are still being "tested", and this merely sporadically and under experimental conditions. To make this normal practice, we need interdisciplinary teams including social scientists, sociologists and labour psychologists. Specific methodical instruments for analysis of activities and the organizational environment are the TBS-GA (16) and the VERA-B (14). Examples for specific communication methods ar described by Kensing (10) and Lanzara et al. (11). What we need, however, is a new continuous projecting methodology.

Cooperation between users and developers demands an ability for partnership from the people involved which does not always exist. A willingness to learn, to accept changes, an ability to conduct sincere communication and to solve conflicts are necessary elements of this process. It assumes the development of such socio-cultural characteristics and at the same time also gives rise to them. Different and unequal demands for such qualification can lead to incompatibilities if one considers that cultural development is based on communitive experience within a society.[17]

System design BY the user brings a new quality into the development of the productive forces, which, because it is a mental activity, does not simply follow the logic of technical and technological processes, but laws of its own. One of the most important prerequisites for the individual is the constitutionally guarantied right to work and to social security.

The interest of the employer under capitalist conditions to make the most efficient use of his capital and employees, causes him to involve the users in system design. However, user-participation has limits set by the conflict of interests between capital and labour and the lack of social security. These limits have to be continually corrected by the struggle of the trade unions to improve working conditions.
Organization within a socialist economic system sets a framework which needs to be used effectively. Although work-design and organization-design can never take place without conflicts (because different individuals have different needs and interests), under socialism there exists a common interest based on the economic and social policies and the right to social security.
These and other prerequisites are the framework for the user-oriented design of information systems and communication systems within a social organization. Different prerequisites in the different economic systems provide different possibilities for realization.
The aim is to accept the human being as a bio-psycho-social entity and make him/her, in his/her creativity, the subject of system design. To the same extent to which we succeed in this, will a conscious dispute begin to take place with the attitudes and motivations of men and women at their places of work.

FOOTNOTES

[1] Dirrheimer, A.: Office automation and training for skilled end-users, in: Briefs, U., Tagg, E. (Hrsg.): Education for System Designer/User Cooperation. Proc. IFIP TC3/TC9 North-Holland, Amsterdam New York Oxford 1985, p.45-55.

[2] Falck, M., Gensior, S.,Manacorda, P., Meier, U., Paetzold, G., Wagner, I., Wenzlaff, R.: Women, Computers and Participation - A Roundtable. In: Docherty, P., Fuchs-Kittowski, K., Kolm, P., Mathiassen, L. (Hrsg): System Design for Human Development and Productivity: Participation and beyond. Elsevier North Holland 1987, p.457 a.f.

[3] Floyd, C., Keil, R.: Integrative Systementwicklung. Ein Ansatz zur Orientierung der Softwaretechnik auf die benutzergerechte Entwicklung rechnergestützter Systeme. - Forschungsbericht DV 84-003 des BMFT Bonn 1984.

[4] Fuchs-Kittowski, K.: Informatik und Organisationstheorie als konzeptioneller theoretisch-methodologischer Bezugsrahmen für die effektive Integration moderner Informationstechnologien in sozialer Organisation. In: IV. Wiss. Kolloquium zur Organisation der Informationsverarbeitung "Information, Organisation und Informationstechnologie" Berlin 1983, p.128-205.

[5] Fuchs-Kittowski, K.: Integrative Participation - A Challenge to the Development of Informatics. In: Docherty, P., Fuchs-Kittowski, K., Kolm, P., Mathiassen, L. (Hrsg.): System Design for Human Development and Productivity: Participation and beyond. Elsevier North Holland 1987, p.3.

[6] Fuchs-Kittowski, K.: Systems and Form, Content and Effects of Information. In: Fuchs-Kittowski, K., Gertenbach, G. (Hrsg.): System Design for Human Development and Productivity: Participation and beyond. AdW der DDR Zentrum für Gesellschaftswiss. Information Supplement Volume of the IFIP-Publication of North Holland Publishing Co. Berlin, DDR 1987.

[7] Hacker, W.: Arbeitspsychologie, Psychische Regulation von Arbeitstätigkeiten Verlag der Wissenschaften Berlin 1986.

[8] Hacker, W.: Psychische Regulation von Arbeitstätigkeiten. Analyse, Bewertung und Gestaltung von Arbeitsprozessen. Kolloquiumsvortrag in der AdW der DDR, gesellschaftswiss. Institute Berlin Oktober 1987.

[9] Keil-Slawik, R.: KOSMOS - Ein Konstruktionsschema zur Modellierung offener Systeme als Hilfsmittel für eine ökologisch orientierte Softwaretechnik. Dissertation TU Berlin-West 1985.

[10] Kensing, F.: Generation of Visions in System Development - A Supplement to the Tool Box. In: Docherty, P., Fuchs-Kittowski, K., Kolm, P., Mathiassen, L.: System Design for Human Development and Productivity: Participation and beyond. Elsevier North Holland 1987, p.285.

[11] Lanzara, G.F., Mathiassen, L.: Mapping Situations Within System Development Project. Information & Management H.8, p.3-20 (1985).

[12] Leontjew, A.: Tätigkeit Bewusstsein Persönlichheit. Beitrage zur Psychologie, Bd.1, Volk und Wissen Berlin 1982, p.103 a.f.

[13] Mumford, E.: Participative Systems Design. In: Th Computer Journal Vol. 27, No. 3, p.283, 1984.

[14] Nullmeier, E., Roedinger, K.-H.: The Limitations of Task Complexity through Information Technologies: Results of a Field Study. - In: Frese, M., Ulich, E., Dzida, W. (eds.): Psychological Issues of Human Compuer Interaction in the Work Place. Elsevier North-Holland, 1987, p. 41-57.

[15] Nygard, K.: Program Development as a Social Activity. In: Fuchs-Kittowski, K., Gertenbach, G. (Hrsg.): System Design for Human Development and Productivity: Participation and beyond. AdW der DDR Zentrum für Gesellschaftswiss. Information Supplement Volume of the IFIP-Publication of North Holland Publishing Co. Berlin, DDR 1987.

[16] Rudolph, E. Schönfelder, E. Hacker, W.: Verfahen zur objektiven Analyse, Bewertung und Gestaltung von Arbeitstätigkeiten mit überwiegend geistigen Anteillen (TBS-GA). - Psychodiagnostisches Zentrum Berlin 1987.

[17] Strützel, D.: Soziokulturelle Implikationen wissenschaftlich-technischen Fortschritts. In: Protokollband zum II. Internationalen Symposium zur Theorie und Geschichte der Soziologie Berlin 1988.

Women, Work and Computerization: Forming New Alliances
K. Tijdens, M. Jennings, I. Wagner, and M. Weggelaar (Editors)
Elsevier Science Publishers B.V. (North-Holland)

COMPUTERIZATION AND WOMEN'S KNOWLEDGE

Lucy Suchman and Brigitte Jordan
System Sciences Laboratory, Xerox Palo Alto Research Center
Palo Alto, California, U.S.A.

In this paper we explore the relationship between authoritative knowledge and the process of technology design, with particular attention to the situation of women. We argue that to design technology relevant to women's concerns requires legitimizing the authority of women's knowledge and competence on their own terms. Two distinctively women's spheres of activity -- childbirth and office work -- provide strikingly similar lessons with respect to technology design in relation to women's knowledge. In both cases, ideological commitments that override the realities of what women know dictate a model of the activity which, when embodied in technologies, enforces the ideology. As a counter-strategy, we recommend an approach based on recent developments in Scandinavian systems design.

INTRODUCTION

Evelyn Fox Keller opens her *Reflections on Gender and Science* with the following quotation from Simone de Beauvoir:

> "Representation of the world, like the world itself, is the work of men; they describe it from their own point of view, which they confuse with the absolute truth. (1985, p. 3)"

While de Beauvoir overstates it for us (it is only in a very particular sense that we are willing to grant the world to be the work of men) we also recognize what it is that she could be talking about. Part of what we are involved in is to see the world as a place populated by women, working in spheres of knowledge and competence rendered invisible by prevailing representations of what competence involves. There is a realm of practice -- embodied, deeply rooted in experience, working from concerns of both heart and mind -- that compels us in our own work and that we want to embody in our designs. Our sense is that it is in the realm of practice that involvement lies, and that it is for lack of tapping that realm that so many women have been alienated from the technological enterprise.

As a way in to this question of involvement in relation to the design and use of technology, we introduce the notion of *"authoritative knowledge"*. Authoritative knowledge is that knowledge taken to be legitimate, consequential, official, worthy of discussion and useful for justifying actions by people engaged in accomplishing a given task (Jordan 1987a, 1987b). We begin from the view that there is an intimate relationship between the setting of human activities, as both physical environment and technology, and the distribution of authoritative knowledge. Changes in the former inevitably have consequences for the latter. Technological change can therefore be a resource for either the expansion of existing forms of authoritative knowledge, or for their transformation.

Both design and use of technology involve *appropriation*. Where technologies are designed at a distance from the situation of their use (as most are), there is an inevitable gap between scenarios of design and circumstances of use. Inevitable, because to some extent technologies have to be designed for unknown users, in unknown circumstances. To get off the ground, designers need to project a scenario of use, the adequacy of which will depend on the adequacy of their understanding of the actual situations of use. However adequate the designers' understanding, the gap will exist and will have to be filled by the user as the technology is interpreted with respect to local concerns and circumstances. In this sense, design is only fully completed in use.

Appropriation can be either a matter of design as we conventionally think of it (that is, an engagement with the technology as such) or of use (that is, an engagement with something else, some other activity, by means of the technology as a tool). In either case, appropriation is ultimately a matter of ownership, of integration of a technology into one's activities in a way that constitutes the basis of one's competence. This is, at least in part, what we mean by involvement, by being engaged with something, and it is the thing that we observe to be noticeably absent from women's relations to computers.

If we put these two concepts -- authoritative knowledge and appropriation or involvement -- together, two radically different, but distinctively women's spheres of activity with which we are familiar -- childbirth and office work -- provide strikingly similar lessons with respect to technology and women's knowledge. In setting these superficially disparate cases side by side we find the relations between them. In both cases, ideological commitments that override the realities of what women know dictate a model of the activity which, when embodied in technologies, enforces the ideology.

THE MAYA CASE

The first case involves cross-cultural work by Brigitte Jordan on childbirth practices or ethno-obstetrics in relation to international development (Jordan 1983). In the Yucatan peninsula of Mexico it is the local collectivity of Maya women who hold authoritative knowledge about birth; that is to say, the knowledge that is considered consequential for making decisions about and managing the event. Childbirth takes place either in the woman's own house or that of her mother, and technologies are familiar household objects. There are unquestionnably experts involved: a midwife, with extensive experience of many births, supported by other women of the family each of whom has her own experience on which to draw. But moment by moment knowledge of the event is produced collectively by the participants and draws centrally on the authority of knowledge of her own body granted to the woman herself.

Into this system, designers of development programs to promote western biomedicine introduce new technologies developed in the context of the high technology hospital. A case in point is the sterile scissors, introduced as an alternative to the local practice of burning the umbilical stump with a candle to prevent infection. The gap between the context of its design and the local conditions of its use (in particular, an absence of stoves) led to a re-interpretation of the technology from a sterile scissors to a pair of scissors dipped briefly into a bowl of hot water. Observation by traditional birth attendants of a subsequent increase in infant tetanus resulted ultimately in their rejection of the scissors in favour of the former, clearly more effective practice of cauterization.

Transported to high technology hospitals, the familiar territory and authoritative knowledge of Maya women are lost to them. As the process of childbirth is transformed into a medical event knowledge becomes privileged, protected and mediated by exotic technologies. In the United States, where childbirth is considered a high-risk pathological process that is best managed with the resources of medical technology, pharmacology and surgery, the person in charge is typically a male doctor, most frequently a doctor with specialist training in obstetrics and gynaecology (which in the US is a surgical specialty). In such births the joint production of information which prevails in traditional births no longer obtains. The woman is transformed from the interactional focus of a collaborative group, an actor, an individual who provides authoritative information about the event, into a mere passive data source as her body is colonized by machines and technical procedures such as stress tests, ultrasounds, and fetal monitors. The artifacts and technologies employed, from hospital gowns to heart monitors, are unfamiliar to her. The machines on whose output the management of her labour is based are inaccessible physically, in the sense that they cannot be touched or manipulated and experientially, in that what the output could possibly mean is beyond her technical competence. Advanced technologies and procedures depend for their interpretation on professionals with specialist training. As a consequence, authoritative knowledge and the attendant power to make decisions in such systems is redistributed.

The case of Maya obstetrics illustrates what we call *the fallacy of the empty vessel*; that is, the belief by those who design new technologies that there is nothing there in advance of their arrival. In the attempts to "upgrade" traditional ways of handling pregnancy and birth in the direction of western biomedical practice, the fact that the new procedures interact with a pre-existing ethno-obstetric system is generally not acknowledged. Time and again, the design and imposition of new technologies that fail to recognize the realities of local conditions leads to a degrading of the local environment and/or rejection of the technology. In the Yucatan, and increasingly in industrialized settings as well, hospital birth is rejected, where possible, in favor of the control afforded by remaining on one's home ground.

THE OFFICE CASE

We can find other examples of the fallacy closer to home. In the case of office automation, the allocation of so-called procedural office work to women in this century, followed by the attempt to delegate it to machines, is made possible by a systematic blindness to the real practice involved in the work (Suchman 1983). Following the tradition of scientific management initiated by Frederick Taylor in 1911, many system developers take office work to be essentially procedural in nature, involving the execution of a prescribed sequence of steps. Taking the meaning of "procedure" in the office to be the same as in the construction of computer programs, analysts have developed models of office work that represent it as a collection of ordered tasks. Office work seems, on this view, to be an activity ready-made for computerization. Interest in artificial intelligence has led to more "plan-based" approaches to office activity which emphasize the goal-directed and contingent nature of actions. More recently, those models have been extended by analogy with distributed computing, people and machines being represented as homogeneous systems of "cooperating agents." In all of these cases, the extent to which office activities remain "informal" or "unstructured" -- that is, resistant to algorithmic representation -- is seen as a problem for system design.

The design of so-called office information systems is based not in an understanding of the practical, situated accomplishment of office work but in a priori, rationalized models of the office as a system of disembodied information flow. In the process, women office workers' knowledge and the technologies on which it rests is rendered invisible.

A detailed study of work in an accounting office suggested that organizational procedures, while oriented to by office workers, in no way provide instructions for what to do next (Suchman 1983). Rather, procedures describe general guide-lines for how things should go and requirements for what they should come to in the end. Those guide-lines and requirements are necessarily underspecified, insofar as what it could mean to, for example, "process an account according to procedure" is contingent on the details of the particular case. The work of the office is precisely to achieve a relationship between the standardized procedures and the unique circumstances of actual cases. It is the accomplishment of this mutual accountability between procedures and real-world events, evidenced in and mediated by the record, that makes up the worker's expertise.

Careful observation of what actually goes on in offices, in other words, reveals that these "unstructured" activities involve both fine-grained structures and, more importantly, a crucial and unacknowledged form of expertise. Like the accomplishment of childbirth, the expertise of office work involves judgment, responsibility, negotiation and knowledge of how to interpret relevant events. The community of co-workers is the resource for that knowledge, bringing information to bear at the time and place that it is needed. An interaction analysis of collaborative work on a single accounting record, for example, shows how while searching files for a missing invoice co-workers continually produce comments that, as assessments of what each is finding, allow the other to monitor their joint work. This monitoring allows for the possibility that either collaborator might "know better" the sense of what the other is finding than the other does herself. And the continual monitoring provides for the possibility that their respective searches might, at any point, develop into concerted attention to any of a number of findings.

Also as in the case of childbirth, workers' knowledge and competence is based in their familiarity with their physical environment. The office workers' environment includes the spatial arrangement of the workplace, the technologies on which she relies, and the others with whom she routinely interacts (Suchman and Wynn 1984). Paper documents, the traditional office technology, are available for physical arrangement in ways that can be read for work to be done and work completed. While the organization of desks, files, etc. is partially inherited from previous owners, workspace arrangements are transformed by each individual in the course of actually doing the work. In this sense territory is appropriated to personal style and individual difference, and the most efficient or satisfactory arrangements are discovered rather than prescribed.

In the office as in other places of women's labour, the introduction of new technologies can bring about disempowerment by transforming previously familiar environments into unfamiliar ground. New computerized systems affect all aspects of the office environment, transforming the physical workplace (including moving some part of it "into the machine"), the technology base and the social organization of office work. To the extent that technology is designed by specialist practitioners working in other settings and imported into the office, the new technology is inherently unfamiliar. And lack of familiarity, we have argued, undermines the basis of competence. This suggests a reinterpretation of the common wisdom that people (women and managers, in particular) are inherently conservative or resistant to innovation. Good reasons for the rejection of new technologies might arise from an implicit recognition of the ways in which models on which the design is based ignore the basis of actual competence. Women are not alone of course in suffering the effects of such models. Our society is specialized to the extent that even managers have technologies imposed on them from a distance, and rationalistic models do no more justice to the realities of managerial work than of clerical work. The crucial difference is that managers can reject new technologies, while women clerical workers are forced to adapt.

What are the lessons we can draw from these cases?

1. There is a politics of technology that comes from the intimate relation of ownership and control of technology to authoritative knowledge and competence. Cross-culturally and historically, women's and men's spheres of influence have in many cases operated as "separate but equal," each holding authoritative knowledge within their culturally allocated arenas of activity. But particularly with the spread of industrial development, the spheres in which men hold influence have systematically been elevated in value and become dominant. In settings where women are primary participants, conversely, the legitimacy of their knowledge has often been subordinated to claims on authoritative knowledge advanced by men.

2. Technological innovation has been a resource for that subordination in two ways. Indirectly, through the representations of knowledge and expertise on which the design of new technology is based. And more directly in the form of ideological commitments manifest in the development and implementation of the artifacts produced. Current models for the design of computer technologies represent activity as rationalized, cerebral and standardized, while suppressing the work's contingent, embodied and differentiated nature.

3. Changes imported from a distance generally are characterized by a design/use gap that must either undermine the pre-existing base of competence, result in work-arounds or adaptations of the technology, more and less satisfactorily, to local conditions or, where users have the power to do so, lead to rejection of the technology. Note that, while they may be required to be users of computer technologies designed by people who do not understand the local conditions and competencies of their lives, women have effectively rejected those technologies en masse.

A STRATEGY FOR SYSTEM DEVELOPMENT

How can we develop new technology in ways that would be sensitive to women's knowledge and concerns, engage their hearts and minds and expand their spheres of competence? In general, we argue that system design must take into account existing technologies and practices, out of which we can identify problems and possibilities for change. One could well imagine a different approach to the design of midwife training courses in the Yucatan, which would take the local ethno-obstetric system seriously, determine which problems it cannot handle, and then design collaboratively with the midwives and women in the community a program that is more sensitive to the material circumstances and cultural realities of childbirth in the region (see Jordan 1987c). In a similar vein, the design of office technologies could be informed by the expertise and work practices of the women who do this kind of work. Only a concerted effort to demystify technology and legitimize the authority of local knowledge will realize that possibility for women.

One implication of our analysis is a radical change in the way that we view the knowledge and skills that go into system design; namely, that we view those of the prospective users of the technology as central, and that we incorporate into the design process as sophisticated an understanding of the social world as of the technology involved. Our own research strategy is to undertake studies of everyday work practice in a range of settings and bring them to bear on the design of appropriate computer systems. Two related methods of inquiry support our research: ethnographic studies of practice and interaction analysis.
Ethnography, the traditional method of social and cultural anthropology, is the intensive analysis of some activity or set of activities in a complex social setting. The goal is what

Anselm Strauss has described as "a conceptually dense theory" (1987, p. 1) articulating the significance of activities and relationships between them in terms of members' beliefs, practices, conflicts, commitments, projects and the like. Such theorizing requires extended participation in the internal life of a setting combined with a critical analytic perspective. Interaction analysis is concerned with detailed empirical investigation of the interaction of human beings with each other and with their physical environment, including technology. Video records constitute the "microscope" for these investigations, where the relevant dimension is time and the requirement that we somehow capture an inherently fleeting phenomenon for systematic review.

Recent challenges to the prevailing systems development process, particularly from Scandinavian computer scientists, suggest the possibility of taking computerization as an occasion to articulate unacknowledged forms of expertise and to take that knowledge seriously as a basis for design (see for example Bjerknes et al 1987, Bodker 1987, Greenbaum 1987, Bodker and Greenbaum this volume, Ehn 1988.) In her article "Outline of a Paradigm Change in Software Engineering," (1987) Christianne Floyd draws on her own experiences as a professor of computer science at the Technical University of Berlin. Floyd characterizes the difference between what she calls "product" and "process-oriented" perspectives on software design as "an on-going controversy between rivaling ideas and attitudes underlying our scientific and technical work in software engineering" (p. 193). As Floyd describes it, the "product" perspective views software as essentially a program and associated documentation, and rests upon the following assumptions:

1. For the design of high-quality computer programs, it is sufficient to consider (formalized) information-processing aspects of the usage context. They can be considered on their own, other aspects being ignored ("abstracted from"). These other aspects include the importance of social communication and cooperation for meaningful work; the role of our body and our senses in dealing with real objects rather than with their computer-simulated counterparts; the environmental conditions under which we perform our work.
2. Human information processing and decision-making is largely equivalent to the functioning of computer programs. The division of work between people and computers is therefore arbitrary and should be determined by technical feasibility. There is no guide-line as to what should and what should not be automated from a human point of view.
3. The suitability of a program in the context of work processes is determined by the information it produces. There is no consideration of how this information can be meaningfully used in communicative action. (p. 199)

In the product perspective, the interaction between the program and its environment is considered to be prescribed by the program's design. The "referent system," or "the part of the real world which we take into account when developing programs" is pre-selected for aspects relevant to the software, while the software development process itself is outside the bounds of the analysis. In the process perspective, in contrast, the referent system is composed of "human work, learning and communication"(p. 194), which are assumed to be subject to continuous development in relation to the software system and vice versa. The interaction between program and environment on this view must continuously change as designers and users change their relation to the technology and as the software is modified to meet new situations and changing work processes. In place of linear "phases" of system development, the work of defining objectives, establishing requirements, specifying functionality, designing, implementing and evaluating comprises different "universes of discourse" that meet through the collaborative interaction of participants in a complex set of activities "interwoven in time." (202)

The words "oriented" and "perspective" are chosen by Floyd to convey the sense not of an opposition or mutually exclusive disjunction, but of prioritization:

"In this paper I have contrasted two perspectives and not two communities of people holding these perspectives. We all argue and act from both of these perspectives. The criticism implied in my argumentation refers to the fact that existing methods and scientific approaches in software engineering embody the product-oriented view almost exclusively. Working towards overcoming this is both highly imperative, considering the role of information technology in the living human world, and also inspiring since it will provide the opportunity for deep insights into the nature of cognitive processes, for richer relations with the people we meet in our work, and thus for personal development and growth while doing high-quality work in technology." (208)

Floyd's call is for a re-prioritization, a kind of affirmative action for the process perspective and its associated concerns. The process perspective focuses designers' attention on the particular settings, activities,interactions between people, and relations between people and machines that Floyd calls the "living human world" (p. 193). More importantly, it directs our attention to the problem of developing a systematic, non-reductionist, experientially based way of designing technology to fit that world.

ACKNOWLEDGMENTS

Our thanks to Paula Goossens for her comments as referee, and to Randy Trigg for helpful comments on an early draft.

REFERENCES

Bjerknes, G., Ehn, P. and Kyng, M. (1987) Computers and Democracy: A Scandinavian Challenge, Aldershot, UK: Avebury.

Bodker, S. (1987) Through the Interface: a Human Activity Approach to User Interface Design, Computer Science Department, Aarhus University, Denmark.

Ehn, P. (1988) Work-Oriented Design of Computer Artifacts. Stockholm: Arbetslivscentrum.

Floyd, C. (1987) Outline of a Paradigm Change in Software Engineering, in Bjerknes et al, Computers and Democracy: A Scandinavian Challenge, Aldershot, UK: Avebury.

Fox Keller, E. (1985) Reflections on Gender and Science. New Haven: Yale University Press.

Greenbaum, J. (1987) The Head and the Heart: Using gender analysis to study the social construction of computer systems. Information and Media Science Department, Aarhus University, Denmark.

Jordan, Brigitte (1983) Birth in Four Cultures. Montreal, Canada: Eden Press.

Jordan, Brigitte (1987a) The Hut and the Hospital: Information, power and symbolism in the aritifacts of birth. Birth: Issues in Perinatal Care and Education 14 (1):36-40.

Jordan, Brigitte (1987b) High Technology: The Case of Obstetrics. World Health Forum 8(3):312-319.

Jordan, Brigitte (1987c) Modes of Teaching and Learning: Questions raised by the training of traditional birth attendants. Technical Report #004, Palo Alto, CA: Institute for Research on Learning.

Strauss, Anselm (1987) Qualitative Analysis for Social Scientists. Cambridge, UK: Cambridge University Press.

Suchman, Lucy (1983) Office Procedures as Practical Action: Models of work and system design. ACM Transactions on Office Information Systems, Vol. 1, No. 4: 320-328.

Suchman, Lucy and Eleanor Wynn (1984) Procedures and Problems in the Office. Office: Technology and People, 2:133-154.

BIOGRAPHICAL NOTE

Drs. Lucy Suchman and Brigitte Jordan are anthropologists at the Xerox Palo Alto Research Center, involved in studies of human activity and interaction in relation to technology design. Brigitte Jordan is also affiliated with the Institute for Research on Learning in Palo Alto, California, a non-profit research institute dedicated to the study of learning in non-school settings.
Correspondence address: System Sciences Laboratory, Xerox Palo Alto Research Center, 3333 Coyote Hill Road, Palo Alto, CA 94304, U.S.A.

Women, Work and Computerization: Forming New Alliances
K. Tijdens, M. Jennings, I. Wagner, and M. Weggelaar (Editors)
Elsevier Science Publishers B.V. (North-Holland)

A FEELING FOR SYSTEMS DEVELOPMENT WORK- DESIGN OF THE ROSA PROJECT

Susanne Bødker and Joan Greenbaum
Department of Computer Science,
Aarhus University, Denmark

This article is based on the design of a research project that will look at intuition, learning processes, language and roles in the development of computer systems. The research project, called ROSA (a Danish acronym for Roles and Cooperation in Systems Development) grew out of our interest in the informal working practices among systems developers, because it is these informal working relationships that are most often overlooked in research about computer science methods and tools.

The project applies a gender perspective to look at the informal work relations of systems developers. The concept of a gender perspective means that we do not intend to look for, or prove, the existence of differences between men and women, but rather to use gender awareness to 'listen to' and get a 'feeling for' how systems developers work together. Our research methods are interdisciplinary and based on action-oriented, participatory methods that help system developers reflect on their own working practices.

The Abacus and the Rose

I, having built a house, reject
the Feud of eye and intellect,
And find in my experience proof
One pleasure runs from root to roof,
One thrust along a streamline arches
The sudden star, the budding larches.

The force that makes the winter growi
ts feathered hexagons of snow,
And drives the bee to match at homet
heir calculated honeycomb,
Is abacus and rose combined.
An icy sweetness fills my mind,

A sense that under thing and wing
Lies, taut yet living, coiled, the spring.

J. Bronowski, *Science and Human Values, p.119.*

As teachers and practitioners of systems development[1] we are, of course, caught up in the whirlwind of debate about how computer specialists and users can better interact with each other. This debate is not new. The spread of personal computers and workstations in the '80s along with the accompanying pre-packaged software applications have clearly brought computer "users" into the forefront of discussion. In fact, the increased knowledge, and to some extent influence of users, has reshaped the lines of the old debate. The issue is no longer whether users should be involved in computer system design, but rather, how their knowledge and experience can be put to good use. And more importantly, from the perspective of computer specialists, the question of how to design systems for, by and with users, has rightfully dominated the debate.[2]

These issues were running through our heads last summer while we were attending a working conference on design and use of information systems. In the midst of the three day conference we were struck by the fact that while the debate had shifted to enfold users' interests, the methods and suggestions were stuck firmly in the historical frame of computer science and its reliance on formal problem-solving. Presentation after presentation addressed the need to find new methods to build better bridges with "the users". But like engineers building bridges, the solutions mostly seemed to focus on quantifiable techniques to *solve* the problem of user-involvement.

The reliance on formal methodology is not new. What was new, however, was our intense desire to climb outside the frame of formal model building and to set up a research project to highlight the *social construction* of computer systems and to pay close attention to the often overlooked *informal* nature of systems development work. The ROSA project,[3] as a small pilot study, was born of that moment. It was one of those points in time when one simply has to stop complaining and "go out and do something" about the situation.

The "something" we chose to do was to construct a research project where we would ask systems developers to reflect on their working practices. The systems developers themselves, with help from researchers, would actively participate in *noting* their ways of working together and seeing if *conscious* awareness of their discussions would provide them with an useful base to influence the way they interact. While we were, and are, keenly interested in the interaction between system developers and users, we thought that our best starting point was within the terrain that we knew the best – the working practices of systems people themselves. This pilot project can, perhaps, serve as a bridge to other studies focusing on the interaction of users and systems developers.

What we present here is first, a discussion of our theoretical base and points of departure for doing the project. It represents for us our 'feeling' for the way systems work is done. And like all feelings, it rests on our past experiences. The second part outlines the

---

1 We will use the terms system and systems development even though these terms as such reflects a certain view of computer applications and their development and use, which has its roots in the rationalistic tradition discussed by Fox-Keller (see below). The reason for using the terms is mainly familiarity. If the reader prefers, the term system can be substituted throughout the text by the term application.

2 Bjerknes, Gro et al., ed.: *Computers and Democracy – a Scandinavian Challenge*, Avebury 1987.

3 The members of the ROSA project are Berit Holmquist, Randi Markussen, Regine Hansen, Lisbeth Rasmussen, and Tage Stephansen, and the authors. The actual research scope of the project is broader than that described in this article. Our focus here uses a gender perspective to look at system development work. Other research perspectives include frames of reference from organizational theory and linguistics. The ROSA project was born out of the research program on computer support for cooperative design and communication at the departments of Information and Media Science, and Computer Science at Aarhus University.

research project itself. While many people note that the emphasis on social interaction and informal working practices reflects women's values, our *focus is not on supporting arguments about the differences between the way men and women work in this field.* Rather, we have chosen to use a *gender perspective* as a way of organizing our thinking and, most importantly, as a pattern for carrying out our research. As Bronowski's poem tells us, we 'reject the feud of eye and intellect', and hope to find our way through the 'honeycomb' to see 'abacus and rose combined'.

OUR STARTING POINT

Evelyn Fox-Keller's *Reflections on Gender and Science*,[1] nicely points out how societally-shaped notions of men and women strongly influence the way that science is done. In her work she shows us, both historically and in practice today, how these notions or myths place women and men on binary poles where their differences are emphasized in the questions asked by scientists and the methods used to answer those questions. These binary poles, or dichotomies, shape the man's world as rational, objective, and quantifiable, while the woman's sphere is painted as emotional, subjective, intuitive and qualitative. As with all myths, their validity is certainly not the issue. What is at stake is the fact that they shape our consciousness and therefore our behaviour.

We quite agree with Fox-Keller's analysis, and, as illustrated in Joan's article entitled *The Head and the Heart* [2] we see clear examples of the ways that these myths effect the development of computer systems. But, and this is an important point of departure for us, we believe that emphasis on the dichotomies gets us stuck within the frame of only analyzing the differences between men's and women's practices. It unfortunately leads us back up the trail where we find ourselves, once again, looking at "women on a pedestal" or "women as victims". A great deal of the literature about women's work behaviour has been trapped into explaining or bemoaning differences using a *sex-role model analysis*.[3] We feel that this was not Fox-Keller's intention, and nor is it ours.

Instead, we look at the *working relationships* for both men and women system developers. In doing this we will focus on four aspects of informal working relations that often are associated with women. These four: intuition (which, for us, is experience-based use of practical understanding) learning processes, language use and roles (ways of acting in different situations), form the themes of our research project. Using these themes as a working definition of informal work relations lets us highlight aspects often kept out of the research spotlight.

While these themes help us focus on working relations that traditionally fall on the woman's side of the dichotomy, we do not do so to claim some pre-conceived pattern. As practitioners of systems development we don't feel that an increase, for example, in the use of intuition as such would make better computer specialists of us all.

We are women, who work as researchers, and whose interests lie in constructing useful computer systems in ways where we can support rather than compete with each other. This statement about *who* we are says a lot about the way we see ourselves in relation to *what* we do. Perhaps, it characterizes the main points of the gender perspective we

---

1 Fox-Keller, Evelyn: *Reflections on Gender and Science,* Yale, University Press, 1985.

2 Greenbaum, Joan: *The Head and The Heart*, DAIMI PB-237, Aarhus University 1987.

3 Kvande, Elin and Bente Rasmussen: *Who Lacks Courage-The Organizations or the Women?*, in *Women Challenge Technology*, Vol. II, European Conference on Women, Natural Sciences and Technology, Elsinore, Denmark, 1986.

apply to our work. Fox-Keller, in describing the research of Barbara McClintock, also describes a research method of 'feeling for', or 'listening to', the subject of the research.[1] Barbara McClintock, a well-known, although un-conventional cytogeneticist, marked her research with a strong belief that one should be "letting the material speak to you". Instead of descending on a research topic with sets of preconceived categories, McClintock choose to think of herself as being part of the material. For her, scientific work was living and being involved in the subject matter, or what she referred to as "a feeling for the organism".

We hope to work in a similar way. Our *gender perspective* starts from our personal involvement in our work and uses this involvement to frame the questions, issues and hunches that we feel are important in the construction of computer systems. In this way, we begin on the "women's side" of the gender dichotomy, but try to use our conscious awareness of this to frame the direction we want to go in. We firmly believe that the material conditions of our work (i.e. that we are *women* who work as researchers in computer systems development) shape our consciousness about our work.[2] All research is, of course, affected in this or similar ways. Our starting point is to use this awareness as a conscious function in the research design and outcomes.

The ROSA project then, develops from this perspective: as women, we highlight the informal, social and supportive aspects of computer system design; as researchers we are constructing a project where the participants speak for themselves and where the research methods reflect their values and opinions; and as systems developers we are interested in ways that people can effectively work together in constructing new applications. In fact our starting point wraps itself around to become our ending point. For our belief is that our perspective and research approach can be useful for systems developers in their work.

Our purpose is not to recommend a new systems development method, or to suggest appropriate tools for better system design. We believe that this emphasis on finding the right solution to the correct problem is too strongly rooted in the technical, scientific (and mainly male) side of the gender dichotomy. It has formed the building blocks of both 'good research' and 'good systems design' for quite some time.[3] The idea that the right solution can be found is a myth that lingers on. Indeed, it was at the conference that we mentioned last summer, that the constant repetition of this myth led us away from the one-solution path. We have chosen not to follow any one, specific path, but rather to find our way through a terrain that, hopefully, bends its way around the poles of gender mythology. By letting systems developers speak for themselves, we hope, that like Barbara McClintock, we can learn to 'listen' and get a 'feeling for' their work.

MYTHS, MYTHS AND MORE MYTHS

The phrase 'women's intuition' has the power to allow women to use and acknowledge their intuition, as well as the pain of stigmatizing women's conclusions as 'hunches' or 'not logical'. In a similar way, the idea that men are rational and objective gives weight to their conclusions, allowing them a positive, persuasive power. Yet, at the same time, it may hinder men in making them feel that they must perform in these predefined rational ways.

---

1 Fox-Keller, Evelyn: *A Feeling for the Organism: The Life and Work of Barbara McClintock*, San Francisco, W. H. Freeman, 1983.

2 Of course the works of Karl Marx on materialism represent a rather well-documented example of this way of thinking.

3 Op. cit. Fox-Keller, Greenbaum.

These and many similar stereotypes are re-enacted at work every day in the ways people relate to each other and reflected in the things that are said. In many situations they are played out as roles, or forms of behaving in different situations. The power of the mythology behind these stereotypes influences much in our daily working lives. The 'good' or 'nice' girl is a role that women are socialized into. Daily, she appears in work situations where, for example, a disagreement threatens a deadline for completing part of a project, and a woman, as systems developer, may be expected to smooth things over by playing this part. The 'capable' or 'clever' guy may be played in the same situation.

There are a variety of myths about women in technical fields that have found their way into computer systems design. Most focus on the fact that women are not interested in technical details, but rather in social interaction. A new version of these old arguments has appeared in the form that women make good (better) project leaders because of their social skills. These stereotypes are often used to *explain* things, like why there are so few women in the computer field, or why women have not advanced very far up the organization/technical ladder, and, more recently, why women should be project leaders. Using socialization and stereotypes as a way of explaining events seems rather like circular reasoning to us.

What interests us here is that these stereotypes push people into believing or behaving on one or the other side of the gender dichotomy. 'Women as better project leaders' plucks women out of the ordinary world and places them right up there on a pedestal – the very place that the women's movement has worked hard to show is not where we want to be. For we all know that the woman on the pedestal can be easily pushed off, if she makes a mistake. And 'women as not interested in technology' moves in the direction of 'woman as victim', where women are denied jobs or promotions because they supposedly lack this interest and skill. In dealing with myths we find it important to consider the difference between the (gender) roles *ascribed* to a person, and the (gender) roles with which the person herself *identifies*:[1] A male manager, for example, may well see (ascribe), a woman as a " sweet girl", but in her work situation, she sees (identifies) herself primarily as a good systems developer. Looking at ways that systems developers identify themselves may help understand how people behave outside of stereotypical roles.

In the ROSA project we will be looking at a specific system development project in a large company where 20% of the system developers and programmers are women. We are focusing on the working relationships among all system developers, keeping in mind the ways that gender may cause possible differences or be an influencing category. We will ask people what they think about these myths and stereotypes, but we can reasonably expect that the women who have been successful in these positions have, themselves, steered a path between gender myths. In our own experience we find that we are both interested in technical details, and in developing the social contacts that keep a project running smoothly. Our personal experiences and those of men and women we know in the field have formed the basis of the ROSA research design.

---

1 Cockburn, Cynthia: *Machinery of Dominance. Women, Men and Technical Know-how*, Pluto Press 1985.

DESCRIPTION OF THE ROSA PROJECT

The ROSA project is set up as a small pilot project to study the work processes of systems development. By looking at the themes of *intuition, learning processes, language and roles,* we set the stage for getting a feeling about informal working practices. These themes are outlined here.In addition to the themes, we have chosen a range of research techniques that bring together a group of inter-disciplinary researchers in order to get a variety of qualitative interpretations. The research methods include *conversation analysis, observation, participant-observation, in-depth qualitative interviews and workshops.* The methods are briefly explained here also.

An important consideration in designing the project in this way, was the fact that we wanted to let the participants speak for themselves. In our double roles as researchers and systems developers we knew that the systems developers we studied should play an active part in reflecting and interacting within the research project. Thus, the use of workshops and participant-observation were intended to let the participants set their own agenda. Researchers, like systems developers, go in and out of a workplace, often leaving behind a wake of 'subjects' whose interests are cut off, as if they were suspended in time. We felt that the project could be seen as a set of action-based themes and activities that would help the participants as well as the researchers, explore the issues from the perspective of their own experiences.

One of our first steps was to outline some central questions about the themes of our work. We illustrate a sample of these here, along with our working definitions of the themes:

* what is the role of *intuition* in the development of computer applications? what does it mean to act according to intuition, or to use experienced-based knowledge in systems development - how does that contrast and supplement the use of formal methods?
* what *learning* takes place during the development process? learning about what (work practice, new methods and other aspects concerning the *process?* and learning about the work of users, the technology and other *product* oriented aspects?) how do systems developers teach each other?
* what types of *languages* are used by systems developers in their daily work and in their meetings? in relation to each other, to management, and to the users.
* what *roles* are played by systems developers as they adapt to work with users, the equipment, and in meetings?

We began our working definitions with the concept of *language*, because we see language as action. Rather than viewing language as just an exchange of information, we see human activity as involving language games[1]. Language games capture the ways we participate and how we make commitments. They are different for different groups and reflect different *roles*. Language then, reflects the relation between individuals, and groups. In systems development, the participants always take part in new language games: the developers deal with users who, under normal conditions, are part of other language games, and have different backgrounds. For this reason it is important to see design as an on-going *learning* process, where the different backgrounds of the participants is an ever-lasting concern, yet, at the same time a source of learning and change.

---

1 The reader may think of Wittgenstein's definition of language games (Wittgenstein, Ludwig: *Philosophical Investigations*, Oxford University Press 1953). For our purpose here, a specific understanding of the term is less important than the notion of situated action and language as action.

Intuition is not a matter of magic, but of practical knowledge or 'know-how'. While intuition is usually associated with some kind of instant insight, we see it as an impression of how people use their experience to react in different situations.[1] And *roles* are the way people adapt as their situations change. Roles usually reflect people's values and assumptions. Like language, learning and intuition, roles point to the overt and informal ways that people react within the formal organization of developing computer systems.

A never-ending characteristic of systems development work is that it always changes. People caught up in the process of designing and constructing computer systems find themselves in new situations, adapting as best they can. Sometimes project deadlines change, often design descriptions change, and most often, decisions are made by others that effect what work is to be done. Having experienced the chaos of constant change ourselves[2], we hope that the ROSA project can help shed some light on it for the participants and for other system developers and researchers.

Our themes have led us in the direction of looking at *situation-based* action and the process of change. Within this framework we will use a broad concept of roles to examine how people adapt. We might expect, for example that traditional 'boy-girl' roles may enable system developers to adapt to some situations, but that in many other areas they use a wide variety of other, non-gender based behaviour. As the following discussion of our research methods shows, the techniques attempt to let the participants identify their own issues.

Our use of multidisciplinary techniques puts together a research toolbox that includes: conversation analysis from linguistics; qualitative interviewing techniques from the humanities; and action-based workshop methods[3] from the social sciences. *Conversation analysis and observation* of different meeting situations will be used in order to focus on language and roles. The in-depth *interviews* with systems developers will be applied to focus on their work life, home life, qualifications, and of course, their feelings about their work. And *participatory workshops,* using techniques like role-playing, diaries, story-telling, and future workshops[4] are intended to help the system developers raise issues that are important to them. The workshops will focus on the questions raised in connections with the project themes. With this wide net of techniques, we hope to learn more about the social interaction of systems development projects; an aspect which is often neglected in systems development methods and theory.

This pilot project will take two research months. Here is an overview of the activities we are involved in:

---

1 See for instance Suchman, Lucy: *Plans and Situated Actions: The problem of human-machine communication*, Xerox ISL-6, 1985, Winograd, Terry and C. Fernando Flores: *Understanding Computers and Cognition: A New Foundation for Design*, Ablex Publishing Comp. 1986.

2 See Greenbaum, Joan and Lars Mathiassen: *Zen and the Art of Teaching Systems Development*, DAIMI PB-238, Aarhus University,1987.

3 Different action oriented approaches similar to ours are described by Freire, Paolo: *The Pedagogy of the Oppressed*, Herder and Herder, New York 1971 Engeström, Yrjö: *Learning by Expanding*, Helsinki 1987. In our version it has roots in the Scandinavian trade union projects (see Ehn, Pelle and Morten Kyng: *The Collective Resource Approach to Systems Design* in Bjerknes, op.cit.).

4 Future workshops are described in Jungk, Robert and Norbert R. Müllert: *Zukunftwerkstätten, Wege sur Wiederbelebung der Demokratie*, 1981 (Danish version *Håndbog i fremtidsværksteder*, Politisk Revy 1984). We are strongly inspired by this way of facilitating brainstorming. We will, however, not apply Jungk and Müllert's method too strictly, but try several different techniques.

1. *Organizational analysis* – a study of organization charts, project group structure and history of project objectives and changes in work organization, including organizational culture, and interviews with project and group leaders.
2. *Conversation analysis and observation* of project group meetings – tape recordings and observation of selected group meetings plus tape recordings and observation of selected work situations.
3. *In-depth interviews* with group leaders and system developers – interviews of approximately 2 hours each with 6-9 people concerning their reflections about the project themes (intuition, learning processes, language use and roles).
4. *Workshops* with system developers – 6 workshops of 2 hours each. The workshops, combining the methods discussed above, will include participants from the in-depth interviews. There will be an equal number of men and women in the workshops, with two workshops where the men and women will be asked to discuss the issues separately.

The company where the pilot study is conducted has its own Systems Development department. The computer system that we focus on aims to provide computer support for financial functions in all branches throughout Denmark (more than 3000 users). The systems development work is being carried out by approximately 30 systems developers, organized in smaller groups with different tasks, and managed by group leaders. The group leaders refer directly to the project manager. Contact with future users is maintained through a reference group, consisting of experienced users selected from the business branches. The systems developers are trained either as business assistants with additional systems development training, or as programmers, engineers, or computer scientists. They equally participate in all aspects of the development and implementation process, with little or no division of work between systems analysis, design and programming.

DISCUSSION AND EXPECTATIONS

The project is intended as only a very limited pilot study. We feel that it is a good place to start, because we need experience with the toolbox of methods that we are using, as well as preliminary feedback from systems developers. Barbara McClintock describes her "feeling for the organism" as an experience where "I wasn't outside, I was down there – I was part of the system".1 We hope that our research methods can help us 'get down into' the experience of systems development work.
Accounts of these kinds of empirical investigations are almost absent in literature about technology. We feel that this should be rectified from both a feminist and a systems development point of view. In the feminist 'world' there are many accounts of women as victims of technology, but few of women as active participants in the shaping of technology. In the systems field there are an almost endless number of books about how systems 'should' be constructed, but not about how work is actually done.

In order to fill this empirical gap, our research questions, focus on three general areas:
1) What are examples of some of the informal working practices used by systems developers in the design of a large system? How do informal practices support or contrast with formal project organization and methods?
2) How does a gender perspective differ from general multidisciplinary research efforts? Does it let us 'listen to' the systems developers, and is this useful for better understanding their working-practices?

---

1 Fox-Keller, Evelyn: *Women, Science and Popular Mythology*, p. 141, in Joan Rothschild, ed.: *Machina Ex Dea,* Pergamon Press 1983.

3) Will our research toolbox be useful to systems developers? Can they use both our research methods and outcomes in their work?

Systems development theory in general, seems to be moving in a direction away from formal methods to situation and experience based action[1], but there is very little empirical work focusing on these aspects of systems development. In fact, as systems development theory moves away from formal, science-based methods and questions, it begins to move into the open, undefined territory between the gender myths. Formal, scientific computer systems methods were clearly associated with the 'male' side of the gender pendulum. As system developers lose some of their reliance on 'the right solution' and move to embrace a wider range of methods and perspectives, the possibility for effective change, can, we believe, increase. We hope that both our toolbox of methods and our use of a gender perspective can be useful to systems developers in their daily work.

REFERENCES

Bjerknes, Gro et al., ed.: *Computers and Democracy – a Scandinavian Challenge*, Avebury 1987.

Bødker, Susanne: *Through the Interface – a Human Activity Approach to User Interface Design*, DAIMI PB-224, Aarhus, University 1987.

Cockburn, Cynthia: *Machinery of Dominance. Women, Men and Technical Know-how*, Pluto Press 1985.

Ehn, Pelle and Morten Kyng: *The Collective Resource Approach to Systems Design,* in Bjerknes.

Ehn, Pelle: *Work-oriented design of Computer Artifacts*, Aarhus forthcoming.

Engeström, Yrjö: *Learning by Expanding*, Helsinki 1987.

Floyd, Christiane: *Outline of a Paradigm Change in Software Engineering*, in Bjerknes.

Fox-Keller, Evelyn: *Reflections on Gender and Science,* Yale, University Press 1985.

Fox-Keller, Evelyn: *A Feeling for the Organism: The Life and Work of Barbara McClintock*, San Francisco, W. H. Freeman 1983.

---

1 Floyd, Christiane: *Outline of a Paradigm Change in Software Engineering*, in Bjerknes, op. cit., Ehn, Pelle: *Work centered design of Computer Artifacts*, Aarhus forthcoming, Bødker, Susanne: *Through the Interface – a Human Activity Approach to User Interface Design*, DAIMI PB-224, Aarhus University 1987, or Winograd and Flores, op. cit.

Fox-Keller, Evelyn: *Women, Science and Popular Mythology*, in Joan Rothschild, ed.: *Machina Ex Dea,* Pergamon Press 1983.

Freire, Paolo: *The Pedagogy of the Oppressed*, Herder and Herder, New York 1971.

Greenbaum, Joan: *The Head and The Heart*, DAIMI PB-237, Aarhus University, 1987.

Greenbaum, Joan and Lars Mathiassen: *Zen and the Art of Teaching Systems Development*, DAIMI PB-238, AarhusUniversity 1987.

Jungk, Robert and Norbert R. Müllert: *Zukunftwerkstätten, Wege sur Wiederbelebung der Demokratie*, 1981.

Kvande, Elin and Bente Rasmussen: *Who Lacks Courage-The Organizations or the Women?*, in *Women Challenge Technology*, Vol. II, European Conference on Women, Natural Sciences and Technology, Elsinore, Denmark 1986.

Suchman, Lucy: *Plans and Situated Actions: The problem of human-machine communication*, Xerox ISL-6 1985.

Winograd, Terry and C. Fernando Flores: *Understanding Computers and Cognition: A New Foundation for Design*, Ablex Publishing Comp. 1986.

Wittgenstein, Ludwig: *Philosophical Investigations,* Oxford University Press 1953

BIOGRAPHICAL NOTE

Susanne Bødker and Joan Greenbaum work at the Department of Computer Science, Aarhus University. Joan Greenbaum is on leave from LaGuardia Community College City University of New York.
Correspondence address: Department of Computer Science, Aarhus University, Ny Munkegade 116, DK-8000 Århus C, Denmark

Women, Work and Computerization: Forming New Alliances
K. Tijdens, M. Jennings, I. Wagner, and M. Weggelaar (Editors)
Elsevier Science Publishers B.V. (North-Holland)

DECENTRALIZED SYSTEM DEVELOPMENT: ADVANTAGEOUS FOR WOMEN?

Leikny Øgrim
Department of Informatics
University of Oslo, Norway

The purpose of this paper is to contribute to a discussion on how decentralized system development can lead to increased influence for women over their own work situation.
Decentralization of organizations and system development may be used by management to weaken the formal influence and power of local trade unions. However, formal trade union influence does not necessarily lead to real employee influence. Beside, women have a weak position in the trade union movement (as in the rest of the society), so decreasing formal union participation may not affect women after all.
In small and informal organized system development projects, the organizational structures to assure influence of the employees do often not exist. The employees may obtain influence over the system development process also through direct user representation. I will argue that informal organization and direct user representation (as opposed to union representation) may be of advantage to women. The sex composition of the project groups are important with regard to the influence of women in the group.

1. INTRODUCTION

For some time I have been interested in the impact from decentralized system development on local trade unions[1] The conclusion so far has been that decentralization often leads to decreased influence from the workers and their local trade unions. While working with these questions, I have often thought of the influence of the **female** workers[2]. We all know there are gender based differences both in society and working life. Females have for instance less tendency than males to participate in trade union work, and female dominated unions are often weak. Is it possible that my earlier belief holds only for unionized workers, or only for the unions themselves, and perhaps not for the individual members? Is it possible that informal organization of system development is advantageous for women?

---

1 Øgrim & Norderhaug(86), Øgrim(87).

2 In the Scandinavian research tradition on system development, described for instance in Bjerknes & Bratteteig(87 and 88) and Ehn & Kyng(87), we have worked with democratization on premises of the trade unions, which in general means white, male production workers.n the Scandinavian research tradition on system development, described for instance in Bjerknes & Bratteteig(87 and 88) and Ehn & Kyng(87), we have worked with democratization on premises of the trade unions, which in general means white, male production workers.

The paper mostly deals with female **users**. Still issues related to influence from women in system development should be of interest to feminist organizations, trade unions and system developers of both sexes.

As a foundation for this paper I have interviewed 10 women about their views and experiences in gender based differences in system development: three system developers, five computer shop stewards, one trade union activist and one feminist activist[1] Instead of presenting my investigation, I have chosen to make a synthesis of their experiences and create a typical example. I give comments to and discuss the example on basis of my earlier work and literature studies (The example is *italicized*, my comments are plain text..

The structure of the paper follows the structure of the argumentation. First the example is introduced, and then one section is dedicated to each of the following statements: 1) Management can use decentralization[2] to weaken local trade unions. 2) But formal influence is not equal to real influence. 3) Beside, women have a weak position in the trade union movement. 4) The employee can obtain influence through user participation. 5) The sex composition of the project groups is also of importance to women.

2. THE FAN ECONOMY DEPARTMENT

FAN ("Find A Name") is a publishing company partly owned by the state. The company mostly publishes school-books, other professional books and de luxe editions. There are approximately 200 employees in the administration, spread in different departments. I will give most attention to the account office in the economy department.

The economy department is divided into three offices, handling different kinds of tasks. The account office has four office workers employed, in addition to the account leader. The office workers are now all women, but earlier there were two male office workers. The account leader and the leader of the economy department are men. Three of the women have children, one of these is a single mother and one has part time work.

The office workers mostly are educated at business colleges, and they have all been working with accounting for at least five years. There is no educational programme in the company. The workers have the responsibility for specific tasks, but can do the work of one of the others in case of illness. Their work consist of some routine tasks, but they also handle special cases.

They use computer systems for accounts, budgeting, cost accounting and order handling. They also use some other computer systems frequently, for instance the stock keeping system.

There are three different unions at FAN, one for the department leaders, one for the employees at the computer department and one for the office workers.

In 1985 management started a decentralization campaign to give more autonomy to the different departments. Many important decisions about trade, plans, production and economy are now made by the local leaders of the departments.

---

[1] I see the scientific problem with the small range of people and the one sided sex composition. Still this is a start to investigate women's experiencies.

[2] The concept decentralized system development is discussed in Øgrim(87). In this context the main characteristic with decentralized system development projects is that they are often informally organized and involve a small number of people, both as participants in the project and as users of the future computer system.

3. MANAGEMENT CAN USE DECENTRALIZATION TO WEAKEN LOCAL TRADE UNIONS

The unions in Scandinavia traditionally have influence through laws and agreements and formal organizational organs for negotiations between management and the unions[1]

Earlier the computer shop steward from the union of the office workers had monthly meetings with the head director and the leader of the computer department to exchange information and ideas about on-going and planned system development projects.

As a consequence of the decentralization campaign, most of the decisions about system development are delegated to the departments, and neither the unions nor management felt need for these meetings any longer. The union now gets information about system development projects after they are planned, and sometimes even after they are started. Moreover, much of the practical project work is organized without any formal representation from the union.

The union leaders now have less control over the union's representatives in the project groups. This problem is increased by the fact that ordinary union members, who usually are user representatives in the system development projects, are less educated in laws and agreements than the central shop stewards. Decentralized system development has made it more difficult for the union club at FAN to keep the overview and control of the system development activities in the company and to pick the right union representatives[2].

The union is however not especially interested in questions concerning computerization or new technology. Such questions have never been discussed at members' meetings, and only briefly on the union board. The computer shop steward thinks that her job was more important earlier, when the workers were more afraid of new technology. Now the office workers feel need for computer systems in their daily work, and the computer shop steward feels that the need for her job has decreased.

This story is quite common. Many computer shop stewards have withdrawn during the last few years.

Decentralization, together with the ideological campaigns to bind the entire organization together in a common company culture, may be used to pacify the members of the organization by giving them a feeling of shared responsibility and participation[3]. This feeling, which does not necessarily mean **real autonomy** for the workers, may be used by management to strengthen the control over the organization[4]..

However, decentralization of system development is often described or interpreted as a part of democratization of the work organizations[5]. User participation is often regarded as the same as union participation both by management and by system developers[6]. The

---

1 See for instance Nygaard & Bergo(74).

2 The same problem is mentioned by Schneider & Ciborra(83) p 248

3 This kind of ideology permeates for instance Carlzon(85)

4 Ehn & Kyng(87) say that ``Direct control and responsible autonomy are complementary strategies for capital accumulation" p 37. Dooreward et al(87) also argue that group organization and integration of tasks go together with taylorism and segmentation. Ciborra(81) and Ehn & Kyng(87) point that these organizational and ideological changes can be used by management to weaken the influence from the trade unions.

5 See for instance Röstlinger & Selldén(83)

6 Ciborra(81)

individual or autonomous group orientation of socio-technique may be seen as a part of this tradition[1].

The view in the trade union movement on decentralization as such, vary very much, perhaps especially as regards decentralized system development. Some hold the view that decentralization is of strategic value for the trade unions, because a greater influence in one area may increase the appetite for more influence in other areas and serve as a training in democracy. Others are of the opinion that decentralization is a threat against the unions, introduced by management to manipulate the workers and to keep them quietly at their posts. Some see the decentralization as a challenge that should be accepted, in order to keep management to their word about cooperation and resource utilization[2].. Most local unions, however, have no conscious attitudes in relation to decentralized system development.

4. BUT FORMAL INFLUENCE IS NOT EQUAL TO REAL INFLUENCE

The earlier monthly discussions turned on economy and technical plans. The computer shop steward had believed she should have influence on what kind of computer system were to be made, but found that the discussions went on the premises of the manager of the computer department. The computer shop steward was quite frustrated over the these meetings. Her knowledge about new technology was too restricted to do a proper job, and neither the local union board nor the other office workers were interested. The local union has no educational programme for their shop stewards.

Now system development in general takes place in close contact with the future users. The local project leaders pick the user representatives to the project groups, maybe after some contact with the local shop steward. At least the shop steward is asked to accept the user representative.

In small and informal organized system development projects, the organizational structures to assure influence from the employees do often not exist. Many representatives for the employees in practice act as hostages in project groups[3]. To avoid this it is necessary that the union/ representatives are well prepared for their tasks and that they have the possibility to discuss the concrete problems with (a group of) their fellow union members.

Participation in system development is a challenge to the local trade unions and to the trade union movement as a whole. Trade Unions are usually concerned with traditional questions like for instance working conditions, night work, shift work arrangements, training and wages. They are not concerned about practical or social work conditions, which is what system development often is about. In addition, the traditional work styles of trade unions are to "fight on the barricades", with formal negotiations, meetings and resolutions and not discussion, investigation and cooperation, which is the way of working in system development[4]..

---

1 The ideas in sociotechnique are described by for instance Mumford(87) and Bostrom & Heinen(77). Ehn & Kyng(87) discusses the sosiotechnical approach (pp 22 - 26), and concludes that it has failed to support local union/ involvement.

2 The union of the trade and office workers in Oslo have worked out a strategy to meet decentralization, see Oslo Handel og Kontor(87)

3 Some of the ``hostage problems" are described in Ehn & Kyng(87) pp 39-40

4 This problem is mentioned in Pape & Thoresen(87), pp 309-310

5. BESIDE, WOMEN HAVE A WEAK POSITION IN THE TRADE UNION MOVEMENT

The female dominated unions are usually weaker than the male dominated ones, with less time and money resources and less prestige and influence both related to management and to the rest of the trade union movement[1]. Between 40 and 50 % of the wage earners in Norway are women[2]J, and about 60 % of these are unionized[3]. Still almost all national unions, and even those that are dominated by women, are led by men.

The trade union activity at the account office at FAN is low. All the office workers, except one with part time work, are members of the union, but only one goes to the meetings and takes part in the union work as a treasurer.

Women have less tendency than men to participate actively in their unions. The trade union movement traditionally have negative attitudes towards part time work, and part time workers are not always offered membership in the trade unions.

The leader of the economy department holds an old fashioned view on organizations. His attitudes towards the decentralization campaign are rather negative, and he seems to be afraid of loosing his prestige if the office workers obtain increasing autonomy. Beside he is overworked and tired. The women at FAN see this as a problem. They want him to retire, but they do not want him to loose prestige or wages. In relation to the top management they protect him. The solution will be to live with him until he retires and then hope for better days.

Many women seem to feel they can not discuss important and difficult matters in their work and struggle inside the union. Instead such discussions take place in for instance feminist groups or different support groups[4]. Maybe some causes can be found in the organizational structure and the formal and hierarchical democracy in the trade union movement.

There are however some tendencies that women's situation in the trade union movement is about to turn. Some of the most important struggles in working life in Norway the last few years have involved typical female workers, like nurses, play-school teachers, airway cabin crew and hotel and restaurant workers[5]..

6. THE EMPLOYEES CAN OBTAIN INFLUENCE THROUGH USER PARTICIPATION

There are the two basic channels for the trade union and the employees to obtain increased influence in system development: **user** participation and **union** participation[6]. In order to obtain as much influence as possible, the workers must use both these channels.

When there is need for a new or modified computer system, the decision is made in the economy department. The computer systems are developed in cooperation between the

---

1 Ericsson(87) pp 63-66, Jensen(86)

2 Jensen(86) p 28

3 Colbjørnsen et al(85) p 31

4 In a playschool strike in 87, the strikers organized groups for discussion and mutual support. In a long lasting conflict about notice of 10 female cleaning workers, organized in a male dominated union, all the discussions about personal finance problems, house loans and the trial are ran in a support group outside the union.

5 See for instance Jensen(86), Palmer(86) and Ericsson(87) pp 129-135 and 138-141

6 Gustavsen(undated)

computer department and the economy department. The leader of the economy department has been participating in all the project groups.

The participating office workers have influence on the projects because of their knowledge, qualifications and experiences. The user representatives are often active and experienced office workers with extensive knowledge, and with attitudes and views that go far beyond the tip of their nooses. By selecting such people, management gains several advantages; and expertise, ideas and creativity are being properly utilized.

A computer system is a part of the work organization context, and is used to carry out some work tasks. The most knowledgeable experts on a work or a work task are those who carry it out. Thus the future users of a computer system possess useful and necessary expertise in relation to the requests to the computer system. This is more and more understood both by management and by system developers[1] .

The new techniques and tools for system development, such as prototyping and the so called 4th generation tools, make it even more difficult than before to run system development projects strictly after the phase models. As the system development process becomes more smooth and iterative, it becomes impossible to insist upon checkpoints and formal negotiations[2].

However, with prototyping techniques, the attention may be more concentrated on influence on the development process, and less on the product, since the product is not well specified[3].

7. THE SEX COMPOSITION OF THE PROJECT GROUPS IS OF IMPORTANCE TO WOMEN

When there were male office workers employed, one of them was usually picked as a user representative together with the leader of the economy department. The female office workers saw this as a result of the leader's old fashioned attitudes towards women. To both of the male user representatives from the economy department the project group worked as a springboard for advancement. Now both of them are employed in the computer department.

In system development project groups men are commonly over-represented. System developers are mostly men, and male workers are in general the easiest to be seen. They are more familiar with and not so afraid of technical innovations as their female colleagues[4], and they are more clever in showing their qualifications. Women are on the other hand often modest on behalf of themselves. When women are asked to be user representatives in a project group they often need to be persuaded.

Reason and impersonal judgment, power and science are associated with masculinity and also with "good" or professional system development, while subjectivity, emotion and

1 The question of **who** the users are, is however not always solved. Friedman & Cornford(87) argue that (with some national restrictions) system developers and management often think that the users are the same as the departmental managers, p 139.

2 Greenbaum(87) describes how microcomputers and the 4th generation laguages have changed the work style and organization of system development, away from strictly division of work in large project groups towards integrated work tasks for the whole project team, p 6. This fits with the Process-Oriented View in software engineering described in Floyd(87)

3 Pape & Thoresen(87), pp 309-310

4 See for instance Lie & Sjøberg(84) on attitudes towards natural science among pupils in Norway.

personal feelings, people and nature are associated with femininity[1]. Even if we know this is ridiculous and even if we know both sets are important in system development, we are influenced by the associations.

About 20 % of the system developers at the computer department are women. Two of them are working with the new budgeting system. They think there are some differences between male and female system developers: Men want to advertise themselves and make fancy systems, while women are better in listening to the future users and in asking when they do not understand. Furthermore they think women are more concerned about teaching and therefore in general better at it.

Both male and female system developers are trained in the rational and objective tradition of computer science. Most women however, possess more social training in "heart work"[2], that is talking about own feelings, listening to other people's personal problems, interpreting unspoken signals and concerning about personal relations. These are important properties in system development.

There also seem to be differences in women's and men's way of cooperation. Female leaders seem to work in a more collective way than male ones. They are used to wriggle through difficult situations. The cooperation and discussion style is more friendly[3]. On the other hand, too much friendliness and no guts for criticizing and for confrontations can be a bit annoying[4].

These points should show the need for female system developers. However, female system developers are no guarantee for influence from women users. Increased influence from women requires consciousness, willingness and fight to strengthen the women's position in working life and the trade unions, willingness to appreciate the "feminine" values and properties and to change the ways of working in system development.

Maybe women workers can obtain influence in system development through "pure" women project groups[5]. Many women will feel confident in a women's group where they will be more outspoken and less reserved. As a research project it could be interesting, but in the real life it is not very realistic. On the other hand, the idea of looking at women as weak creatures who need to be hidden away from the ugly men is not appealing. The sexual undertones and flirting between men and women working together can be annoying and disturbing, especially when you are one single woman in a group of men, but it may also be inspiring and fun.

In my opinion there is need for both sets of values and for both sexes in system development[6]. In fact I don't belive that women have all the "good" and men have all the "bad" values and work styles. As I hopefully have argued, I also think there is need for some of the "hard" or "masculine" values in system development.

---

1 The examples are stolen from Greenbaum(87). Hacker(87) also empasize these myth, pp 183-185.

2 Greenbaum(87)

3 Novak(86) claims that women have the necessary patience and trust and therefore have good contact with end users.

4 Bjerknes(87) suggests confrontations to promote understanding and communication between different groups of people in system development. She argues that most of the known techniques in system development may be used to promote confrontations if the process is planned with confrontation in mind

5 Microsyster(86) show some experiencies in system development for feminist groups in England

6 The title of Greenbaum(87), ``The Head and the Heart" emhasize the need for both sets of values, for both ``head work" and ``heart work" in system development.

8. FINAL REMARKS

As a feminist I will emphasize the position of women in the trade union movement. Computer shop stewards must have proper training and education, they must not work alone, but they should have a group of available fellow members around. More work should be done to pick female user representatives to the project groups and there should be educated more female technologists.

As a computer scientist I will emphasize the fact that computer systems are based on a harmonic model of parts of reality. It is important to be aware of the real conflicts to see what parts of the reality that is modeled away[1]. Males and females are representatives of different kinds of experiences, and it is of importance to have both these realities represented in the project group. Both sets of values and work styles are important in system development and to computer scientists and professional system developers.

Management can use decentralization to weaken local trade unions as they did at FAN. But formal influence is not equal to real influence. Beside, women have a weak position in the trade union movement. To obtain influence over system development and thereby over their daily work conditions the employees must use both channels of influence: user **and** union participation. Typical ways of working in system development lead to both cooperation **and** conflict. To combine the strategies has become even more important since new tools in system development make it more difficult to rely on formal checkpoints and negotiations. Informal ways of organizing system development allow several ways of working and thus give women the possibility of working the way they are socially trained to.

The discussion in the working group showed that the condition both for women and for trade unions differ a lot from country to country[2]. I would like to give my acknowledgements to all the participants, and especially to the referee Titia Bos and the chair person Judith Gregory.

Despite of the national differences, I have become stronger in my belief that decentralized system development and informal work organization in system development may be of advantage to women, both as users and system developers.

REFERENCES

Bjerknes, Gro(1987): "Kommunikasjon: konfrontasjon eller manipulasjon." (Communication: confrontation or manipulation), pp 97-108 in Sydpol Working Group 1: **Arbejdsformer sat på dagsordenen.** (Work styles on the agenda)

Bjerknes, Gro & Tone Bratteteig(1987): "Florence in Wonderland. System development with Nurses." pp 279-295 in Bjerknes et al (eds.): **Computers and Democracy** Gower Publishing Company, Vermont

Bjerknes, Gro & Tone Bratteteig(1988): "Computers --- Utensils or Epaulets? The Application Perspective Revisited.", Paper presented at **Symposium Angepaßte Technologie,** January 18-22, Technische Universität Wien

1 Bjerknes & Bratteteig(88) point out how the Application Perspective among other things can be used to select aspects and constraints in the area of the system development project and thereby to avoid a lot of conflicts, p 10

2 Some of the differencies are shown in O'Brien &Ramioul(88)

Bostrom, Robert P. & J. Stephen Heinen(1977): "MIS Problems and Failures: A Socio-Technical Perspective. Part II: The application of Socio-Technical Theory", pp 11-28 in **MIS Quarterly, december 1977**

Carlzon, Jan(1985): **Riv pyramidene!** (Tear down the pyramids!,), Gyldendal, Oslo

Ciborra, Claudio (1981): "Information Systems and Transaction Architechture", pp. 305-324 in **Journal of Policy Analysis and Information Systems,** Vol. 5, No. 4,

Colbjørnsen, Tom, Arvid Fennefoss & Gudmund Hernes(1985): **Så samles vi på valen.... ?** (Then comrades, come really.... ? (these are the first words in the refrain of the workers' song "The International", and not a word by word translation)), FAFO, Oslo

Dooreward, Hans, Wil Martens, Harrie Regtering & Bernard Riesewijk(1987): "Dance Macabre: The Fortunes of **Integration and Segmentation", pp 231-244 in Bjerknes et al (eds.):** Computers and Democracy., Gower Publishing Company, Vermont

Ehn, Pelle & Morten Kyng(1987): "The Collective Resource Approach to Systems design", pp 17-57 in Bjerknes et al (eds.): **Computers and Democracy.**, Gower Publishing Company, Vermont

Ericsson, Kjersti(1987): **Søstre, kamerater!** (Sisters, comrades!), Forlaget Oktober, Oslo

Floyd, Christiane(1987): "Outline of a Paradigm Change in Software Engineering", pp 191-210 in Bjerknes et al (eds.): **Computers and Democracy.**, Gower Publishing Company, Vermont

Greenbaum, Joan(1987): **The Head and the Heart. Using gender analysis to study the social construction of computer systems**, Aarhus University, Computer Science Department, Aarhus

Gustavsen, Bjørn(undated): "The Demand for Flexibility and Evolving Patterns of Enterprise Organisation.", Draft for an article in **International Labour Review**

Hacker, Sally(1987): "Feminist Perspectives on Computer Based Systems and democracy at the Workplace", pp 177-190 in Bjerknes et al (eds.): **Computers and Democracy.**, Gower Publishing Company, Vermon

Jensen, Siri(1986): "Hun har bare sine lenker å miste" (She has only her chains to loose), pp 27-39 in Ericsson (ed.): **Kvinnekamp. Vi eier morgendagen.** (Women's struggle. We own the future), Forlaget Oktober, Oslo

Lie, Svein & Svein Sjøberg(1984): **"Myke" jenter i "harde" fag** ("Soft" girls in "hard" trades), Universitetsforlaget, Oslo

Microsyster(1986): "Computerizing women's groups: Microsyster's experiences from April 1984 to March 1986", Presented at The IFIP Conference **Women, Work and Computerization,** August 27-29, Dublin

Mumford, Enid(1987): "Sociotechnical Systems Design: Evolving Theory and Practice", pp 59-76 in Bjerknes et al (eds.): **Computers and Democracy.**, Gower Publishing Company, Vermont

Novak, Jasminka(1986): "Women, Work and Computerization --- the Yugoslav Way", Presented at The IFIP Conference **Women, Work and Computerization,** August 27-29, Dublin

Nygaard, Kristen and Olav Terje Bergo(1974): **Planlegging, styring og databehandling.** ("Planning, management and com ization"), Tiden Norsk Forlag, Oslo

Oslo Handel og Kontor(1987): **Klubbstrategi ved oppsplitting av bedrifter, en håndbok for klubbtillitsvalgte.** (Union club strategy on splitting of enterprices, a handbook for shop stewards, written by a working group in Oslo office- and trade-worker's federation)

Palmer, Marion(1986): "Vi tøffe damer i kvinnfolkjobber" (We tough ladies in women's jobs), pp 69-79 in Ericsson (ed.): **Kvinnekamp. Vi eier morgendagen.** (Women's struggle. We own the future), Forlaget Oktober, Oslo

Pape, Tom Chr. & Kari Thoresen(1987): "Development of Common Systems by prototyping" pp 297-311 in Bjerknes et al (eds.): **Computers and Democracy.**, Gower Publishing Company, Vermont

Röstlinger, Annie & Jan Selldén(1983): "How to Perform an Empirical Study on a Change Analysis Method", pp 281-305 in: **Report of the 6th IRIS 1983.**

Schneider, L. & C. Ciborra(1983): "Technology Bargaining in Norway.", pp 243-250 in Briefs et al (eds.): **Systems design For, With and By the Users.**, North Holland Publishing Company, Amsterdam

Øgrim, Leikny & Erik Norderhaug(1986): **Desentralisert systemutvikling og fagbevegelsen** (Decentralized system development and the trade union movement), Master Thesis, Department of Informatics, University of Oslo

Øgrim, Leikny(1987): "Decentralized System Development and Local Trade Unions.",in Järvinen (ed): **Report of the 10th IRIS 1987.**

BIOGRAPHICAL NOTE

I work as a research assistant in the System Development Group at the Department of Informatics, University of Oslo, Norway. I teach in "computerization and society", in addition to be working on my Ph.D. on conflicts and cooperation in system development. My main interests have been related to the influence of workers and their trade unions on system development and the relation between power structures, organisation development and system development.

Address: P.O. Box 1080 Blindern, N-0316 Oslo 3, Norway. Tel: + 472 45 43 31. Net Address: LEIKNY @ IFI.UIO.NO.

Women, Work and Computerization: Forming New Alliances
K. Tijdens, M. Jennings, I. Wagner, and M. Weggelaar (Editors)
Elsevier Science Publishers B.V. (North-Holland)

GIVEN THE LIMITATIONS OF TECHNOLOGY AGREEMENTS FOR WOMEN WORKERS: HOW CAN WOMEN BETTER DEFEND THEIR INTERESTS IN THE FACE OF TECHNOLOGICAL CHANGE

Anne P. O'Brien and Monique Ramioul,
University of Leuven, Belgium

The impact of technology agreements is limited for female workers, because they are peripheral groups within the workplace and because trade unions do not take into account the sexual division of labour. A trade union policy that can defend women's interests must be worked out at centralized level. Secondly, female workers should seek structural improvement of their position in trade unions. Third, women can increase their influence by creating their own channels of information exchange.

1. THE IMPACT OF NEW TECHNOLOGY AGREEMENTS ON WOMEN WORKERS

1.1 Introduction

The focus of this paper concerns technology agreements and their implications for women workers, primarily the defence of the interests of women workers in the face of technological change. The concern is therefore with the collective bargaining process and the resulting technology agreements rather than with the effects of new technology on women's work. Much work has already been carried out in the latter area with forecasts of the 'factorisation' of service sector employment, large scale redundancies, dequalification, deskilling ... If these forecasts are true, how then, can women workers best defend their interests in the face of such change?

This paper is presented in two parts: the first part is based on research[1] carried out 1986-1987 in France, Federal Republic of Germany and the United Kingdom. The methodology incorporated meetings with a network of national experts drawn from research institutes, universities, trade unions, management organizations and women's

1 The research was carried out in 1986-1987 at the Sociological Research Institute, Catholic University at Leuven, Belgium on behalf of DGV, Commission of the European Communities.

organizations. In addition, two case studies of best practice for women workers were carried out in Belgium.[1]
The first part of this paper covers the main conclusions of the research using national examples as appropriate. The second part of the paper covers strategies for the future so that women workers can better defend their interests in the face of technological change.

1.2 New Technology Agreements

One of the earliest conclusions of our research was that the drawing up of new technology agreements is not normal practice in any of the countries studied. The introduction of technology is being dealt with under many different headings such as: Rationalization Protection Agreements (FRG) Health and Safety Agreements (U.K.) Job Classification Agreements (France) ... What appeared to hold true for each of the countries studied is that trade unions face major problems in trying to cope with the introduction of new technology mainly in adapting existing bargaining methods to suit technology. The information they receive from management is often in adequate, or indeed they do not have the technical expertise to act on such information and thus simply 'react' to management rather than propose alternatives.
As TURU[2] points out "influence requires expertise" which the trade unions have not yet been able to acquire. The single underlying casual factor seems to be the weakened position of the trade union movement in Europe in general, this has been attributed to the world economic recession. The result being that, when faced with new technology, trade unions act in a responsive (to management) rather than innovative fashion, in a defensive manner attempting to protect their workers against quantifiable negative effects.

It is not surprising that resulting technology agreements tend to concentrate on quantitative aspects such as hours of work with visual display units (VDU's), number of 'voluntary' redundancies ... and not with such areas as job design. Few trade unions have a tradition of bargaining in the latter area, it is more often viewed as 'management's area'.

In addition trade unions do not always present a 'united front' in the face of technological change. In France the trade union movement is divided on ideological grounds, in the U.K. because technology cuts across traditional demarcations, white/blue collar trade unions are thrown into a situation of potential conflict.

Contrary to expectations, legislation to increase information and consultation rights does not always improve the bargaining position of the trade unions. Clauses such as "appropriate time" "(FRG 1972 Works Constitution Act) or "important new technology" (France - Auroux Laws 1982) can be used by employers as escape clauses so that trade union representatives do not have adequate time and/or information with which to act. Trade Unions are often forced to use whatever avenues are open to them to deal with the introduction of new technology. Mercier[3] reports on a case study of the introduction of

1 O'Brien, A.P.: The Repercussions for Women Workers of Collective Agreements Between the Social Partners Concerning New Technology, Phase 1, Part 1 - The feasibility Studies, 1987.
O'Brien A.P. and Van Hootgem, G.: The The Repercussions for Women Workers of Collective Agreements Between the Social Partners Concerning New Technology, Phase 1, Part 2 - The Analysis of the Case Studies and Methodology, 1987.

2 Trade Union Research Unit (TURU): New Technology Agreements: Five Years On. TURU Occasional Paper No. 95, Oxford, Trade Union Research Unit, 1985.

3 Mercier, N.: 'Second Etape de L'informatisation: "Une Banque" (Second Stage of Informatisation: "A Bank"), in Reynaud, J.D. (ed.): Dispositions et Problèmes Concernant la Diffusion et L'Utilisation de L'information des Entreprises: Le cas de la France. Rapport pour la Commission des Communautés Européennes, Paris, (Rules and Problems concerning the Disclosure

technology in a bank in which the trade unions attacked the issue solely on the basis of the training of staff.

1.3 The impact of the new technology agreements on women workers

Women workers, like their male counterparts suffer from the inadequacies of resulting technology agreements. However, women often find themselves in an even more vulnerable position. Williams[1] points out that managements need for worker consent varies with a major factor being centrality of the group to the production process. By and large agreements afford greater protection to 'core' workers. Unfortunately women workers are more predominant in peripheral groups. For example, in the U.K. women constitute 89,6% of all part-time workers.[2]
Williams[3] reports on a case study of the introduction of new technology in a cigarette making firm ..." managements design of jobs with the new equipment was shaped to ensure that there was an occupational niche for each of (what management felt were) the 'key' groups amongst the existing workforce". Thereby the "low skilled female workers received far less attention".

However, often it is the indirect, secondary, spin off effects of the agreements that are most discriminatory, for example:

* Training Provisions Training is usually limited to the skills necessary to operate the new technology and to the users of that technology. If women are not part of this group their exclusion from potential new jobs, which are usually technical, can be reinforced.
* Redundancy Clauses Agreements which include redundancy stipulating, 'last in first out' can mitigate against women workers who have interrupted their career for childrearing purposes. Agreements stipulating 'natural wastage' or even 'voluntary redundancies' may, on the surface, appear to be strictly egalitarian but as Williams points out 'natural wastage' rates are significantly higher for (younger) women workers".[4]

With regard to 'voluntary redundancies'. Maruani and Nicole[5] describe a case in point where management were looking for 'voluntary redundancies' from the workforce. They introduced shiftwork in the factory (the introduction of new technology often facilitates the move to shiftwork so that maximum use can be made of the equipment). They offered the unions a reduction in daily working from 8 to 7 hours. 110 workers resigned, 80 % were women workers. Compared to the decision of male workers - based on their work, the decision of the women workers was for 'family reasons'. Under shift work conditions it is impossible to combine work and family life - the 'dual career' of many women workers.

---

and Use of Company Information: the Case of France. Report for the Commission of European Communities), 1985.

1 Williams, R.: "The British Case" in Berta, G.: Industrial Relations in Information Society: A European Survey. Rome, Adriano Ollivetti Foundation, 1986a, p. 474.

2 TURU, Women and Trade Unions: Trade Unions and Women. TURU Technical Note No. 100, Oxford, Trade Union Research Unit, (referring to the OECD Employment Outlook), 1986.

3 Williams, R., ibidem.

4 Williams, R.: Analysis of the Content of New Technology Agreements and their Provisions Relating to Women Workers. Paper presented at the one day meeting on New Technology Agreements and their Implications for Women Workers at TURU, Oxford, 1st December, 1986b.

5 Maruani M. and Nicole, C.: "Quelques réserves sur l'armée de réserve": Droit à l'emploi et politiques de gestion de la main d'oevre féminine", (Some remarks on the 'Reserve Army'. The right to Work and Women Labour Market Policy), in Revue Française des Affaires Sociales, Avril-Juin, 1985, p. 30.

The experts consulted as part of this research seemed to agree that the agreements concerning new technology reflected the deep seated sexual division of labour in the marketplace. Although the situation in each country differs from an industrial relations or collective bargaining point of view, the situation of women workers is similar: at the lower levels of the qualification structure in the enterprise, in peripheral jobs, in jobs in the frontline of technological advance ... et they, who above all need to be able to defend their interests in face of the technological change, are under-represented in trade unions and at the bargaining table.

1.4 Women workers and trade unions

Focusing on new technology agreements did help expose the inadequacy of the level of representation of women workers in trade unions and at the bargaining table. In the FRG, female membership of trade unions was 22,5% in 1982[1] but their participation at decision making levels of the DGB (the central trade union congress) was less than 10%. Similarly for the U.K., where women form 34,2% of total trade union membership[2] but are not comparably represented at decision making levels. Where women do participate it is often in traditional 'women's areas' such as welfare.

Women workers could be described as 'invisible social actors', they are not recognized partners at the bargaining table. This has been partly attributed to the fact that trade unions, according to Rogerat[3] not take the sexual division of labour into account and thereby do not recognize that the social effects of the introduction of new technologies are different according to whether they are applied to a male or female working population. Trade unions (notable the CGT in a policy statement quoted in Rogerat[4] see the concerns of male and female workers as 'not differing fundamentally'.

A TUC report on Equality for Women within Trade Unions[5] reported that the main barriers to the active participation by women members are: dual commitments, lack of childcare facilities, the timing of meetings, a lack of confidence and "the sexist attitudes of male colleagues". Even more fundamentally, Cockburn[6] points out that in the U.K. trade unions are organized on an occupational basis reflecting strong craft traditions "these were men's unions framed for men's purposes and they continue to reflect this origin both in their agendas and in their procedures". Similarly, in the FRG, the Steel, Metal and Printing Unions have formed the 'strong arm' of the industrial relations system with membership traditionally drawn from male skilled labour. Perhaps the trade union movement is not the most appropriate vehicle for women workers to defend their interests.
So, the question remains - how can women workers defend their interests in the face of technological change? How can they become 'visible' social actors? What are the strategies for the future?

---

1 Eurostat, Trade Union Membership. Luxembourg, Commission of European Communities, 1985, p. 43.

2 TURU, Women and Trade Unions: Trade Unions and Women. TURU Technical Note No. 100, Oxford, Trade Union Research Unit, (referring to the OECD Employment Outlook), 1986, p. 13.

3 Rogerat, C.: "Nouvelles des Technologies, Syndicats et Division Sexuelle du Travail", (New Technologies, Trade Unions and the Sexual Division of Labour), Les Temps Modernes, Mars, 1986, p. 103.

4 Ibidem, p. 107.

5 TUC, Women and New Technology. London, Trade Union Congress, 1984b.

6 Cockburn, C.: Women and Technology, Workers Educational Association Studies for Trade Unionists, 1984b, Vol. 11 No. 41, p. 17.

2. STRATEGIES FOR THE DEFENCE OF WOMEN WORKERS' INTERESTS

2.1 Introduction

Strategies to make the effective influence of women possible must, of course, be directed to all the critical points cited above. The highly threatened position of women on the job with the present technological rationalization no longer permits the development of any single strategy.
A structural improvement in the weak position of women in the labour market is the primary objective, certainly in the context of technological innovation. Positive actions must focus explicitly on the job situation of women workers and must be more than the demand of a quantitative presence of women. This problem is dealt with in more detail elsewhere in this conference.
A greater impact and a long-term policy of the trade union with respect to technological innovation is another essential measure. In sum: the specific situation of women must be taken into account in any collective bargaining and not simply in technology agreements. Specific clauses concerning job security on the introduction of new technology can also for example be included in the regularly recurring wage agreements.[1] We come up to this point further on.
Firstly we need to focus on strategies that can present women workers with opportunities to defend their interests. A better defence of their job situation must be accomplished at the level of the work place as well as beyond it. Only a mobilization of the women involved, in or out of the context of a trade union policy that is explicitly oriented to women and that takes account of the barriers noted above, offers a chance of improving the position of women. The trade union movement, by the mobilization of women employees, can rest on a broader base in terms of members and experience which will help them to achieve a real technology policy.

2.2 Mobilizing trade unions for women's concerns?

The explicit orientation of the trade unions towards women workers is important for the position of women workers as well as for the empowering of the trade union movement itself. However, the mobilization of women to participate as active partners in trade union activities is only possible when trade unions themselves can be mobilized for specific women's concerns. They must focus on all existing problems of women workers before they can expect an increase in the participation of women workers: for example, in the recurrent negotiations on conditions of work the specific defence of women in flexible working times is necessary; the demand for childcare facilities is essential for a real participation of women on the labour market... These are only few but important examples whereby the trade unions can better take into account women's concerns in their policy. However, the move towards a change in orientation of trade unions cannot be brought about on a decentralized basis. History has taught us that it is not successful or efficient to leave such important concerns to shop floor trade unionism.

Indeed, it would be naive to place all one's hopes in the readiness of male union members to stand up for (or with) women. Rather, we believe that competitive relations between men and women on the labour market will prevail over any feelings of solidarity, especially at a company level. In the situation of large scale unemployment there is a labour market surplus, which leads to tense relations between the various

---

1 O'Brien A.P. and Van Hootgem, G., op. cit., 1987.

groups on that market. In addition, the consequences of technological innovation are different for men and women. Thus, men sometimes see a clear improvement of their job situation, while women see their jobs becoming even more precarious by the introduction of flexible labour. In this sharpening of opposition and increasing segmentation, women may not allow the defence of their interests to depend only on readiness and good will, especially when a condition for support from male union colleagues is the discernment of common demands.[1] Only problems that are felt as problems by both male and female employees seem to generate solidarity from the men at company level. So it is clear that it is necessary to work out general strategies and policies about the working situation of women at a central and sectoral level. Only a explicit policy towards the improvement of the position of women workers, worked out at central level can generate a better defence of interests of women workers at shop floor level.

A structural improvement in the participation of women members in the trade union movement is a second important way of enhancing the defence of women workers interests. Therefore it must also be an on-going objective of women employees themselves to search for support for their demands from their male colleagues as well as to acquire an equivalent place in all the branches and at all the levels of the trade union organizations. One strategy could be the implementation of so-called networks of women workers at all levels of the trade unions. The organization of specific channels for women within trade unions in order to generate direct communication and permanent contacts between female trade unionists is an essential factor for co-ordination of demands and strategies. Or indeed, a trade union specifically for women, as is the case in Denmark (a union of unskilled women workers) and in the Netherlands (the women's union FNV), women's congresses, and women's demand packages becoming part of national policy. Perhaps there are some lessons to be learnt from strongly organized women's unions and the basis on which their strength lies.
However, to be effective it is important that such channels, or 'women affairs' or 'equal opportunity offices' within the trade union movement do not remain isolated from the regular trade union activities and policies.
The development of these strategies is a long term process. On the short term, the bargaining about new technologies can be a 'back door' to bring in again the problems of working conditions of women workers: where former equal opportunity plans failed, the negotiations about changes in the working place that occur with the introduction of new technologies can be an opportunity to pick up again specific women's concerns.
In short, positive action in trade unions and mobilization for specific women's concerns is necessary but not sufficient. Women workers have also to look for their own means of exerting pressure, their own platform for their demands out of the context of trade unions.

2.3 Mobilization at a decentralized level and the conditions for an effective defence of interests on the job

Women must first be organized collectively in order to search for the ways to defend their interests, irrespective of whether the fight is conducted in or out of the union. One could seek specific organizational forms that are explicitly oriented to women who are facing the consequences of new technology. Organizing women beyond the company level has been shown to be important in the attempt to share information and experience about the consequences of new technologies.

---

[1] Pisa, W. and Van der Avoird, M.: "Opkomen voor de schemerzone in ons arbeidsbestel, De Vrouwenbond FNV" (Fighting for the Twilight Zone in the Labour Market, The Womens Trade Union), in Personeelsbeleid 21, 1985, nr. 11, p. 375.

One example is the French "Reseau Femmes-Informatisation-Bureaux". This is a network of women who are confronted with office automation. It's objective is to democratize and transmit information that can be used as power by women who work in offices. It is based on women's natural affinity for democratic relations. The aim is to encourage women to play an active role in their own destiny, which is being increasingly shaped by new office technology. The network permits their own information to circulate in order to empower their position in the enterprise. This is possible by establishing links between women whose work is affected by office automation systems.[1]
However, it seems clear that the motivation of women workers to defend their jobs in an active way is most likely at a decentralized, shop floor level. Evidence for this is found in a very interesting study about the action potential of women workers. This study, carried out by the German Institut zur Erforschung Sozialer Chancen of Cologne, came to some very interesting conclusions regarding the mobilization of women workers for the defence of their interests.[2]

The researchers concluded that the condition for mobilization is the devoting of sufficient attention to the specific situation of women workers and to their uniqueness as women. According to the researchers, the dual role of women in the production and reproduction process, as employee and as housewife, confronts them with conflicting demands, subjects them to two diametrically opposed behaviour logics. These opposing qualification requirements and expectations, however, can be complementary for the woman's personality: a "transfer" of experiences can occur between the one sphere and the other.[3]
In the social and emotional qualities that women develop as wife and mother, therefore, are opportunities to realize a real political commitment in the job situation. These qualities are:

* the strong personal involvement of women;
* the great capacity of women for conflict resolution and for accepting responsibility for the well-being of people near them, which permits real commitment;
* the priority women give to mutual understanding instead of to individual striving, which displays a stronger aptitude for solidarity.

Also, women workers want to be recognized in their totality: the problems of the working environment and those of the home belong to the same life world. The "frontier transcending" experiences and the cited qualities contain relevant political potentials that require specific forms of application to be realized.[4] It is difficult to convert these potentialities into action in the professionalized, formal union organs, which are far removed from the life world of women.

Women do feel attracted to participating in unions or other institutions under the following conditions:

* when meetings are held at times that are realistic for women: during the working day or when childcare can be provided;

---

1 Da Costa I. and Glogau, M.: "Women and Computerized Office Information Technology", paper presented at the conference Women challenge Technology, 15-19 November 1986, Elsinare, Denmark.

2 Frerichs, P., M. Morschhäuser and M. Steinruecke: "Interessenvertretung von Frauen und Neue Technologien - Ein Projektbericht" (Defence of Interests of Women and New technologies - Overview of the Research Project), IFG Frauenforschung, 1986, No. 3, pp. 46-58.

3 Frerichs, P., M. Morschhäuser and M. Steinruecke: "Weibliches Arbeitsvermögen und Politikzugänge von Frauen. Am beispiel betrieblicher Interessenvertretung von Frauen" (Female Working Capacities and Avenues for Politics. An example of Defence of Interests at Shop Floor Level) in Fricke, W. (ed.): Jahrbuch Arbeit und Technik in Nordrhein-Westfalen 1986, Bonn, Verlag Neue Gesellschaft GmbH, 1986, p. 484.

4 Ibidem, p. 484.

* instead of meetings in the official, institutionalized union and consultation organs, the meetings must first be held on the decentralized, departmental level;
* there must be opportunities to express every-day problems at work instead of concentrating on abstract and long-term problems;
* face-to-face communication instead of representation; informal contacts instead of procedures, and the approach to problems in their totality instead of being purely work oriented.[1]

The existence of, for example, consultation groups, therefore, offers a greater chance for more direct involvement by women. In France, too, the experience of the so-called 'groupes d'expression' has shown that these succeed better in mobilizing women and that they are more appropriate channels for the expression of desires and dissatisfactions. This is because of their informal character.[2]

2.4 Conclusion

Realistically, the opening of the unions to the equal participation of women will be a long term process. Women cannot rely on it exclusively. Women remain still strongly dependent on the attitudes and behaviour of male trade unionists for the protection of their job situation in the face of technological change.
Therefore, attention must be given to fostering awareness among male trade unionists to the specific problems of women both on the job and in the unions as a whole by a means of centralized trade union policy towards women. Secondly, the search for their own channels for opposition must continue in order to increase the chance for the defence of women workers' interests and their position in the labour market.

ACKNOWLEDGEMENTS

We wish to thank all participants of the working group 'Strategies for system development - can trade unions influence the design process for the benefit of women' as well as Judith Gregory, who was the chair person, Titia Bos, who was the referee and all other participants for the discussion and useful remarks on our paper.

BIOGRAPHICAL NOTE

Anne O'Brien, M. Soc. Sc., is working as a consultant in the private sector on 'end-user computing'; the human and organizational issues involved in introducing new computer systems. Correspondence address: System Dynamics Limited, 72, Merrion Square, Dublin 2, Ireland.
Monique Ramioul, Lic. Soc. Sc., is an assistant at the Department of Sociology of the Katholieke Universiteit Leuven, Section Sociology of Work and Organization.
Correspondence address: K.U. Leuven, Sociologisch Onderzoeksinstituut, E. van Evenstraat 2c, 3000 Leuven, Belgium

---

1 Ibidem, p. 488-491.
2 O'Brien, A.P., op. cit., 1987, p. 46.

Women, Work and Computerization: Forming New Alliances
K. Tijdens, M. Jennings, I. Wagner, and M. Weggelaar (Editors)
Elsevier Science Publishers B.V. (North-Holland)

PART 3

TRAINING: NEW OPPORTUNITIES FOR WOMEN?

Margaret Weggelaar
University of Leiden, The netherlands

INTRODUCTION

In general it can be said that women's position in organizations and on the labour market is weak. Women very often find employment in low paid jobs and in gender based occupations.
Moreover, their position is threatened by developments accompanying technological innovations, such as the loss of jobs in the services and in the administrative sectors and the deskilling of jobs.
Training and education in our days become more and more important in order to keep one's job or to improve one's position. Training and education are especially important for women. Not only for women who are already employed, but also for women who want to re-enter the labour market.
The first category of women, the women who are employed, are confronted in many cases with changing job-requirements, or the threat of losing their jobs. In order to maintain their positions, they have to adjust to new requirements or to apply for other types of jobs. The second category of women, those who want to re-enter the labour market, face the fact that rapid changes have been taking place over the last decennium and many women find that their skills and training no longer fit the new requirements.
Women in general face a number of problems regarding education and training. Very often they are employed in positions, where admittance to training is restricted. Also, they are confronted with the myth 'women and technology don't go together', which in many cases implies that for scarce training possibilities in organizations and firms women are excluded. The only training they get is very often restricted to 'how to push the buttons'. They are not made familiar with the computerized systems.
For women re-entering the labour market specific needs have to be met, like child-care, part-time education and compensation for admittance fees.
In this section some of the problems and opportunities regarding training and education for women, especially in relation to technology, are explored.
Two examples of specific training programmes for women re-entering the labour market are offered. S. van Hoek in 'Strategies and Education' describes the case of Women's training schools. These schools in the Netherlands were established as a result of an initiative of women in the trade unions and are financed by the government.
J. Verhoef and H. Voorhaar present in 'Realizing Training Programmes for Women on informatics' the results of a vocational training for women, also in the Netherlands. In

both articles the importance of supplementary conditions for women with household and family obligations, in order to ensure success of participation, are stressed.
When we discuss the relation women and technology, we have to take into account that (for instance) we have to deal with an exclusion of women from the technical culture, and exclusion, which is reproduced in the working place ànd in the family. About this phenomenon G. Doniol-Shaw reports in 'Technical Culture, Training and Female Employment'. The myth that women are not interested in technology or in technical aspects of computing is attacked by B. Elkjaer in 'Myth and Reality about Women and Technology'.
Another interesting aspect of the relation women and training - in view of the success of training programmes - is the idea of gender based differences in learning. D. Brecher argues in 'Gender and Learning, do Women learn differently?' that curriculum and content of training programmes have to take into account that the gaining of in depth understanding of computers is essential to women before application programmes are used. This is part of the idea that women have a holistic learning style, whereas men learn more linearly.
On the changing skill requirements of clerical occupations in relation to technical developments and the implications for training-programmes we find some interesting remarks from the research of C. Pullman and S. Szymanski. They stress in 'The Impact of Office Technology on Clerical Worker Skills in the Banking, Insurance and Legal Industries in New York City: Implications for Training' that we do not only have to take into account the training of technical skills, but that the new developments in integrative computerized information systems also will bring the need for the recognition and training of abstract and interactive skills.
But we even may go further than stating the fact that it is of great importance to establish specific vocational training programmes and training schools for women. Regarding the reorganizations to come in the economics of the Western world, according to new technologies, new structures of work organizations will emerge. The changes to come again stress the need for recurrent education. S. Gensior, L. Hohenberger and H. Schicke in 'Women, recurrent Education and new Technologies' report on the situation in the Federal Republic of Germany.

Women, Work and Computerization: Forming New Alliances
K. Tijdens, M. Jennings, I. Wagner, and M. Weggelaar (Editors)
Elsevier Science Publishers B.V. (North-Holland)

STRATEGIES AND EDUCATION

Saskia van Hoek and Els Rienstra,
Women's training centres: Alida de Jongschool,
Utrecht, The Netherlands

1. GENERAL INFORMATION ON WOMEN, THE LABOUR MARKET AND INFORMATICS

As compared to other West European countries women's participation on the labour-market is still very low in The Netherlands. Reasons for this may be found in the Calvinistic background of the Dutch where a women's place is in the home, but also in the enormous shortage of child-care facilities in this country. Already in 1900 the percentage of female industrial workers was low compared to the surrounding countries. Especially the percentage of married women with a paid job was low. The late industrialization did not provide enough work so that there was already a very large male reserve-army.
After the second world war, the labour market position of women changed. The employment of women during the first and especially during the last world war decreased the prejudice against working women. Women in the trade unions played a prominent part in this.[1] But still there is a lot to change.
There is an actual need for 200,000 children to be looked after during working-hours: 100,000 of them are on waiting-lists for a place in a day nursery.[2] According to estimations 60,000 women don't work at all or less hours than they would like to because of this. During a recent national conference, The Netherlands was said to be "a Third World country in Europe" concerning childcare. Recently our Minister of Social Affairs and Employment (including women's emancipation) stated that "childcare" should not be a government issue. Therefore he did not intend to reserve any money for it. In spite of these bottle-necks, more and more Dutch women try and succeed in finding paid employment.

Table 1: Labour-market participation of women, The Netherlands compared to 8 other countries (1985)[3]

Sweden	79.7	Finland	74.9
United States	65.5	United Kingdom	61.0
Japan	57.2	France	56.4
Germany	50.4	The Netherlands	40.8
Italy	40.5		

As from 1971, women's labour market participation increased from 30 to 44 %. Yet there is a long way to go. The international women's movement has made it visible that the difference in power between men and women is unmistakably linked with the traditional division of labour: Women doing the unpaid work at home and men going out into the world to earn a living. Economic independence and an equal position at the

1 H. Stasse and others: Staat en seksenstrijd, (State and gender struggle), p. 57-58.
2 Ministry of Social Affairs and Employment, Labour Market Review, 1987
3 Ibidem,

labour-market - important women's demands - are conditions to alter this division. In The Netherlands, these conditions are still far from being met.
The participation of women is increasing but so is unemployment!! In 1987, the official unemployment of women is 17 percent. The figure concerning men is 12 percent. Moreover, if we take into consideration that there are about 280,000 women who want to re-enter the labour-market but who are not registered at the employment agencies, the figures are doubled. Women's employment is concentrated in low-paid jobs. Women are more than men limited to a small number of occupations. Women's jobs can mainly be found in the field of services e.g. sales, administration and general care. In The Netherlands the number of temporary and flexible labour contracts in 1987 was estimated to be 11 to 12 % of the total number of employees, more than 90% of them concerning women. More than half of the working women work part-time. The number of female part-time workers has increased the last ten years by about 60 percent and thus counts for the majority of the new women's jobs.

Table 2: Labour-market Information 1987[1]

* Participation of women (15-64 years)	ca. 44 %	
(in 1971	only 30 %)	
* Unemployment	men 12 %	
	women 17 %	(250,000 women)
* Hidden unemployment		
women who want to re-enter the labour-market	17 %	(280,000 women)
* Temporary and flexible labour-contracts	11 - 12 %	(500,000 people)
* Part-time jobs	men 8.9 %	
	women 54 %	
* Growth of female part-time workers over the last ten years	60 %	

Let us now consider the situation in the computer field. The general conclusion is, that women play a small role in this sector. The women's share even decreased, from 9.4 % in 1976 to 6.1 % in 1986[2], a decrease of no less then 35 % . 6.1% means about 4000 jobs for women. Except this quantitative under-representation, women also appear to be working in a limited part of the automation sector: they occupy predominantly the lower jobs. Segmentation is also at stake here. Managing, technical and designing jobs are occupied by men while the processing and administrative jobs are mainly occupied by women. Those jobs in particular are at risk in the years to come, because of automation. In automation too, we see a segmentation of jobs. Women end up in the so-called "dead end"-functions and men in what we call "growth" functions. When we consider the figures concerning higher qualified jobs, we see a deterioration in the last 4 years whereas the position of women is concerned, from 3.5 to 1.9 percent.

Table 3: Filling of automation jobs.[3]

	men	women
System analysts	94.8 %	5.2 %
Programmers	91.7 %	8.3 %
Data typists	5.7 %	94.3 %
Operators	85.7 %	14.1 %

Table 4: Percentage women in higher qualified automation jobs.[4]

	1982	1986
Head of programming) network controller) project adviser)	3.5 %	1.9 %

1 Ibidem.
2 Research automation market 1987 by Inter/View, in Computable 27-2-87.
3 National bureau of Statistics, The Hague, 1983 and 'Girls in the ISI-project', PCBB, August 1987, p. 11.
4 Berenschot 1986

Why do fewer women than men work in automation jobs? There is no objective reason why women can not do these jobs. It is a rather new profession: physical strength are not required and the conduct of work can look very ladylike. Yet there are some explanations. Automation already has a male image and an own culture and circuit with a complicated jargon that does not make it easy for women to enter. Socialization of women also makes their attitudes to computers different then those of boys and men, who play with the computer more often and have one at home at their own disposal more often. Most automation jobs are still fulfilled full-time and employers do have outspoken or hidden preference for male employees. It is hard to imagine that they might also have had a woman in mind if one looks at advertisements for automation jobs.

Let us now turn to training and education. In the Dutch discussion on how to solve unemployment, training and education are considered to be a very important instruments. Regular educational institutions are pressed to direct their courses more and more to the demands of the labour-market.
The increase in courses in informatics has been explosive. The slow reaction of regular educational institutes has caused an increase in the private educational institutions. 70 percent of trained personnel in automation never had their training in regular education but through in-company training, re-training projects or otherwise.[1] However, women's participation in these kinds of education is still very low.

Table 5: Women's participation in informatics education.[2]

University/Polytechnic	10 - 11 %
I.S.I. (a large scale school-leavers project)	ca. 30 %
Retraining projects by the employments agencies (programmers/system analysts)	10 - 35 %

In The Netherlands the percentage of women amongst the students attending information theory at a university or polytechnic level is about 10 - 11 %. In the lower level e.g. the I.S.I.-project (a large scale government supported school-leavers' project) girls do not fill the 30 % . And here too, the boys take the high road of programming while the girls take the low road of word-processing.
Projects are financed by the Ministry of Social Affairs and Employment which have as their most important goal to solve the bottle-necks in the labour-market and to make it possible for unemployed to (re)gain a place there. Again, within these projects the participation of women is low; it varies between 10 to 35 percent. Little is done to have women enroll in these projects.
Another large impediment for women are the entry requirements. To join many (re)training courses one has to be qualified at university or polytechnic level or to have mathematics as a examination subject in secondary school. One should not be older than 35. The employment bureaus tend to support the mechanism to jack up entry requirements, so that it becomes more difficult for categories of people to enter the labour market.

The largest part of training for automation jobs takes place within the companies themselves and in private institutions. The latter are usually too expensive for women to enroll. If women have a job they often work at companies which do not have. The average age of employees entering in-company training is between 20 and 30 years. Training opportunities by companies are bigger for young people than for older ones. Women are hardly stimulated to join automation courses. Part-time work is a definite disadvantage. Many managers still think that investment in training part-timers does not pay off. The best chances for in-company training are an age below 35, a stable labour contract, full-time employment and the proper education. Therefore women have virtually no possibilities.

---

1 Information in preparation for this speech, collected by P. Biemans, 1988

2 'Girls in the ISI-project', PCBB, August 1987, p. 9.

The conclusion we, women in the trade unions, reached at when faced with this situation was - that if women were to have a chance- they should have to do it themselves. Informatics provides new and sometimes challenging job opportunities and women should try and obtain a substantial part of this employment.

2. WOMEN'S TRAINING CENTRES

2.1 History

The Netherlands is quite unique in having a separate union of housewives. This Vrou–wenbond (Women's Union) is affiliated to the Dutch confederation of trade unions FNV. Comparatively speaking, there are a lot of housewives in The Netherlands. Directly after the second world war the then NVV confederation founded a union for housewives as a support organization. Obviously, women were needed to take care of many obligations of the NVV unions (like caring for the sick, old and disabled, organizing fund raising, moral support during strikes etc.). Later on the Vrouwenbond NVV became influenced by the women's movement and started more and more to do something for the women themselves: organizing support centres for women doing paid homework, getting involved in social security rights for women etc.

It was only in the eighties that this union finally, after a long struggle, obtained voting rights in the FNV confederation board. The other unions had to be convinced that non-employed women could also be union members and wanted and needed a say in union policy. Since then, the FNV unions have women's rights more integrated in their own words and actions. Still, there is a long way to go.

In 1982 the Vrouwenbond FVV decided that more professional facilities for women should be created, because in this country we do not have suitable facilities for women. For instance hardly any women are trained in the Technical Training Centres for Adults. These centres are now under the obligation to try and get 5 % women instead of the 3 % they reach now. Of course, we in The Netherlands have equal rights laws and action programmes (sometimes very elaborate works showing the multitude of things that needs to be done) but these are mostly words though and very little action or money. With the support of the trade unions we convinced the government to support financially the start of a women's training centre. We also got some financial support of the European Social Fund. In September 1984, we were able to open the doors of this first centre: the Alida de Jongschool in Utrecht, right in the heart of The Netherlands. Alida de Jong was an almost forgotten female trade union official, who made an enormous contribution to the social position of women.

2.2 Aim of the project

We as organizers came from this housewives union. That is why we had knowledge of the target group. Secondly, we had knowledge of the labour market: employment possibilities and quality of working life e.g.. In the setting up of the project no existing educational institutions were involved, so there were no institutional interests. The aim of the project was to train women who wanted to re-enter the labour market in jobs with good perspectives. The idea of creating training centres just for women was not based on the premises that all training for women should take place in separate institutions. Separate centres were needed to show what possibilities were available: to set an example to all regular institutions; to show that women are interested also in traditionally non-female jobs and in high technology jobs -if they can fix it on their own conditions, if they are supported and met on their own terms-; to show that selection before and during courses is not so natural or necessary; that many women are anxious to get jobs and can reach a high level in a relatively short span of time. In a sense we were tired of always defending women and criticizing policy. We wanted to focus on the positive aspects and to show what could be done.

2.3 Target group

It is estimated that 350,000 Dutch women want to re-enter the labour market. They have many reasons: financial necessity, economic independence, the wish to be useful to society, escape from isolation, personal development,or appreciation from society. In fact women face many problems: overall high unemployment, lack of child care facilities, prejudices to working married women ("shouldn't the family go first"), insecurity due to 'sitting' at home for some time, no recent work experiences, no up-to-date or relevant education or training and . part-time availability When setting up a trai ning project for women who want to re-enter the labour market, one has to consider all these factors.

3. THE CONCEPT OF THE WOMEN'S TRAINING CENTRES

The concept of the women's training centres consist of:

A. Material conditions.

These are very important, especially to enable women to participate in training. On our courses we have realized the following conditions: short termed courses, part-time courses, child care within the premises of the centres, no financial obstructions: cheap fee, allowances for traveling costs, exams and books, training hours to coincide with school hours, within reach by public transportation, low formal diploma demands; mathematics is not a prerequisite, no obligation to be a registered unemployed, no maximum age and no "school" atmosphere

B. Choice of professions.

The choice for which professions the women should be trained is determined by shortage on the labour market and number of vacancies. We carry out our own research on labour market developments. It is also determined by traditionally non-female jobs or jobs with career perspectives and our own ideas about interesting jobs for women

The school offers courses in the following areas:

* informatics (only 6 % of the jobs in this field is held by women), such as
    * computer use in administrative settings
    * adviser or counsellor in the use of software and hardware
    * programming and systems control
* entrepreneurship (only 16 % of the Dutch entrepreneurs is female)
* management (3 % women)
* commerce (20 % women)
* computer use in the printing industry (17 % women)
* computer numerical control (cnc) in the metal industry (1 % women)

C. Content of the courses.

Of course it is very important that women get professional knowledge. We offer courses on various levels, depending on previous education and experience. A broad training is important, not only preparing for today's jobs but also for tomorrows'. A lot of attention is paid to new technology, not only to make women know how to handle the computer (software as well as hardware like repairing small defaults etc.), but also to know that it can be used in many different ways and to stimulate a critical awareness of those uses. We work mostly with women teachers especially for the more traditional male subjects, because it helps women to get a positive identification.

D. Recruitment and intake of women students.

Traditional women's magazines and free newspapers wrote about the courses, because of our publicity campaign. We also informed the unions, women's organizations, social security offices, community centres etc. Right from the start hundreds of women applied. We organized information meetings and asked all participating women to return for an individual intake interview. Then, we talked to each women in order to know whether she was in our target group, motivated, orientated towards a paid job and suitable for the training. In effect, far too many women proved good candidates. Thus in the end we had to draw lots. Maybe because of our union background, we decided that of course there should be a positive action policy for those women with the least chances e.g. black women or women with lower education. 15% of the women at the centres belong to ethnic minority groups. We tried to show all those we could not place whatever other limited training possibilities were available.

E. Supervision and training.

Counseling and supervision of women are important parts of our courses. We have is both individual supervisions and group trainings. Women need time to put into words the processes of change that take place when they start again to live a life of their own. Important topics for discussion and training are:

* learning to learn, to study again
* to develop self-confidence and inspire women to aim higher
* to combine study and responsibilities at home
* to re-appreciate domestic work
* to prepare re-entering the labour market, for example practise job interviews and applications

Part of this training and supervision is done by our trainers and much is also done by colleague-students: telephoning each other; keeping up the good spirit; daring to aspire for better jobs; comforting each other when things are going wrong, so having their own networks during the training and there-after.

F. Ways to have 'our' women employed

Competition on the labour market is heavier, especially at the lower end of the informatics jobs where our students have to make their start. So obviously we need to have a very active policy on how to have women employed. What do we organize to realize this? First of all, our own officials have working on contacts with companies. Excursions to companies and institutions are organized. Throughout all courses practical work is used so that employers can get used to re-entering women and to the capacities of our students. This proves to be a way to re-adjust students to paid work and to re-estimate her own labour market value. For our teachers and counsellors this is a way to make contact with companies and learn more about vacancies and current job-demands. We organize separate meetings with personnel departments of companies and mail material on our courses regularly. We also actively try to get all kinds of government grants and subsidies from labour agencies. In all our contacts we stress that these women are not a pitiful group but have a lot to offer. Of course the training, counseling and network-activities that we mentioned are essential parts of our strategy to get women employed. When the course has ended, women return to the centres for updating their knowledge, for improving our courses, for their own network-activities.

4. OUR STUDENT-RESULTS UP TO NOW

Drop-out rates during courses are very low compared to other adult training activities. Indeed, 10 to 15 % of our students drop out. When they do, this is mostly because of: personal of family reasons (such as health or relation problems) and the wrong choice of

profession or training The average drop-out of similar mixed courses varies from 25 up to 40 %.
One effect of the training centres is that women learn to appreciate their own qualities and find out that they are better students then they ever thought about. A comparatively high percentage of women decides to go on studying after the course.
Of course, most important is the number of women actually finding a job. On average, more than 80 % of the students finishing our courses does so, or decides to attend further education.

Table 8: Results of the ALIDA de JONGschool

	'84/85	'85/86	'86/87
Women who found a job	84 %	75 %	88 %
Women who attend further education	6 %	15 %	7 %
Women seeking employment	10 %	10 %	5 %

As we said before, students do return to our centres after finalizing their course. They tell us how their first year of being employed cost a lot of sweat and tears, but they are never sorry. Most women are very proud to be financially independent and find their lives are more interesting now.
What we hoped but did not dare to expect, is that the majority of women has found a part-time job. Some started at 40 hours a week but they are now working less hours. Many of them go to in-company training courses as well and it seems that many of them still want to go a long way in this respect. Quite a number of women who re-entered as programmers are now employed as system analysts, information analysts, teachers or project-leaders.
Those who attended the lower administrative courses have much more difficulties to find steady and rewarding jobs. On behalf of these women we recently started our experiment with evening classes.

5. CONCLUSIONS ABOUT BROADER RESULTS AND EFFECTS

5.1 More centres started

In 1984 we started with one training centre and 90 students. Nowadays, centres are operating in 11 towns, with 800 students every year. So the growth rate is quite substantial. We found that the idea of women's training quickly gained popularity amongst politicians, local employment agencies, women's union- groups. Also for many women themselves the centres were eye-openers: it is still possible to catch a second chance, to start a new.

5.2 Eurotechnet

The women's training centres were acclaimed to be so renewing that they were submitted to a program by the E.C. called 'Eurotechnet': a European Technology Network of over 70 innovation projects on technology training and education. We have attended a few Eurotechnet Conferences already. In spite of a quite long and sometimes very effective EEC-policy in favour of women, we met very few women there. Only 5 out of 70 projects focus on women. Most other projects supposedly aim at people regardless of gender; in fact, most focus on men and many more than 5 are exclusively for men. Women and technology seem miles apart on the European level as well. Needless to say that the most prestigious of the Eurotechnet projects, where real innovation is supposed to take place (here we speak of competing with the US and Japan),are NOT THE WOMEN'S PROJECTS. Women tend to focus on how technology can be made more accessible to all, and this can be realized at comparatively cheap rates.

It is always very stimulating for us to participate at these conferences, because it tells us to aim higher, be more daring and demanding. What is being spent on women is really peanuts, in spite of what our government suggests. Our courses are in fact cheap, effective and successful. Nevertheless, we get the burden to prove ourselves over and over again.

5.3 Future of the schools

The Dutch government forced us to sign a declaration of good intentions and co-operation, because they wanted us to agree that in the long run separate training centres for women would not be necessary, that our centres have to help others and to show them how to work effectively for and with women. In return the government promised to do its best to create more training facilities for women in all training and educational centres. Thus, we are now searching for ways to transfer our knowledge to others, preferably by creating mutual projects. Furthermore, we try to push institutions and companies to: organize special recruitment of women, examine their selection and testing methods, improve their material conditions and accessibility, offer not only a technical course but to broaden the content of courses, organize special training and counseling facilities for women, recognize the importance of courses guided by women teachers

This proves to be a long way for existing institutions. Some are trying, but usually they only try one or two conditions from our concept. It is not enough to have an excellent course but forget the special recruitment. It is not enough to start a course with 50 % women if a majority cannot finish it or can find a good job . One wonders whether this will work.
Therefore we believe separate training courses and centres for women will be a necessity for a long time to come. They prove to be more renumerative, because of less drop-outs and higher placement rates. This implies lessons for government-funded training as well as for in-company training, especially if one considers the waste of human potential, time and money in companies due to the way their female employees are treated.

5.4 Final remarks

Of course, we realize that training alone is not enough. What is the use of training when employment is lacking? Or when employers are not willing to hire women, even if they have had the best possible training? Therefore, an active involvement with the quality of working life is needed as well: an employment policy and positive action programmes. We cannot leave this to individual employers, the government or even the trade unions: women themselves must get involved.
Sure, we shall meet set-backs at this road. But it does help to know that a lot of women all over the world -and some men too- are striving for the same goals, like that is shown at this conference.

BIOGRAPHICAL NOTE.

At the time of the conference Saskia van Hoek was director of the Alida de Jongschool. A month later, Els Rienstra succeeded her in this capacity.
Correspondence address: Women's training centres, c/o Alida de Jongschool, Mariaplaats 4a, 3511 LH Utrecht, The Netherlands.

Women, Work and Computerization: Forming New Alliances
K. Tijdens, M. Jennings, I. Wagner, and M. Weggelaar (Editors)
Elsevier Science Publishers B.V. (North-Holland)

MYTH AND REALITY ABOUT WOMEN AND TECHNOLOGY

Bente Elkjaer,
Institute of Informatics, Royal Danish School of Educational Studies,
Copenhagen, Denmark.

In this article I argue for the necessity of developing a general qualified education in the humanities and social sciences within the technical educations. I present some of my own research dealing with gender and computing in the 8th grade of compulsory schooling.
These results suggest, however, that the confidence we as feminist scholars exhibit to young girls and women in order to insist on the general aspect of computing rests upon of fragile and perhaps transitional foundation. I suggest that this faith in girls might be reserved for middle-class girls (and boys), elucidating the social distinctions in the labour market and in the decision-making processes. I argue, however, that these research results are overshadowed by the concrete male dominance of the classroom and the generalized rejection of the peculiarities of women's values. I offer an explanation of this devaluation of women and women's values. The final part of the article argues that it is necessary to keep insisting on integration of general and technical education. We must, in terms of making strategies, begin with be an assessment of the political and economical situation in general, and policy towards women in particular. Doing that we can take our point of departure in the fact that information technology is more women friendly than earlier forms of technology. However, we must be aware of the fact that if women are to attain more power this comprises a threat to both men and women.

1. MYTHS ABOUT WOMEN AND TECHNOLOGY

In general, one can speak of two opposing myths about women and technology existing side by side. On the other hand there is the myth that women and technology are incompatible and unnatural. On the other hand there is the myth that says that women have the potentials to liberate the development of technology. Women are by nature and culture (through socialization) solicitous towards people, society and nature and foster a technology with a human face.
The first myth is predominant in public opinion. The second myth thrives within the community of feminist scholars as a way of insisting upon those women's values that are not recognized, let alone rewarded. The concept of women's values is controversial. It seems to be randomly accentuated or denied according to whether it is expedient to use the term. I use the term women's values to connote the traditional (stereotyped) choices women make in working life where women's occupational choice seems to lie with the human and social element. I regard this as a positive trait in a society which denies humanity in order to celebrate technology. This might be changing because

gender roles are perhaps decomposing, and as I shall suggest later in this article women's values is that they have a very low priority in society in terms of power and prestige. This is because a lot of the necessary work women do in society, e.g. in the social- and health care sectors, is very low paid. In stating this, it is evident that the first myth of women and technology is the most powerful. This myth contains the notion of men as the norm against which women's qualifications are measured and found inadequate. It is women as lacking in qualifications that forms the predominant picture of women in society of a whole. Women are regarded as the subjects who must change. They are not regarded as subjects who might have something special to contribute. And women do not regard themselves as such persons.

The outline of these two myths about women and technology is my departure in my research on gender and computing in secondary schools. My aim is to draw a more balanced and true picture of women and technology and to contribute to an understanding of why these myths exist.

2. CHANGING WOMEN......

By now you might think that I have overlooked the many changes in women's position in society that has taken place within the last 20 years or more. Have not women's participation in the labour force, in politics and in the decision-making processes changed the picture of women and technology?
There has been a change in ideologies and policies directed towards gender, implying an ideal of equality between gender. Today there also exist an ideal of gender as more integrated instead of segregated. And some actual changes have taken place. Women and men meet each other in a lot more employment and social situations than they did before. Today's ideal is not the suppresses women and the oppressing man.
Yet I wonder whether these ideals have led to the abolishment of the power-relations between gender of if rather patriarchy becomes invisible and hidden instead of visible and open.[1]

3. ....CHANGING TECHNOLOGY.

Another changing framework that must be considered when discussing women and technology is that today's technology is information-technology. This technology is to some degree more fit for the traditional cultural role of women.[2] Women have traditionally chosen jobs - and in the family had tasks - that implied knowledge and skills in information, communication, mediation and provision of service.[3] More and more jobs demand these qualifications today. In the information-technological society, many jobs involve creating contacts and networks something demonstrated by the growth of the service-sector. At the same time there is a decline in traditional male dominated, heavy trades (agriculture and fishing, construction and industry).[4]

---

1 See also: Haavind, Hanne: Förändringar i förh llandet mellan kvinnor och män. (Changes in the relation between women and men). In: Kvinnovetenskaplig Tidskrift no. 3, 1985.

2 I argued theoretically for this in: Elkjaer, Bente: The New Technology is Women's Opportunity.In: Women Challenge Technology. European Conference on Women, Natural Sciences and Technology. Contributions, vol. 1, Elsinore, Denmark 1986.

3 Balbo, Laura: The servicing work of women and the capitalist state. In: Political Power and So cial Theory, vol. 3, 1982.

4 See also: Schurmann, Franz and Close, Sandy: Kvinder soger efter respekt i jobs uden for hjemmet. (Women are looking for respect in jobs outside the home). In: Information 27th of August 1987. And: Henningsen, Inge: Mor er Major. (Mother is major general). In: Naturkampen no. 44,1987.

When looked upon separately it appears on the one hand as if women will get their chance to really break-through into qualified and powerful positions in the labour market and that their historical functions in the family will be acknowledged as necessary and valuable prerequisites in the high-technology society. On the other hand, the opposite could also happen, i.e., that men will become more powerful or that no real shifting in power-relations between gender will take place. The creation of information-technology has its base in formal logic, e.g. in mathematics. It is not within areas where formal logic dominates in its purest form that we find most women. On the contrary, these areas are very male dominated.[1]

I argue that in the very development of technology we have a sort of potential gender-based struggle for defining the essential content of the development of technology. Is it technology's connection with information and communication that has the priority, or its connection with the formal logic? Is it the connection with the humanities and social sciences that becomes predominant, or with the natural sciences?

4. INTRODUCING RESEARCH ON GENDER AND COMPUTING.

The purpose of my research is to go beyond the myth about women and technology. In order to draw a more contemporary picture of the combination of women and information-technology, I have chosen to look at technology and the learning processes involved in an 8th grade computing class. The reason for choosing computing is that the content of this subject in principle consists of technical and general aspects. The subject involves both working on and with the computer and discussions of general aspects of applications and implications of information-technology for individuals and for society as a whole. One could say that the subject of computing serves as an example of whether it is the women's of men's traditional professional interests that have priority in the learning-processes.

From quantitative and gender-neutral research, we know that pupils in computing prefer the technical part of the subject.[2] I wanted to see whether this was true when preferences were examined gender-specifically. I also wanted to obtain a deeper understanding of why the pupils had their specific preferences in their learning-processes of computing. This led me to do qualitative research.

I have made classroom-observations in two schools, one in a working-class setting and another in a middle-class area. On the basis of these observations I made interviews with some of the pupils going into more depth about why they chose the subject, what they prefer and why.[3]

In order to measure the changing gender roles I began with two stereotyped theses:

1) Girls wish to place the technical part of computing in a human and social context in order to experience the subject as meaningful.
   - They have chosen computing because of its utility-value and they distinguish the technical part of the subject from its utility-value.
2) Boys are fascinated solely by the technical part of computing.
   - They want knowledge for fun.

"Technical" is to be understood as the technical part of computing, while the human and social context is the general part of the subject. My assumptions were that girls would

---

1 Foged, Brita et al: Kvinderne kan ikke lide lugten i bageriet. (Women do not like the smell in the bakery). In: Nordisk DATAnytt no. 1, 1986.

2 Dorf, Hans: Teknologiorientering og datalëre -burgernes vurderinger. (Technology-orientation and computing - "the users" evaluation). In: Pëdagogisk Orientering no. 1, 1986.

3 For a more detailed description of my research design, see: Elkjaer, Bente: Are the girls the problem? Gender and Computerscience in the Danish Secondary School-system. In: Contributions to the 4th GASAT Conference. Vol. I, University of Michigan, Ann Arbor, MI, USA, 1987.

not find computing meaningful (useful to learn) unless they also got the chance to talk about their feelings and worries about the technological development. Unless they got the chance to air their considerations for humans and society in the light of the technological development. They feel nurturant towards humanity and society and will show it in their behaviour in the subject and in their reasons for choosing it. They will participate in discussions of the applications and implications of technology more actively than boys and will place priority on this aspect over the technical part. On the other hand, the boys seem not to care about the general part of computing. They just want to play with the(ir) computers. In other words I claim that the interdisciplinary syntheses as laid down in the description of the subject of computing, cannot be established unless the preferences of the girls find room in class.

5. PRESENTATION OF RESULTS FROM RESEARCH ON GENDER AND COMPUTING.

Space does not permit me to go into depth about the research results, so I shall only mention some conclusions.
The before mentioned have been both exploded and confirmed in classroom-observations. Both sexes apparently prefer the technical part of computing, but there are some relations indicating quantitative (boys talk more than girls) and qualitative (boys set the content of the classroom-discussions) dominance within gender in school which make shared discussions of the general part of computing very difficult. The girls' attempt to contribute to discussion are killed by noise in the school in the working-class area. In the school in the middle-class area the means used are ridicule and invisibility of the girls' attempts. An equal and shared discussion cannot be achieved because of these relations of dominance among gender toward the conquest of the classroom. The boys set the terms in the classroom either by public obstruction or by creating sub-cultures for boys only in class. The measurement of the theses is therefore difficult to assess solely upon the class-room observations.

I have picked out 4 interview-cases for thorough preliminary analysis of the interviews. An academically and technically weak girl and strong boy from the school in the working-class area and the opposite from the school in the middle-class area.

The girls' interviews are stereotyped when questioned about why they chose computing. They chose it out of necessity. It will be useful for them to know computing because of the competition on the labour-market. They differ in the evaluation of the necessity of defining computing as more than a technical subject. They both define computing as a technical subject, but the girl from the school in the middle-class area thinks that the class discussions of applications and implications are necessary in order to understand the general part of technological development, even if it is boring. The girl from the school in the working-class area does not see this necessity. To her it is necessary to learn how to use the computer and not anything in general about technological development. They are both, however, aware of the dominance of gender. The girl from the school in the working-class area is not aware until I mention it to her and then she resigns towards it. The girl from the school in the middle-class area brings it up spontaneously several times. She blames that for her being not very active in class-discussions. And she is very aggressive towards the power-relations of gender:

> "I'm feeling too insecure (in class-discussions), I think. And I don't want to be that. (...) It's because the boys from our class laugh, when we say something wrong. (...) It's OK for the boys to say something wrong because the boys don't put each other down, they put the girls down."
> (girl from the school in the middle-class area).

I think that these differences among the two girls represent both the academically and technically strong and weak, province and capital and social differences. The academically and technically weak girl from the school in the working-class area thinks it is sufficient to learn the technical part of the subject, whereas the academically and technically strong girl from the school in the middle-class area also is aware of the general part of the subject. The girl from the school in the working-class area from the province knows that the struggle to get a job is difficult where she comes from, whereas the girl from the school in the middle-class area lives in a setting with sufficient educational- and vocational possibilities. The social differences also reflect that the girl from the school in the working-class area does not see herself as influential in decisions about technological development. The girl from the school in the middle-class area does. She wants to participate actively in democracy. Perhaps these social differences are also mirrored in the girls' view of the power-relations between gender as, respectively, resignation and aggressiveness?

The two boys' interview-cases are different. On the one hand the boy from the school in the working-class is stereotyped. He defines computing as a technical subject and has chosen the subject in order to learn how to do programming. The boy from the school in the middle-class area is difficult to categorize. On the one hand he places himself within the stereotype by saying that programming maybe is the most fun, but he is academically and technically weak even if he does not like to admit it. Perhaps this is why he also places priority on the general part of computing.

> "If we had more work on the computer, I don't think we would understand it properly or understand it as well." (boy from the school in the middle-class area).

Perhaps he has these (middle-class) priorities in computing because he lives in an area where people see themselves as being important in the democratic decision-making process. The boy from the school in the working-class area defines his chances on the labour-market on the basis of technical computer-knowledge, but this has to be more advanced than what he learns in school. This is one reason why he quits the subject.

6. CONCLUSIONS.

Neither the myth about women's inferiority nor that of their potentials with regard to technology were confirmed. The unilateral technical male fascination was not firmly confirmed either. The male dominance in the shared classroom was, however, confirmed. And it was suggested that this dominance in the class could be the main reason for the suppression of the girls' potentials. When questioned about reasons for choosing computing and especially about the definition of computing, my preliminary interview analysis suggests that we also have to consider the issue of class - perhaps not only in terms of the single pupil or person but also in terms of the general setting of the scene.

I shall return to possible explanations of why the myth about woman and technology exists at all. First I think that it is necessary to emphasize that the traditional middle-class women's interests seem to have very few chances of survival if we consider these girls to be an expression for the whole area of technical education. I do not think we can or should do that, but perhaps we can use the results to put our faith in the middle-class girls' active protests against pure technical education into perspective.

At the same time I also wish to emphasize that the display of women's potentials are masked by male dominance. Perhaps the priorities of the technical part of computing is also due to the possibilities for girls of creating rooms for girls only around the work on the computer. That independent computerwork in single-sex pairs within the classroom changes the relation of gender in favour of the girls' dominance. Girls are in this way

able to create fully accepted subcultures for girls only in class, as the boys do in classroom-discussions. The computerwork thus becomes a retreat from male dominance.
One could ask if there is any reason at all to give priority to the general part of computing when all pupils apparently more or less prefer the technical part of the subject? I think there is. First information-technological development demands other qualifications than the technical. The aim of information-technology is the creation of information and communication even if the development of the actual technology is based upon formal logic. We have today a technology that more than ever requires qualifications which can close the gap between technical and general qualifications. Second I do not think that we should underestimate both the concrete male dominance in class (and other places where technical education takes place) and the general male dominance in terms of the development of society and especially in strategies directed towards women in the labour-market. I am referring to the politically and economically strongly supported strategy of channeling women into male dominated (technical) areas. That of course has an impact on girls' (and boys') preferences in technical education. Third it is important to bear in mind the actual work with the technical part of computing. Especially the working-methods are here strongly stereotyped along the lines we know of when talking gender. They are confirmed in my research, e.g. that the girls cooperate and the boys do not. The boys are strongly competitive and feel very insecure with relation to any technical weakness in comparison with their work-mate when there happens to be differences. Finally I think that we cannot overestimate the suppression of women's values in society and in women's own interpretation of the appropriateness of adapting to male society and the predominant male values when she wants to take part in that. Especially in times of great attention towards the needs of the trades and industries that results in the strategy of channeling women into male dominated areas as I mentioned earlier. I do not claim that this is a very conscious and conspiratorial suppression. On the contrary I think that suppression of women's values is very unconscious for women - and men - and that it takes place on a deeper psychological level. But in order to unveil this we have to be aware of the gender struggle and we have to agree upon a technological development, education and creation of jobs that restore and preserve what we think and have to think in order to have hope that somewhere there is the potential to develop technology that is life-preserving instead of death-ridden.

One way to do this is to insist on a qualified content of general knowledge about humans, nature and society (psychology and social sciences) on all levels of technical education. Another and cooperative way is to develop new job profiles suiting women's traditional choices and new information technology. Potentials are not just inherent in biology or culture, especially not in a changing society. Not all women know of or acknowledge the existence of women's values. They tend to represent undervalued and rejected qualifications that are not immediately regarded as adequate in a high technological society - or in any society. Our tasks as feminists is to lay claim to that which valuable in women's values and to claim that women's values get room in schools, and in the content of technical educations in times where these are given high priorities.

7. WHAT CAN BE DONE?

I think that we need to re-assess the political situation in order to analyse the possibilities and limitations of feminist struggle. In that analysis we need to be constantly aware of suppressed women's values. The strategy must be to develop new educational- and job-profiles that are politically justifiable and are grounded in a knowledge of why the suppression of women's values is so massive. I shall try and give a concrete idea of what that means in terms of actions to be taken.

8. ATTEMPT TO ANALYSE THE POLITICAL SITUATION.....

10-15 years ago in the good old days there was a form of consensus in the new women's movement that the struggle for women's liberation also implied the abolition of capitalism. At that time we - women in the Women's Liberation Movement and students of feminist scholarship - were fully aware that neither capitalism nor male society.
Years passed and a good deal of women from Women's Lib found themselves in a situation where they had completed their higher education. Meanwhile women's liberation had become women's equality (with men).

Women's equality was now acceptable to most people (read: men) because of the aura of justice that radiated from the mere concept of equality. Some of the well-educated women of the Women's Lib even landed in jobs whose purpose was to carry out the official recognized and adopted Equal Rights policy for women.

During the same period of time the governments of the Western European countries have changed from the social-democratic to conservatives. The labour-parties are in opposition and in deep ideological and political crises in most of these countries. Even the socialistic world has introduced capitalistic (market-economy) oriented principles.
The political situation has on the one hand become more right-wing. On the other hand we have a more or less accepted gender-equality policy and we have feminists placed in the public sector executing the equal rights policy.

9. ....AND THE SUPPRESSION OF WOMEN....

I earlier suggested that the suppression of women in this context is expressed through the maintenance of the technical predominance in education. I also suggested that this suppression needed to be understood on a deeper psychological level because on the surface, women and men are appearing to become more equal, at least as an ideal.
Dorothy Dinnerstein[1] has something to contribute with in terms of understanding the suppression of women and women's values.
Dinnerstein suggests a connection between, on the one hand, the fact that it is through a mother - a women - all children experience early childhood and, on the other hand, the suppression of women. The mother-dominated early childhood leads to the unconscious feeling that women's will and power are omnipotent (compellent). The woman is the person who made the child feel impotent (powerless). That leads to an attempt to seek compensation later on in life for this impotency.

Humans do that through sexuality and creativity (enterprise). This is gender-specific as girls later on feel the sense of guilt towards the mother, the woman, for having deserted her love for the father, the man. The sense of guilt means that the girl later on will feel nurturant towards women and nature because women are perceived as nature because we experience the love of mother before we can separate ourselves from the outside world (nature).

Men do not have this sense of guilt as they keep mother (woman) as prime-targets for love. Neither does he feel any identification with her. He thus does not feel the necessity

---

1 Dinnerstein, Dorothy: The Mermaid and the Minotaur. Sexual arragements and human malaise. Harper and Row, N.Y. USA 1976. See also: Elkjaer, Bente: Patriarkatet som en flugt fra modermagte - en prësentation af Dorothy Dinnerstein. (Patriarchy as an escape from the power of mother - a presentation of Dorothy Dinnerstein). In: Patriarkatsteori - en antologi. The University of Roskilde, Denmark 1984. And: Elkjaer, Bente: Vi frygter alle kvinders vilje og magt. - et bud pÜ en anden mÜde at tënke teknopatriarkat. (We all fear women's will and power. - A suggestion for another way of thinking about technipatriarchate). In: Forum for Kvindeforskning nr. 4, 1985.

to be nurturant to woman and nature. On the contrary he tries to regain the feeling of omnipotent by controlling women and nature. He replaces the mysterious - women and nature - with something he can control with machines - with computers - with the control of technological development.
This is a very crude summary of Dinnerstein's ideas about gender. I think though that it is possible to use Dinnerstein's theories to help us focus the power of women instead of the powerlessness that we are so familiar with when we think about the suppression of women. Women and women's values are being suppressed because they represent women's power that needs to be controlled by men but also by women because both men and women feel the ambivalence toward feminity. On the other hand the woman is necessary and deer to us and on the other hand her power and will constantly reminds us of our childish feeling of impotence.

10. ....AND ACTIONS.

Focusing on the power of women and on the possibilities and limitations within the general political and equal rights policy while bearing the qualitative potentials of the information-technological development in mind, should give us some clues as to concrete feminist struggle concerning women and the labour-market. In my opinion this means that we must be aware of the need for women's values in the development of qualification-patterns and how we can suggest strategies along these lines knowing in advance that it will be a struggle. The reluctance to acknowledge that women could possess anything that men (and women) cannot control, has deep roots and will be denied as being gender-based.

Instead of going into areas where we have no fore-mothers (role-models) that is in male-dominated areas of the labour-market, than we should insist on creating new areas combining what we want to develop (the general aspect of education and vocational training) with that which is unavoidable considering the political - and economic - climate (the technical part of information-technology).

In the area of office-automation I could thus conceive of completely new job-profiles where office-workers received more technical education on the one hand and on the other hand developed education and knowledge of the decision-making process in order to attain access to more qualified and powerful jobs. The same could be done in the welfare-sector and in all other sectors where information-technology can be used. Further education and training along these lines could help prevent women from them ending in so-called dead-end male dominated areas of the labour market.

BIOGRAPHICAL NOTE

About the author: 34 years old, no children. M.A. in social science and geography and at the moment working on a dissertation in educational studies. The article is part of this. I have been working in the area of feminist research since I began my studies in 1976. During the last 6 years on the subject of women and technology. In shorter periods I have worked as a grammar school teacher of social science and geography.
Correspondence address: Bente Elkjaer, Institute of Informatics, Royal Danish School of Educational Studies, Emdrupjvej 115B, DK-2400 Copenhagen NV, Denmark.

Women, Work and Computerization: Forming New Alliances
K. Tijdens, M. Jennings, I. Wagner, and M. Weggelaar (Editors)
Elsevier Science Publishers B.V. (North-Holland)

GENDER AND LEARNING: DO WOMEN LEARN DIFFERENTLY?

Deborah Brecher,
P.O. Box 68, Jenner,
CA 95450, USA
(707) 632-5763

GENDER AND LEARNING, DO WOMEN LEARN DIFFERENTLY?

There has been increasing concern about integrating women into the technology-based workplace. This has resulted in a variety of affirmative action efforts including the creation of women's re-entry programs devoted to getting women into science and math college programs, math anxiety workshops, special women as role-model programs designed to attract high school girls into technology-based careers, etc.
These efforts although laudable, miss the real point, which is there are gender-based learning differences. Since school curriculums are designed for the dominant learning style (male), they favour boys and discriminate against girls. Whether the curriculum is delivered by a male or female teacher is therefore irrelevant. It is the method of delivery that is the problem.

The theory presented in this paper are based on 20 years of observations. During that period I have taught computer skills in various environments to various classes of learners. These include seminars for faculty members in California State College, in-staff training for employees of a Medical school and in my own school, which offers small women-only seminars.

LEARNING STYLES

Based on these experiences, I have come to say that there are two different styles of learning. One style of learning is rule-based. That is, the material is presented linearly. The student follows the rules with no understanding why the rules are valid. Understanding is gained through experience. Usually, this means the student learns by accidentally violating the rules and discovering the consequences for this. Learning is by trial and error.

The other side of learning is holistic. This means an overall understanding is attained before the detailed rules are encountered. The holistic style of learner uses the generalized understanding to provide a framework within which the rules are organized. Here understanding the process by which the various components inter-relate is the necessary precedent for learning which keys to press.

GENDER-RELATIONSHIP

These disparate learning styles are particularly obvious when teaching computers. Equally obvious is the gender relationship. These differences are seen quite dramatically when working with new computer users. A man who has never touched a machine, hits

some keys and waits what happens. A women hesitates to touch the computer, afraid of the consequences of doing something wrong.
Similarly, most men find it relatively easy to follow a set of procedures even if they don't understand them. They are willing to learn by trial and error, figuring that with experimentation they'll discover how and why a computer works.
Women, on the other hand, generally need to understand first what they are doing -- and why -- before they feel comfortable "playing around with the machine". Until they have that understanding they are reluctant to experiment, often explaining they are afraid they will break the machine, or destroy important data. Once they have a holistic sense of the reasons behind the rules, they are just as eager to try out the computer. This is not to say that either women nor men do better, or worse -- merely that they approach the subject differently.

BIOLOGY OR ENVIRONMENT?

When I talk about my theories of gender-based learning differences I am always asked if I think these differences are genetic or cultural. I only have to look around me to see the overwhelming effect culture has.

Most obvious is the effect on games on children's perceptions., Boy's games are rule-centered (baseball, football, etc.), while girls games are process-centered (dolls, playing house, etc.). Thus, girls become accustomed to expecting a scenario or framework to fit isolated events into. Boys, on the other hand, become accustomed to playing by rules -- rules which exist in isolation with no rational foundation. I believe this explains why so many of my woman students have been unable to learn to use the computers in their work and home environments after their male co-workers or husbands "showed them what to do". These teachers didn't realize that demonstrating a rule is not the same as explaining it so that it makes sense.

Boys games give men another advantage as well -- the experience of failure. In sports, a player's performance evolves over time. Repeated attempts produce improvement and the message the player gets is that a single failure is not a disaster. Instead the child get the opposite message. He sees that repeated attempts will produce success.

Becoming proficient with computers requires precisely this attitude. In fact, most rule-based instructions assume this approach. It is merely most obvious in computers. For example, learning a new program requires the user to read the instructions and try to follow the directions. A dialogue then follow with the user receiving acknowledgements of successful attempts and error messages when attempts are unsuccessful. The error messages prompt the user into trying a new approach. Eventually the user succeeds. Individuals who are uncomfortable with this approach are at a disadvantage. A single error message may cause them to give up or seek help.

Computer training is not the only field that demands this learn-by-failure-and-retry pattern, it is merely the most obvious. However, in all cases it is the female-style learner who suffers. Teachers working with math-anxious women report similar stories. Teaching math skills to these women required a new curriculum, one that recognized the sources of the anxiety. One basis for anxiety was the misconception on the part of the students that there was only one correct answer. These women had no experience with trying an approach to a problem, then discarding it for a better solution. They could not validate such a process, and only experienced a sense of failure when abandoning their initial attempt. Symptomatic of this was the lack of understanding of approximating an answer and then refining until the correct answer was arrived at.

Spots are not the only sex-linked games that affect learning styles. Model building is another. In past years boys built airplanes or cars. Now they build very intricate plastic robots. These are very definitely boys toys -- girls do not play with them.

What does this have to do with learning styles? I became aware of model building after noticing that many of my women friends had difficulty setting up their newly purchased VCRs (video recorders). They complained that the instructions were poorly written. But the written instructions weren't really meant to be read. Instead, the schematic drawing of the required connections was really the key to understanding how to hook-up the VCR.

Women have little experience with schematics. Men on the other hand have once been boys building plastic models. The instructions for putting these models together are similar to the VCR schematic -- they have no words. Instead, the plan for the models refer to various components by number. Each step in the model building process is dependent on the previous step. By following plans and building models, boys get patterned themselves -- they are introduced to both linear processing and schematic representation. They feel secure that by carefully following the instructions they will achieve their desired result even if the process will make no immediate sense to them.

Another obvious cultural difference is our household roles. Although women's role in the workplace has changed dramatically, our jobs at home have stayed the same. We continue to be responsible for cleaning up the messes, both on the material and emotional levels. What does that have to do with computers or learning styles? As the nurturers in their home, women are much more aware of the consequences of creating 'messes', on the toll it takes to remedy them. Therefore I propose, women are much more apt to consider the consequences of their actions, and are much less cavalier about creating problems that someone else will have to deal with. This concern inhibits the experimental-style of learning, since fear of consequences does not promote risk taking (i.e. trying out some keystrokes to see what will happen).

The rule-based learning style presupposes a certain level of comfort with risk-taking. The learner tries out the rules, encounters allows the user to determine the parameters. By trial and error, the new user 'learns'.
However, such experimentation is a necessary ingredient in working with computers. Adjusting the curriculum so the student understands how the machine works, before using it, enables the holistic learner to experiment since she is not taking blind risks. By understanding the environment she is working within, she can determine for herself that she will not be harming anything. She understands the consequences of making mistakes and recognizes that they are responsible.

MEN AS FEMALE-STYLE LEARNERS

Although I refer to learning style differences as gender-based, women are not the only holistic style learners. Many men are also 'female-style' learners, that is, people who have a hard time learning procedures without understanding the reasons behind them. Often these are the same people who had difficulty in math and science.
Most math and science classes use the linear rule-based approach to teach their subject-matter. Those uncomfortable with that approach self-define themselves as poor in math, or poor in science. While the majority of those labeling themselves in this way are women, many men also have difficulties with these subjects. I believe these men are also holistic learners.

By self-selecting out of these subjects, and going into process-oriented fields such as the arts, writing, sociology, etc., these learners are choosing to learn via their 'natural' styles.
As the level of schooling progresses, more and more girls drop out of these rule-based subjects. That's quite obvious. What isn't so obvious is that many boys also drop out. Math and science courses have the reputation of being 'difficult'. But, maybe the problem isn't with the subject. Perhaps the problem is with the method of teaching.

A MODEST PROPOSAL

I therefore propose that the entire school curriculum needs to be revised. Instead of rewarding rule-based learners by teaching in their native style, and penalizing process-learners, I suggest that alternative curriculums be designed to offer both modes of learning.
I have done exactly this in developing a 2-day computer change. Rather that teaching computer applications by simply having the student interact with the programs, in-depth understanding is gained first -- before the student uses the application programs.
The results have been quite dramatic. My change, The Women's Computer Literacy Training, has now been offered to over 4,000 women in small groups of 10. Each training has been characterized by repeated expressions of delight on the part of the student, coupled with complaints about how traditional education had failed them. However, this paper is not about the details of that curriculum handbook which is available in a book.[1] Rather, this paper is about the underpinnings of the method, why it's necessary and why it works.
Our curriculum Recognizes these gender-based learning differences and uses a holistic approach. For example, instead of the reader learning how to use a computer by rote, she learns how the computer works. Then she doesn't have to memorize rules, they make sense and become obvious.

WHY CURRICULUM REFORM IS NECESSARY?

Assuming that the gender-based learning style issues as discussed above are accurate, what are the implications of the current teaching methods, and why is curriculum reform so necessary?

Currently, process-oriented learners self-select out of technical fields. This means that our new technology-based society is being designed by one group -- only rule-based learners. This group may not be the best suited to design our new society.
Will redesign of the curriculum encourage these 'soft' learners to re-direct their careers into the 'hard' sciences. I believe this to be quite true. One only has to look at the marketing efforts of Apple to see the effects of their redesign.
The Apple Macintosh employs a more process-oriented user interface -- the so-called 'desktop'. By using a less rule-oriented interface than the IBM 'command line', a whole new population of users is being attracted to personal computers. The Macintosh is being used by precisely that group that I have termed 'female learners' -- artists, architects, writers, etc. I can only assume that similar results will be obtained if the academic interface is redesigned, i.e., curriculum revision.

---

1 The Women's Computer Literacy Handbook, NAL, 1985, New York and Run, Stop, Go, Orlanda Frauenverlag, 1988, Berlin.

Women, Work and Computerization: Forming New Alliances
K. Tijdens, M. Jennings, I. Wagner, and M. Weggelaar (Editors)
Elsevier Science Publishers B.V. (North-Holland)

REALIZING TRAINING PROGRAMMES FOR WOMEN ON INFORMATICS

Drs.J.Verhoef and Ir.H.Voorhaar
Vocational Training Centre on Informatics for Women
Nijmegen, The Netherlands.

The paper presents the results of a vocational training for women who want to return to the labour market after having children. Those trainings require not just practical, organizational adjustments of a school, but also measures concerning selection and the curriculum and training programme. An excellent combination of labour market demands and women's job wishes is the training for 'Supervisor of personal computer use'. Women, especially those with a career as housewife and mother, have first class experience with understanding and problem solving. Qualifications which are very important for a supervisor. Thus, in this training programme two at first sight completely different interests have been combined. Labour and skills of women have been used, which have been neglected for too long.
In this paper the following aspects will be presented: the initiative: women want paid jobs; the practical demands of education programmes for women; the specific terms of training contents; an example: supervisor of personal computer use; conclusion

1. THE INITIATIVE: WOMEN WANT PAID JOBS

1.1 Introduction

The Vocational Training Centre on Informatics for Women in Nijmegen was set up in October 1985 by a small group of women. Almost three years later, we know there is a lot more to say about running a women's school. Meanwhile we developed a new training programme and new educational concepts, responding to changing demands on the dynamic labour market. Before we go into this new programme we would like to take a short look back at the initial start of the project.

1.2 Women want paid jobs.

The very high unemployment rate among women in this region lead to an action in February 1985 of women's organizations. The action aimed at revealing the hidden unemployment especially of women with children. Although many of them are not officiall registered as unemployed, they are looking for a job.
In 1985 in Nijmegen 60 % of 350 women questioned were not registered; they are so-called "discouraged workers".

Table 1. Unemployment in Nijmegen, February 1985:

	unemployed people	% of the dependent professional population
men	17.463	26.3 %
women	8.893	28.7 % *
total	26.326	27.1 %

* for The Netherlands the figure was 17.9 %

1.3 The position of women on the labour market.

Women have a weak position on the labour market. They work mostly in traditional women's jobs in lower position with relatively low wages, little opportunities for in-house training, job enrichment and promotion. Job segregation severely limits women's prospects in different sectors. In addition to these facts we should mention the exceptionally low degree of women's participation on the labour market. Compared to other countries in Northern and Western Europe we stay way behind. Wages, taxes and the social security system benefit the bread-winner and discourages women to earn their own living. As far as the necessary facilities such as child care and maternity leave are concerned, we must admit they are very poor indeed. Due to the economic recession, the Dutch Government cut down expenses on education, public health services, child-care, care of the old aged and family welfare. Women were hard hit by these measures while many of them held jobs in these sectors. At the same time the large-scale introduction of office automation led to structural changes on the labour market.
The unemployment rate among women rose sharply. The situation deteriorated still further as a result of a growing number of married women wanting to re-enter the job market. A serious impediment for those women is their low level of education, often combined with a lack of vocational training and recent job experience. Especially women above thirty find it hard to compete with younger workers who are better qualified with regard to the new technologies.

1.4 Education programme on informatics

From the very beginning we were determined to develop a project offering realistic opportunities to women re-entering the labour market. We took good notice of the regional job market. As a result of our telephonic inquiry among personnel managers of 50 companies (small and medium sized) we predicted a shortage of well trained employees skilled in office automation and informatics. The introduction and implementation of new technologies leads to a disparity between the old and the new skills for office workers. Women need to be retrained to qualify for these new jobs. There was for sure a demand for people familiar with microcomputers. So an appropriate vocational training should combine the regular training for office work with a special training on using computers. Such a programme was set up for two groups of sixteen women from 25 to 45 years of age. The entrance qualification for this course was junior general secondary education ("MAVO"- level). To higher educated women, more interested in informatics, a course for system controller was offered. Eighteen women enrolled after passing the selection. Our plans for a special Training Centre for women were accepted and subsidized by the Labour Office owing to the lack of opportunities for women at the existing training institutes.

2. PRACTICAL DEMANDS FOR A SCHOOL FOR WOMEN

We analyzed the position of women who intend working again after a break of several years. We took account of the female double role and its implications. A training programme stands little chance of success unless a range of conditions are met and facilities provided for. Class hours and holidays should be adjusted to those of primary schools in the region. Childcare facilities are a necessity to enable women with children under the age of four to participate. The location of the Training Centre must be in easy reach of public transport since most women don't have a car at their disposal. As for the duration of the vocational training programme we think it should be limited to 1 year or 1 1/2 years to keep women motivated. In order to overcome difficulties in the process of retraining, additional guidance, support and lessons on job orientation are required. And last but not least: the costs. Many women who want paid jobs are willing to study for one year in order to get one. However they often depend on lower family incomes and are not able to pay the full price of a professional training programme on informatics.

3. SPECIFIC TERMS OF TRAINING CONTENTS

3.1 Analyzing the problems

Training women for the labour market not only requires practical, organizational adjustments of a school. It also asks for measures concerning the selection and the curriculum of the training programme. During the first weeks of the project we found a women enrolled in the course on office automation to be a "whiz woman". She switched to the course on system control and finished as the top student of the informatics group. This was considered an exception and in the next project switching from one programme to the other wasn't possible because of timetables. Nevertheless the same thing was noticed: some women in the administrative group would fit in better into the informatics group and vice versa. Women over thirty often turn out to have had a school-career considerably under their learning capacities. In their youth it was hardly seen necessary for a girl to study. Preparing for housekeeping and bringing up children seemed a better basis for their destined career as a housewife and mother. On the other hand, housekeeping and bringing up children are also valuable assets in a mature women's life. They learn to organize and manage their small-scale business. Present-day parents find studying more important for their children. Many women assist their children with their homework. In doing so they not only guide their children in the learning process, but benefit by it themselves. Hence selecting a professional training for women in terms of their previous formal education might be the wrong thing , because their schooling would often not correspond to their present abilities.

Women who for a long period of time have not participated in a learning or working process need more than just a professional training before re-entering the labour market. First of all they will have to learn to combine their studies with their tasks at home. Even with a part-time training programme they can only spend half of their previous time taking care of the house and family. Secondly they will need to refresh and update their working experience. Office work has changed considerably during the last few years. When applying for a job, recent working experience is an absolute must. A period of two months of practical work should be included in the programme. During this training-on-the-job women get a chance to prove their competence and overcome their uncertainty and anxiety. By taking this important barrier away with support from teachers and employer means a much better start of their first paid job .

The informatics programme in the first project was developed by a Software House, experienced in training working people. The programme contained a great deal of programming in Basic and Cobol and very little knowledge of standard software. For many women this abstract, theoretical training programme was very difficult. When they started applying for jobs, trade and industry in the region made clear, they were more in need of people who were able to work with standard software and to instruct fellow-workers in its use. These critical remarks from both students and employers resulted in the development of a new training programme: Supervisor of personal computer use.

3.2 Introduction period

At our Training Centre a more accurate judgement of women's knowledge and capacities is obtained during a six week introduction programme for all students, whatever their previous education has been. About one third of this newly developed programme consists of informatics. For many women this is their first encounter with a computer. In addition to this introduction course in basic informatics the students are given keyboard instruction, Dutch and English grammar, basic economics and administration, and, simultaneously, labour orientation.
In the first lessons on labour orientation the students learn how to combine their studies (as well as their future job) with family responsibilities. After that, the counsellor pays a lot of attention to explaining the most significant aspects of their future profession.
At the end of this introduction period both the students and the staff will have obtained a better idea of a student's learning capacities. The students know what informatics and working with a computer is all about, including the job prospects.

The staff assists the student in choosing one of the two vocational training programmes: the one mainly dealing with administration subjects or the other with more informatics. If a student decides not to study at our Training Centre we show her the way to alternative courses at other institutes. Ten percent of the women does indeed switch after the six week introduction period. Thus this pre-training period primarily performs a screening function. It gives a good indication of future career opportunities. As such it has proven to be a much better selective instrument than entrance qualifications based on formal education or working experience.

3.3 Support in learning and job finding

The complementary lessons on labour orientation are quite necessary for a successful learning process. As we mentioned above, labour orientation starts tackling the practical problems of the women's double role. This is followed by a survey on today's labour market, including knowledge of corporate organization, labour unions and the effects of automation on job contents. In an early stage, visits are paid to companies and bureaus. The last term is spent on training in job searching and application. Labour orientation thus enables the students to obtain a clear view of their future working possibilities and rediscover their skills and interests. We are convinced that support and guidance form a crucial factor and essential element in the training programme.

3.4 Practical training on the job

To gain recent working experience, the last two months of the training programme are spend on training-on-the-job in various businesses and enterprises. The staff tries to find the best possible combination of student and company for this training period. The counsellor who takes care of the lessons labour orientation and is acquainted with the student's capacities and interests, is responsible for this matching. The overall positive appreciation by students as well as by employers underlines the importance of training-on-the-job as an essential part of women's vocational training. An additional effect of this period on the students is the confirmation of their knowledge and skills. To many it means a rediscovery of the value of work and a strengthening of their motivation.
For many employers the trainee is a first acquaintance with women re-entering the labour market. Existing prejudice against elderly female workers may be eliminated in a relatively safe, unrestrained situation. Such a prejudice tends to turn into the opposite after the training period. Employers discover that these well trained and skilled women are particularly reliable, motivated and loyal workers.

3.5 New opportunities on the labour market

Trying to deliver specialists in the field of informatics after a part-time one year training is not a realistic approach. Therefore the Training Centre has chosen for a basic training programme in informatics in which a broad field of subjects is presented. This programme gives the target group, usually not acquainted with the working field, an overall knowledge. For the employers they are a trained group, qualified for a range of tasks. By means of a thorough evaluation of the training period the Training Centre obtains a clear idea about the current demands on the labour market. These evaluations added to information from labour offices and trade unions influence the curriculum of the next programme. Year after year the programme is adjusted to the changes on the labour market. This flexibility in programming gives the students a greater job finding guarantee. It also provides the labour market with up-to-date trained employees.

3.6 Implications of a Training Centre for women.

A Training Centre for women only, equipped with an almost completely female staff, appears to function as a testing ground and frame of reference.
Fellow students and most staff members have to combine housekeeping and studying or working. Both the students and the staff have to find solutions for problems resulting from this combination. During the course the student will experience that no teacher can learn her a general strategy to deal with the technical, practical and social problems. The personal constraints of every student is unique. Therefore a range of solutions has to be considered. Working and studying merely with women provides a basis for discussing all these problems and find individual solutions. The creativity with which the solutions are found is very stimulating to all concerned. As a result women which are looking for a part-time job, had no problem in accepting a 30- to 40- hour working week by the time they left the Training Centre.

4. AN EXAMPLE : SUPERVISOR OF PERSONAL COMPUTER USE

The labour market demands and women's job aspirations are both perfectly met by the training programme for "Supervisor of personal computer use". This course was developed in close cooperation between the Training Centre and several employers. The latter feel the need to bridge the gap between everyday users of computer hardware and software and the people developing computer systems. Users and specialists often find themselves at the far ends of the automation process. Most users know perfectly well which aspects of their work can be improved. However they have difficulties in translating their wishes in such a way that it enables specialists to take appropriate action. On the other hand, specialists know in which way the organization can benefit to the most from computerization. However their lack of insight into the every day working situation often rules out participation of users and turns computerization into a long and troublesome process.
This is where the Supervisor comes in. She is well acquainted with micro computer systems and has learned to work with several standard software applications, such as word processing, accountancy, spread sheets and data filing. Her training programme also includes communication and instruction, reporting, organizational structures, system development methods and problem solving techniques. The supervisor is able to solve the basic problems of users. She has learned to think at user level and she knows how to translate user problems and preferences into solvable questions for programmers and technicians.
Women, especially those with a career as housewife and mother, have first class experience in understanding and problem solving. They are competent at small scale managing, instructing, organizing and budgeting and in creating an atmosphere in which

others are stimulated to an optimal use of their talents. That is what housekeeping and bringing up children is all about.
Our training programme provides women with an opportunity to make full use of all their talents, knowledge and skills and apply them successfully in a professional setting. In the first place the students get a theoretical and technical training in a broad field of automation subjects: basic informatics, office automation, programming, standard software applications and tools. Next they are acquainted with organization theories and their social skills are stimulated for solving technical and communicative problems. This kind of training is for many women the key to the field of automation and business. When they get their first job they have to study further and maybe reorient themselves about their place in the automation business.
In July 1986 the first group of 15 Supervisors completed their training programme and started their search for a job. By October 11 of them had actually found a job in a wide range of companies and bureaus, varying from large multinational corporations to very small software houses and schools. No matter how different their environments may be, nearly all of them have found a job in which instructing and guiding fellow workers accounts for the greater part of their activities. The year 1987 about 75% of the students found a job according to the end terms of the vocational training centre. Evaluation of each course is very important for the development of the programmes for the forthcoming years.

5. CONCLUSION

The development of this training programme is an outstanding example of combining the latent labour resource of woman and the current job vacancies in automation. Women want paid jobs and are eager to be trained in order to qualify for newly created jobs offering opportunities for a career in new technologies.
Following the local labour market by development of special training programmes for woman gives the guarantee to establish the success of the training programmes. A training centre for women has to met important practical demands, such as childcare facilities and free school holidays. Re-entering women need above all support and individual counseling in this transition stage from homework to job. The target group of women that have to be retrained has to reach also the group of working women to consolidate and up date their knowledge and skills for the security of women's jobs. Employers are thus able to fill vacancies and utilize human resources and capacities of women re-entering the job market. Efficient use of their labour and skills may contribute to further economic development in the region of Nijmegen.

LITERATURE

Opleidingscentrum Informatica voor vrouwen Nijmegen: Project verslag 1985-1987 (Report 1985-1987 of the Vocational Training Centre on Informatics for Women in Nijmegen, The Netherlands)

BIOGRAPHICAL NOTE

Joke Verhoef is manager director of the Vocational Training Centre on Informatics for Women in Nijmegen, The Netherlands. Riëtte Voorhaar is member of the board of the Training Centre. Correspondence address: Opleidingscentrum Informatica voor Vrouwen (Vocational Training Centre on Informatics for Women), Postbox 1187, 6501 BD Nijmegen, The Netherlands.

Women, Work and Computerization: Forming New Alliances
K. Tijdens, M. Jennings, I. Wagner, and M. Weggelaar (Editors)
Elsevier Science Publishers B.V. (North-Holland)
IFIP, 1989

TECHNICAL CULTURE, TRAINING AND FEMALE EMPLOYMENT

Ghislaine DONIOL-SHAW,
GEDISST-CNRS,
Paris, France

This paper is based on field work the author has done on the work of blue-collar women and on training of women for highly technical and traditionally male crafts. It questions the role played by technical culture and re-training on the entry or exclusion of women from certain types of jobs.
If the level of qualification of women's employment evolves favourable in the economic sectors where women constitute the majority of the labour force -- and in the service sector in particular -- such is not the case in "traditionally male" sectors, especially when women's entry into those sectors takes place at low levels of qualification.
Thus, if it is possible to argue that technical culture should become part of all kinds of training attended by women, one should not forget that serious obstacles still remain, since the workplace is permeated with gender relations.

1. INTRODUCTION

Among the studies we have carried on at the GEDISST*, several indirectly question the relationship of women to technology. Two of these studies deal with the issue of the qualification of blue-collar women workers[1], and another with the problem of the training and entry of women in electronics and industrial computerization occupations.[2] We have observed that access or barriers to technical training and culture were crucial stakes for the control women could have over their work as well as their access to certain apparently "non-technical" jobs. We will use throughout this paper Philippe Roqueplo's definition[3] of technical culture as "the possession of the kind of knowledge and know-how which can potentially grant a minimum of personal control over both our environment and the activity of those whose skills prove indispensable. The general lack of this type of culture constitutes a major cause of generalized alienation. Its development is therefore necessary in order to abolish this alienation as much as possible." The fact that technical culture is in general too scarcely developed, makes users dependent on

* The GEDISST is a research group within the CNRS -- The french national research centre -- which specializes in the study of the division of labour along social and gender lines.

1 G. DONIOL-SHAW, Qualification des ouvrières et déqualification du travail,, rapport GEDISST-CNRS, 1985. G. DONIOL-SHAW, "Ergonomie, qualification et division sexuelle du travail", in Division sexuelle des techniques et qualification, rapport collectif GEDISST-CNRS,1987.

2 G. DONIOL-SHAW, Formation et insertion professionnelle des femmes dans les métiers de l' électronique et de l'informatique industrielle, rapport GEDISST-CNRS, 1986.

3 P. ROQUEPLO, Penser la technique, Ed. du Seuil, Paris 1983.

users dependent on specialists or experts. To question this kind of dependence is for both men and women to question the hierarchical power relations that prevail in society in general; but, for women in particular, it also means, at the same time, to question the power relations involved in gender relations.

We have already showed in a recent communication[1] how women were excluded from technical culture and the role this exclusion played in the definition of social relations between men and women; furthermore, the studies of the GEDISST we have already mentioned, show that the ways through which women, and particularly working class women, are excluded from this type of culture are constantly produced and reproduced at the workplace as well as within the family. In this paper I would like to stress that women do want to have access to technical culture, to control their environment and their work. Their high demand for technical training is one way in which this is expressed. It meets all kinds of barriers, which has important consequences at the workplace as we will see. I will also show that, furthermore, in order to increase the qualification of women's jobs, it is important to introduce specifically technical topics in all kinds of training, given the growing importance of new technologies, which are changing the work environment in all sectors of activity

2. WOMEN WORKERS AND TECHNOLOGY

One often hears that the reason why women do not fill technical positions in industry is that they're not interested in technology. But then what does it mean to be interested in technology? And how is this "lack of interest" of working-women for technology actually produced by their environment?
As far as we are concerned, during the interviews we made of a number of women workers, we have listened time and again to their multiple demands directly connected to the mastering of their activity and, as such, referring to a need for technical training in its broad definition.
However, the choice for working-women is rather meagre; the employer's point of view is that they don't have any technical training to start with, and are therefore given jobs somehow designed according to this "lack of training". Thus, according to the employers, they fill positions which do not require any technical training. This situation makes it difficult or even impossible to ask the hierarchy for technical training. The demand in nonetheless real, as we have outlined, and expresses itself in several ways. It is related either to a formal need for technical training -- most often at an elementary level though indispensable to master the work -- or to a more diffuse demand which would then be more precisely related to a kind of technical culture.

2.1 A demand for elementary technical training

We have observed this type of demand during a inquiry on women who worked in a plant producing safes and had had some training on locksmithing, having to do in particular with the assembly of the doors of the safes. The training took four hours a week for a period of 6 months and it was attended by six women workers, all of them unskilled, and by four male workers, three of whom were skilled assembly-men and one who was unskilled.
The interviews we made of the women workers made it possible to evaluate the impact of the training on the control of their activity but, above all, also showed the initial effects of the lack of such training when the women workers were first hired. The

1 G. DONIOL-SHAW, "Le rapport des ouvrières la technique: une construction sociale à analyser", a communication for the international workshop on gender relations, Rapports sociaux de sexe, Paris, november 1987.

following excerpt from the interview of one of the women workers is significant of the opinion of all the women and it explains the role and the status of "technology" in their work. The woman we quote was 55 and had 35 years of professional experience, 15 of which had been spent in that plant.

> "I used to call on the men more often than now... The young ones had the scheme and I would copy what they were doing... The assembly men stayed in the same team for a long time; they didn't really like to show what they knew... I would have liked to have had this training before working in the assembly of the doors; I did not know how to use a screwing or a boring machine. I didn't know any of the machines; I didn't know how to change the drills... I would have liked to go and help the others from time to time to be able to use the tools, make some rivet holes... When one arrives in a men's workshop there is plenty of things one doesn't dare to ask. They say it's easy, but a women is not used either to a drill or an electric screw- driver... I would have liked to learn how to use them before starting on the job."

This interview, which expresses the same problem the other women told us, shows the elements of the constitution of a "technical lack of skill" due to the simple fact the women worker's have no relation to the tools before entering on the job. However, as far as learning is concerned, it is rather "easy" and little time-consuming to know and be able to manipulate "simple" tools such as a drill or an electric screw-driver, as is mentioned in the quote. Moreover, if women are excluded from such a learning process, some studies show that young male workers when they are new comers in the shops benefit from the experience of their elders in the form of on- the-job training.
This example shows the conditions under which one of the forms of the *appropriation* of technology by men takes place, or rather is "socially instituted". The access of women workers to technology requires that we unveil what is really at stake, for the relationship between men and women, in the appropriation of technology by men.
However, the barriers are multiple and the example we gave is not conducive to optimism. Actually, of the seven women workers who benefited from the training and who all expressed positive feelings about its impact over their work, only two were still working in the assembly of the safe-doors three months after their training. Four of them had been transferred and were doing less qualified tasks which did not enable them to use, and above all to develop through experience, what they had learned during their period of training. Only one was shifted to a more qualified task and took the advantage to get enrolled in an electricity CAP (Certificate of Professional Capacities, which is a French national diploma delivered to students having attended short technical courses, mostly before high school).
The stakes involved in technical training thus appear to go beyond the simple issue of productivity; in fact, all forms of social relations can be upset by the access of women to the mastering of technology.
As Chantal Rogerat writes[1], one must "elaborate a critique of the 'obvious facts' used to 'explain' the inferior status of women as well as of the attempts to dissociate this status from the structuring of a society which pretends to guarantee them equal treatment, particularly thanks to education."

2.2 A demand for technical culture

What follows is based on another study we have done, an ergonomic evaluation of working conditions in the workshop of a firm producing water-heaters and gas-boilers. There were 20 women workers in that workshop and they were all unskilled.

---

1 C. ROGERAT, "Le rapport des ouvrires la formation: un enjeu de classe et de sexe", a communication for the conference on Technical Culture and Training, Paris, december 1987.

The job of these workers is to assemble the gas-part of both types of appliances. It consists in putting together a series of pieces on a frame called the central part. They use certain mechanical tools, particularly a pneumatic screw-driver and a blower, in order to clean the dust and chips out of the pieces, and they also have small manual tools. There are several models of the gas-part according to the size of the water-heaters and the country they are to be used in. Since the workers are not specialized in a precise model, they must adapt themselves to each change. If one considers the "seemingly" technical dimension of the assembly tasks, one may think that no technical training is required to master the job. The women workers have, however, clearly expressed certain needs which the hierarchy refused to attend although it had partially favoured their emergence. What happened was that the women workers had been offered, sometime before our intervention in the plant, a training course of three half-days. This training was to be common to different workshops, and to deal with the knowledge of materials, the overall process of water-heating producing and the general methods of production management.

Actually, *all* the women workers responded positively to this proposal. Facing such a massive demand, the hierarchy did not try to organize any training for this specific workshop but, on the contrary, discouraged the majority of the women workers to attend by explaining to them that there were so many demands pouring in that it would take two to three years to satisfy them all! The "nonsensical" explanation given to the women who had showed their interest in the training proposal made to them had the foreseeable "demobilization effect", but at the same time the women denounced the incoherence of the situation. One of them thus explained to me that:

> "up until now they only proposed training courses having nothing to do with our work, wallpapering and tapestry ... and for once that they proposed something interesting they say we'll have to wait two years...".

The fact that the women workers responded so massively to an offer of general training concerning the product they were making, its design, and the production and management processes, constitutes a demand that comes under the sphere of technical culture. The justifications they made in terms of control over their work *"in order to better understand what we are doing"*, and over the quality of the product *" so that we don't assemble bad pieces"*, well situates the stake of technical culture at the level of an increased mastering of the environment through the acquisition of higher capabilities for autonomy and action. In this sense, it is hardly surprising that the strictly male hierarchy, having implicitly measured the stakes of a collectively attended technical training for women workers, thus "resisted" the demand they had themselves initiated and which would lead, if they had allowed it, to a weakening of their hierarchical position and their power over the women workers.

2.3 Different training needs and responses in different branches of the economy

The evolution of the training and qualification of women workers in France varies according to their sector of activity. In fact, in certain sectors the level of qualification of the women workers increases at the same time that their total number decreases. Such is the case in electronics and more generally in the whole metalworking sector.[1] By contrast, the industrial food sector, which is going through rapid technological changes, seems nowadays to increase women's work both quantitatively and qualitatively. The clauses on professional equality negotiated in that branch really seem to include better

1 A. LEROLLE, "Ouvriers et ouvrières de l'électronique face aux nouvelles politiques de gestion des ressources humaines", communication for the international workshop on gender relations, Rapports sociaux de sexe, Paris, november 1987.

conditions for the workers.[1] We don't have enough data to make any conclusions yet, but it does seem that the industries in which the rate of female to male workers (rate of "feminity") grew the most between 1970 and 1985, in parallel with an evolution of their producing process towards increasing automation, are the most likely to offer their women workers real chances for a higher qualification of their jobs. The coexistence within a branch of a significant rate of women and dynamic employment policies seem to be the necessary conditions for the evolution of the "quality" of the jobs held by women in those sectors. In this sense, the industrial food sector, whose total employment grew at the same time as there was an increase in its rate of feminity[2] -- which went from 30 to 34.4% between 1979 and 1984 -- is a particularly important sector, and should be the object of further qualitative and quantitative studies.

3. NEW TECHNOLOGIES AND TECHNICAL CULTURE

The study we made of the professional training and entry of women in electronics and industrial computerization crafts[3], showed how the acquisition of a form of technical training could be a determining factor both for getting a job and the progress of a future career in apparently non technical jobs, such as different levels of office work for example, or jobs with a profile that does not correspond to the one women get from their technical training. We would like here to discuss this result rather than the obstacles, still quite numerous, women who had that sort of training met in trying to integrate such jobs as robotics or automation technicians.
The French Ministry of the Rights of Women took the initiative to reserve for women several long term (1200 hours) training courses aimed at the integration of women in the sectors of automation, industrial computerization, and electronics and micro-computers maintenance; these courses were financed by the Ministry and the participants were remunerated. We have done a qualitative assessment of this type of training through interviews of the participants as well as of the instructors in a sample of the courses. The choice of the participants in the training courses was done in quite different ways. Thus they took in women whose previous education could have been scientific and/or technical, literary or commercial, with levels going from the CAP (see page 3) to the licence (3rd year of the university). The professional experience of the participants was also quite diverse, some had been qualified technicians, others had occupied several low qualification jobs, and others still had yet to enter the active population. It is among the participants having had initial education in literature, commerce, or science in general, which did not go beyond, in the latter case, the level of the high school diploma, that we have observed the unforeseeable consequences of a scientific and technical training on their job research. They constituted about a third of the population concerned by the training courses.

Actually, the training proved itself insufficient to enable these trainees to get employment as technicians as was initially intended, but several among them have integrated jobs in offices or executive positions in a high level technical environment. Our estimate is that such positions touched 10% of all the trainees and 30% of those we mention above.
Some examples of the type of employment they entered in will clarify what we mean:

* A trainee with a literary high school diploma and speaking fluent English was hired as marketing-assistant in a firm making robots. She defines the content of her job as follows: *"in charge of all relations with the press, expositions, robot leasing,*

---

1 C. ALEZRA et coll., La mixité du travail, une stratégie pour l'entreprise, cahiers Entreprises n.3, La Documentation Française, 1987.

2 B. BELLOC et coll., "Les femmes et l'emploi" in Données sociales 1987, INSEE.

3 G. DONIOL-SHAW, 1986, op. cit.

'open-door' events, documentation and marketing tasks". It must be noted that she was the one getting the highest salary at the end of the training. When compared to students of BTS or DUT (college level technical diplomas), or to women who had previous qualified professional experiences, this example becomes all the more significant.

* A trainee having a literary high school diploma, then having studied architecture for a year, and several years later getting a CAP (see page 5) in electronics, was hired at the agency of an automobile firm specialized in robot design; she organizes and manages the planning department there.
* A trainee holding a high school diploma in experimental sciences (bac D), was hired as commercial secretary in a printing house specialized in the publishing of scientific works.
* Another trainee with the same diploma became the secretary of the technical manager of a firm producing high technological level medical materials.

In all these cases the capacity to integrate a universe highly marked by a scientific and technical culture, thanks to the mastering of a vocabulary common to engineers and technicians, was determinant for the hiring. Therefore, the acquisition of a technical culture, even if it was based on the formal learning of scientific and technical questions, is what made possible here these forms of professional integration. Such forms of training, which not only favour the professional integration of women but also enabled them to start on a job at a higher level -- which necessarily has positive consequences for the unfolding of their careers -- must of course be developed.

However, one must also think about what kind of training conditions can be better suited to the characteristics of this population. In fact, we have chosen but successful examples of professional integration, whereas several of the trainees who had the same profile as the ones we've mentioned were not able to find a job at the end of the training period. The fact that the instructors orient their teaching according to the entrance requirements of strictly technical jobs, does not facilitate the trainees' development of alternative and apparently divergent projects. This is true even for the trainees whose participation in the courses was based mostly on some sort of intellectual interest, having to do more with abstract knowledge than with manipulation. There was a desire not to be left out of what looks like a major evolution of society, without necessarily aiming at a specific craft. However, the courses do not hint at any possibilities other than technical crafts, while, at the same time, they offer too short and insufficient a training for the participants to be able to get the necessary level to really master such a future technical craft.

In this sense, there is work to be done in order to develop training actions specifically aimed at the acquisition of a technical culture to facilitate the integration of certain professional milieus. The acquisition of such a culture must no doubt encompass specifically scientific and technical questions, but it must also valorize temporary in-plant practice (stages en enterprise) and include a number of concrete case studies. The assessment of the training courses we have done actually shows that a successful period of in- plant practice -- depending on the type of training this could last between 4 to 8 weeks -- was a determining factor for being hired afterwards regardless of the initial profile of the trainees and the characteristics of their future jobs.

If training on "technical culture" should always be part of *all* kinds of initial education courses, the accumulated backwardness in this field requires make-up measures in order to facilitate the professional integration of women and an increase in the level of their qualification within the job hierarchy.

It must be stressed that we do not consider computer related work as "traditionally male jobs". On the contrary, computerization seems to really offer training and employment opportunities for women having low levels of initial education. Moreover, it also tends to guarantee a qualitative evolution of those occupations. In fact, as we will suggest below, the massive introduction of computers in the service sector is most likely to have favoured the general increase of women's qualification in this sector in which they constitute the majority of the workforce.

4. TRAINING POLICIES: ORIENTATIONS TO FOLLOW OR TO CURTAIL

We can indeed already observe notable positive changes in the qualification of women's employment in the service sectors since "the concentration of women in certain occupations, particularly lower qualification ones, tends to be reduced: women have increased access to qualified categories of employment, especially in the occupations and industries were they are a majority of the labour force".[1]
Data on women's employment and professional integration also shows that concerning the lower levels of qualification, particularly the CAP, the girls' professional integration "chances"are higher if their training is aimed at the service rather than at the manufacturing sectors: in 1983, 50% of the young girls trained for industrial crafts were unemployed; the figure was 60% for those having a CAP in the clothing sector. In the service sectors only 40% of the young girls trained were unemployed. However, it is important to add that among the young girls having some sort of training at the CAP level, 34% worked in the craft they had been trained for when their CAP was in the service sectors, whereas this was the case for only 12% of those trained in construction and 17% of those trained in garment making. This unusually high discrepancy between training and employment, concerning young girls trained for the industrial sectors, tends to outline that the policies of diversification of female employment to enable them to enter traditionally male crafts, must be developed with a view on the higher levels of qualification where the chances for integration seem to be the most favourable.[2]

ACKNOWLEDGEMENTS

The author would like to acknowledge her referees Els Rienstra and Cydney Pullman.

BIOGRAPHICAL NOTE

Ghislaine DONIOL-SHAW is an ergonomics engineer. Together with Danielle Chabaud, she is in charge of CNRS (National Centre for Scientific Research) research team called GEDISST (Research Group on the Division of Labour along Social and Gender Lines). She is currently studying the impact of technological change on the working conditions of women blue-collar workers and the advancement of job integration.
Correspondence address: Ghislaine DONIOL-SHAW, GEDISST-CNRS, 59, rue Pouchet 75849, Paris CEDEX 17, France

---

1 M. HUET, "L'ampleur du chômage féminin: un indicateur d'une moindre reconnaissance du droit l'emploi des femmes" to be published in Les Temps Modernes.

2 C. MARRY, "La scolarisation des jeunes filles dans l'enseignement supérieur scientifique et technique", communication for the conference on Technical Culture and Training, Paris, december 1987.

Women, Work and Computerization: Forming New Alliances
K. Tijdens, M. Jennings, I. Wagner, and M. Weggelaar (Editors)
Elsevier Science Publishers B.V. (North-Holland)

THE IMPACT OF OFFICE TECHNOLOGY ON CLERICAL WORKER SKILLS IN THE BANKING, INSURANCE AND LEGAL INDUSTRIES IN NEW YORK CITY: IMPLICATIONS FOR TRAINING

C. Pullman and S. Szymanski
The Labor Institute, 853 Broadway Rm. 2014
New York, New York 10003, U.S.A.

The authors summarize the results of a research project which investigated the impact of office automation on clerical skills in the banking, insurance and legal industries in New York City, and outline a pilot training program which incorporates some of the research findings. Both the research study and the training program emphasize the importance of technical, problem-solving and communication skills as critical to clerical work in an automated environment.

INTRODUCTION

This paper is adopted from the final report of the same title. The study was conducted in 1986 by the Labor Institute for the Private Industry Council of New York City** Four hundred New York City clerical workers in computer-related jobs in four banks, two insurance companies and three law firms answered 160 questions about their background, job content, training, opportunities for advancement, work organization, job satisfaction, work environment and productivity. Over 40 managers from the work units of the surveyed clerical workers were interviewed on the same topics..[1]

---

* The Labor Institute is a non-profit organization which conducts research and develops educational materials and training programs on workplace issues for unions, community organizations and government agencies. The Private Industry Council is a federally funded employment and training agency for underprivileged individuals in New York City. Similar councils exist in other major U.S. cities.

1 In this study computer-related clerical jobs are those involving the "new" office technologies -- wordprocessors, personal computers, and other microcomputers -- relative to mainframes and batch processing.
Overall the clerical workers who participated in this study are young (80% under 40 years), high school educated females with a fairly extensive work history, several years tenure with their current employer, and with earnings clustering in the $15 - 18,000 range. The sample is 48% white, 30% black, 12% hispanic, 5% asian and 5% other, with white workers clustering in the legal industry and minority workers in banking.

Two major questions guided the study:

1. How are office technologies altering the skill requirements of clerical occupations and what are the attendant effects on work organization, job mobility, job satisfaction and work environment?[1]
2. What are the implications of the new skill requirements for training programs, particularly for pre-employment programs directed toward entry level workers?

These questions represent important concerns for our society and its policy makers for several reasons. The jobs most affected by office technology are clerical jobs. Nineteen million clerical workers, 80 percent of whom are women, comprise the largest occupational category in the United States Labor market. This means that nearly one out of every three working women is in a job that is susceptible to technological change. Minority women are being placed in particular jeopardy by office technology. Minority women first began entering clerical work in the late 1960s and today comprise about 15 percent of all clerical workers. This percentage is expected to increase as minority women replace white women in the lower-paying clerical positions as white women move into higher paying clerical jobs or management. However, the direction in automation is to eliminate the type of jobs in which minority women and less-educated urban white women have recently entered.[2] Those without the skills associated with computer-related work will be left out of the clerical labor market entirely or restricted to the most menial jobs.

CLERICAL EMPLOYMENT IN NEW YORK CITY

In the United States the banking, insurance and legal industries are among the main employers of clerical workers. Similarly, these three industries play an important role in the New York City (NYC) economy and are critical to future job opportunities for clerical workers.[3] Growth in the FIRE sector (finance, insurance, and real estate) and the Service sector (including legal services) along with the concomitant decline in manufacturing are resulting in occupational shifts in NYC toward "white collar" workers, particularly clerical workers. In 1970 clerical workers and production workers constituted the two largest occupational categories -- each representing 26 percent of the City's total employment. By 1983 a radical change occurred. Clerical workers increased their share of total city employment to 30 percent -- becoming the largest occupational category, leaving production workers in second place with a 20 percent share of total city employment.[4]

---

1 Although the focus of this paper and the final report is on the identification of the skills clericals use in computer-related jobs, the survey and interviews included questions on other aspects of job quality (work organization, training, mobility etc.). These categories were included because we believe that office automation involves the transformation of work that goes beyond the introduction of new machines and mechanical procedures. The same machine, for example, will generate various skill and training needs when introduced into different offices, dependent on other workplace factors and social relations.

2 Stafford, Walter W. 1985. Closed Labor Markets: Underrepresentation of Blacks, Hispanics and Women in New York City's Core Industries and Jobs. Community Service of New York.

3 The stock market crash of October 19, 1988 is expected to have a detrimental effect on the employment of clerical workers in New York City, particularly those in the finance and legal industries as well as in advertising and accounting. Matthew Drennan, an economist at New York University, has developed an econometric model which predicts that 75 percent of the 71,000 estimated jobs lost in New York City will be office positions (this includes professionals and managers as well as clericals).

4 Haren, Robert. 1986. New York City Department of Labor. Unpublished study on New York City.

Clerical workers in the banking, insurance and legal industries account for almost 6 percent of total NYC employment[1] and represent over 50 percent of the total workforce in all three industries.[2] Unlike the insurance industry, the numbers of clerical workers in banking and legal services increased. In all three industries the percentage of clerical workers is declined while the percentage of professionals, officers and managers is rising.[3] The increasing use of office technology can account to some extent for these chancing staffing ratios.

In New York City, 45 percent of all white women; 48 percent of black women and 41 percent of hispanic women are in clerical occupations.[4] Although banking and insurance are the most significant employers of clericals of all races, the location by industry of NYC clerical workers differs by race. Almost one-quarter of black women clericals, 22 percent of hispanic women clericals and 13 percent of white women clericals are in banking. The insurance industry accounts for 10 percent of both black and hispanic female clericals and 8 percent of white female clericals. Three percent of white women; 2 percent of hispanic women and almost 2 percent of black women clericals are in the legal industry.[5]

These three industries -- banking, insurance and legal -- are already highly automated. As these industries move from less reliance on Mainframe and batch processing to more dependence on micro computers and networking, new forms of work organization and new types of jobs are evolving. Clerical skills and training needs are shifting in response to the new technologies and changing work environments.

JOB CONTENT

In these study, our analyses of the effects of technology on clerical tasks and skills was made within the conceptual framework represented by the term, "job content." Job content has two aspects:

1. A set of operations -- the tasks an employee can be seen to be engaged in over the course of a work period, such as keyboarding, filing, answering the phone. This aspect is usually the basis for job descriptions.
2. The ability to perform -- the technical, interpersonal, cognitive and experiential skills that underlie the manipulation of the tasks. This aspect includes the "hidden or invisible" skills associated with many "women's jobs."

In order to perform any job successfully, all workers must incorporate these two aspects into their jobs. Both of these aspects tend to be recognized more -- in terms of pay and status -- in professional and executive positions. jobs that are considered "lesser skilled" and/or associated with "women's work", such as clerical jobs, tend to be characterized and compensated only in terms of the first aspect --as routine tasks. There is growing recognition that this singular, task-oriented definition of clerical work results in the underestimation of the type and amount of skills necessary to perform in

---

1 Ibid.

2 New York State Dept. of Labor, Occupational Patterns in Selected Non-Manufacturing Industries, April-June 1975; Employment by Occupation--Finance, Insurance and Real Estate, May-June 1981.

3 Ibid.v

4 Calculated from N.Y. State Dept. of Labor, Bureau of Labor Market Information, Unpublished Insured Employment Data, 1985 and Employment Review, State of N.Y. Dept. of Labor, Aug., 1985, Vol. 38 No. 8, Tbl. B-3, p.20.

5 Stafford. Op Cit. Appendix -- Table 1A.

computer-related clerical jobs.[1] To the extent that these non-quantifiable aspects of clerical work (the second part of job content) are ignored, computer systems and the people using them will be less effective in performing their jobs.

Office technology continues to be used to fragment and deskill clerical work.[2] However, research indicates that clericals working with computers are using new skills as work becomes more decentralized and as more of the routine, lesser skilled entry level jobs are eliminated.[3] The new skills include analytic, problem-solving and interactive abilities, as well as good technical training. We observed four trends which suggests the need for these types of skills.

1. Clerical work is increasingly characterized as computer related.

Working with a computer requires more mental effort and analytic capabilities than performing the same task manually. Clericals enter information into a system and manipulate the data through a new, symbolic, technical language which is based on precise and sequential procedures.[4]

2. Tasks are being reintegrated resulting in multifunctional work.

Tasks that were once performed as discrete jobs by several different individuals are now integrated and performed by one clerical with the aid of a computer. For example, clericals in a bank who handle foreign exchange transactions perform a variety of data input tasks as well as balance books, correct errors and receive funds. Interspersed with these computer tasks is phone interaction with traders who give them instructions about money market transactions. According to managers, such multifunctional work requires clericals who can grasp many procedures, work under pressure, and have a broad understanding of the operation of their firm and industry -- its specific language and labor processes.

3. Data is directly entered into the computer without the mediation of a supervisor.

Increasingly, clericals who have direct contact with customers and clients enter the transaction into the system without additional verification or authorization by another employee. Because automation allows the almost instantaneous transmission of information throughout the system, the inputting of correct data is crucial. As labor processes

---

1 Office of Technology Assessment, United States Congress. 1985. Automation of America's Offices. Washington, D.C.: U.S. Congress.

2 Murphree, Mary. 1984. Brave New Office: The Changing Role of the Legal Secretary. In Karen Brodkin Sacks and Dorothy Remy (eds.) My Troubles Are Going To Have Trouble With Me: Women's Toils and Triumphs at the Workplace. New jersey: Rutgers University Press.

3 Appelbaum, Eileen. 1985. Technology and Work Organization in the Insurance Industry. Labor Relations Review. June.
Adler, Paul. 1983. Rethinking Skill Requirements of New Technologies. Working Paper 9-784-027, Graduate School of Business Administration. Boston: Harvard University.
Some research indicates that even the more routine data entry type positions require increased cognitive skills. See for example, the Women's Bureau, U.S. Dept. of Labor. 1985. Women and Office Automation: Issues for the Decade Ahead.

4 Zubhoff, Shoshanah. 1982. New Worlds of Computer-Mediated Work. Harvard Business Review. September/October.

become more integrated with the use of computers, each data entry mistake can result in larger consequences. A knowledge of the organization, social skills, independent judgement, accuracy and responsibility for the data are some of the skills critical for this type of clerical work.

4. Office technology is creating new clerical jobs.

The necessity to control and verify data accompanies the diffusion of computers and the example, more secretaries are receiving their own personal computers or terminals. Frequently, these secretaries are scattered throughout an organization, often on different floors. A "Data-base Management" or "Systems Control Operation" type position has been created to maintain and control the automated systems -- control access to the mainframe, archive work and answer computer related questions. In addition to excellent technical skills, a knowledge of the whole production process, the ability to explain procedures and to deal with non-routine situations.

PROFILE OF CLERICAL TASKS

Table 1 presents a rank ordering of all the tasks which are performed by the clerical sample "using a computer most of the time" and which are "very or quite important to their job." Given the selected sample, we would expect data and text entry to head the list, followed by input and word processing related tasks. The ranking toward the end of the list of such tasks as, "create a filling system," and "directing mail and phone messages," suggests that clericals in our sample are not in jobs or firms which utilize the more sophisticated technologies such as electronic mail or computer networking. Almost 93 percent of our entire sample performs at least one of the top six ranked tasks.

Table 1: Ranking of Computer-Dependent Tasks by Clericals in NYC Banking, Insurance and legal industries

1.	Data Entry	13.	Statistical Computation
2.	Text Entry	14.	Money Handling
3.	Process Records	15.	Support Services
4.	Write Standard Materials	16.	Spreadsheet Applications
5.	Write Original Materials	17.	Read Letters & Inquiries
6.	Proof/Edit	18.	Read Professional Journals
7.	Prepare charts	19.	Create Filing System
8.	Billing	20.	Filling
9.	Information Gathering	21.	Bookkeeping
10.	Equipment Maintenance	22.	Plan Meetings
11.	Develop Forms	23.	Direct Mail/Phone Messages
12.	Fill in Forms	24.	Receptionist Duties

Computer-dependent tasks were also ranked by industry. Data or text entry tasks are more important in the banking and insurance industries than in the legal industry. Sixty percent of the sampled clericals in banking perform data entry tasks exclusive, while the percentages are 56 percent in insurance and 46 percent in legal services. Text entry is more significant in the legal industry. Forty percent of the clericals in the legal sample perform text entry to the exclusion of data entry. In banking, only 18 percent do text entry tasks and no data entry. For insurance the percentage drops to 14 percent who do text but not data entry.

PROFILE OF CLERICAL SKILLS

Table 2 represents the skills ranked as most important by the entire clerical sample. Table 3 includes the ranking of skills by clericals in each of the three industries. All the skills were categorized as either "technical," "interactive", or "abstract." The technical skills reflect the abilities more associated with the first aspect of job content as previously discussed; while interactive and abstract skills reflect the second aspect. The abstract and interactive skills head the ranking both for the entire sample and for the three ranked skills in each industry. Of the top six skills, four skills cover half of the sample population in the legal industry, while three skills cover half the insurance industry and two skills cover over half the banking industry sample.

Table 2: Ranking of Skills Most Important to Clerical Workby Clericals in NYC Banking, Insurance and Legal Industries

1.	Accuracy (A)*	14.	Sense of timing (A)
2.	Cooperation (I)	15.	Explain procedures (I)
3.	Concentration (A)	16.	Mathematics (T)
4.	Communication (I)	17.	Sense of proper dress (A)
5.	Eye for detail (A)	18.	Formatting of charts, tables (T)
6.	Memory (A)	19.	Take directions from more than one person (I)
7.	Spelling (T)		
8.	Read Handwriting (A)	20.	Deal with people outside own work unit (I)
9.	Grammar (T)		
10.	Coordinate work (I)	21.	Creativity (A)
11.	Problem-solving (A)	22.	Sense of mistakes hurting firm (A)
12.	Tact/Diplomacy (I)	23.	Sense of mistakes hurting self (A)
13.	Knowledge of company (I)		

*A=Abstract skills; I=Interaction skills; T=Technical skills. Note: The labeling of skills as either abstract, interaction or technical is somewhat arbitrary because each skill can be characterized as a combination of these three categorizations. However, we find the distinctions useful in our effort to isolate and identify clerical tasks and skills for training purposes.

Table 3: Top Ten Ranked Skills By Clericals in NYC Banking, Insurance an Legal Industries

Banking	Insurance	Legal
1. Accuracy (A)*	1. Accuracy (A)	1. Accuracy (A)
2. Cooperation (A)	2. Eye for Detail (A)	2. Cooperation (I)
3. Concentration (A)	3. Cooperation (I)	3. Spelling (T)
4. Communication (I)	4. Communication (I)	4. Concentration (A)
5. Eye for detail (A)	5. Concentration (A)	5. Read Handwriting (A)
6. Spelling (T)	6. Memory (A)	6. Grammar (T)
7. Grammar (T)	7. Problem-solving (A)	7. Communication (I)
8. Read Handwriting (A)	8. Coordinate Work (I)	8. Memory (A)
9. Memory (A)	9. Read Handwriting (A)	9. Eye for Detail (A)
10. Knowledge of firm (I)	10. Grammar (T)	10. Coordinate Work (I)

*A=Abstract skills; I=Interaction skills, T=Technical skills.

Table 4 lists the skills managers indicated as most important for clerical work.

Table 4: Managers' Listing of the 15 Skills Most Important to Clerical Work

1.	Knowledge of the industry	8.	Computer abilities
2.	Intelligence	9.	Congeniality
3.	Organization	10.	Analytic abilities
4.	Match	11.	Flexibility
5.	Communication	12.	Keyboarding
6.	Efficiency under pressure	13.	Accuracy
7.	Reading/Writing/Spelling	14.	Problem-solving
		15.	Work independently

As Tables 1-4 show, managers and clerical workers agree that a combination of technical, interactive and abstract skills are absolutely critical for entry into clerical work, the interactive and abstract skills are key to maintaining jobs in an increasingly automated office environment. As on a manager stated, "I would not look as much for the technical keyboarding skills, but for the interpersonal skills because this opens up the possibility for the employee to move up within the organization."

TRAINING

Both clericals and managers feel that the training clericals receive is not adequate to the demands of the increasingly automated office. In the initial stages of office automation, vendor-provided training was used as a major form of introduction to computer systems. The orientation to computers provided by vendors was adapted by trial and error to the needs of work units and often the vendor-trained employee (s) became the informal trainers (s) for co-workers. Today, the increased diffusion and complexity of computer systems is placing new demands on training needs for clericals that goes beyond vendor based programs.

Although more companies are establishing in-house computer training programs, indicating a recognition that the maintenance of an internal training capacity is essential for responding to changing skill requirements, clericals tend to have limited access to formal company computer training. A national survey indicates that clericals receive the least amount of formal company training compared with professionals or managers -- clericals receive, on average, 11 hours per year while professionals receive 27 hours and managers 33 hours per year.[1] Almost half of the clericals in our study -- 48% --indicate that no formal training was available to them in their companies. In the several instances where formal training is available it tends to take the form of approximately 5 half day sessions and the clericals often are required to read the entire manual prior to classes.

The main from training for clericals is on-the-job (OJT) training. In our study, 94% of the clericals receive on-the-job training for their computer related work. The OJT is usually conducted by supervisors and experienced employees. A large proportion of the clerical respondents -- 82% -- found their OJT useful. But almost a quarter of the clerical sample said that the computer training they received in their companies was inadequate for their jobs. The most useful training cited by the clericals in our sample is informal training for their co-workers. Fifty-seven percent of the clericals indicate that they turn to each other to learn more about the computer systems on which they work. This is in spite of the fact that the majority of managers we interviewed indicate that employees are instructed to ask their supervisors, not co-workers, about computer related problems. When asked "who has been the most useful in teaching you about computers?" the response, "supervisors," received the lowest percentage (11%) compared with the responses "your co-workers" (38%), "yourself"(18%), "instructors" (18%) and "other" (14%).

Although co-worker training cannot and should not be a substitute for a comprehensive training program, it is underestimated and undervalued as a critical form of computer learning. In fact, our regression analysis reveals that informal training has a negative effect on the earnings of those clericals who must rely on training from their co-workers. This suggests that formal training is very beneficial to management because employees are "picking up the tab" for providing training that is essential to performing their jobs.

---

1 Training Magazine's Industry Report, Training. October, 1984.

The managers we interviewed were particularly concerned about the lack of a generic and conception orientation to computer systems which is essential to employees' ability to adapt rapidly to new and changing computer tasks and environments. Also, managers -- especially in data entry intensive work -- were concerned about the need to avoid errors in computer-mediated work. As the labor process becomes more complex and integrated and as clericals input information directly on-line, the chance for costly errors increases. Managers feel that errors could be avoided if training programs provided a contextual overview of the functions of the work unit and the industry. furthermore, managers listed such skills as independent judgement, communication, responsibility, and organization as critical to clerical work in an automated office. According to the managers, these types of skills are not integrated into any new technology training programs.

RECOMMENDATIONS

Training is not a panacea -- it cannot immediately impact such macroeconomic problems as unemployment or illiteracy. But training can begin to influence some of the problems associated with computer-related work for individual clerical workers -- such as high turnover, limited mobility and rapidly changing skill requirements. In our final report we outline a variety of recommendations for clerical training programs. The recommendations elaborate on the need for a broad-based approach represented by the phrase, "training for tractability." By this phrase we mean helping workers develop skills that will enable them to be flexible and creative in response to the workplace restructuring and reorganization characteristic of automated offices. This type of training includes a technical component which emphasizes the concepts of computer use, not merely keyboarding; and components which focus on developing the interactive and abstract skills previously discussed.

"OFFICE OF THE FUTURE" -- AN EXPERIMENTAL TRAINING PROGRAM

The Labor Institute is currently involved in an experimental training program, the "Office of the Future," which incorporates some of the findings from the research study. The training program is sponsored by the Private Industry Council of New York City (PIC) for unemployed workers. The sixth month program has four components (three taught by PIC staff and one contracted out to the Labor Institute): 1) a technical curriculum which focuses on a generic introduction to computers, instruction in word processing, spreadsheeting, data base management and typing; 2) self-paced instruction in math and English; 3) job readiness which includes discussions about work expectations, how to write resumes and conduct job interviews; and 4) the Labor Institute's curriculum on technology, work and the economy. The Institute's curriculum, "Office Work and Technology in our Changing Economy," looks at the relationship between clerical work, technology and the economy as the context in which to develop problem-solving, analytic and communication skills. The development of these types of skills is implicit in the teaching method which is based on a small group approach whereby students participate in problem-solving activities and the instructor acts as a facilitate rather than a lecturer. Activities are constructed to analyze topics such as: the definition of technology -- its advantages and disadvantages; the history of office work -- how it became female dominated; what is the service sector; the effects of technology on clerical employment; the definition of skill and the identification of clerical skills; health problems associated with computerized office work; and, the relationship between job design and computer use.

Most computer training programs in the United States, particularly employer and government sponsored ones, are limited to technical, job specific training. Unions also sponsor technically oriented training programs for their members but usually incorporate discussions about the effects of technology on employment, skills, and health in other types of education courses or during preparation for negotiations. Within this context the scope of the Office of the Future curriculum is rather unique in so far as it includes a component about technology, work and economy. However, the program is limited to the extent that its federal and corporate support is determined by how many students are placed in office jobs. Thus, the emphasis of the program is on the "short term" in the sense of recruiting students who are considered "marketable" and airing on the side of turning out technically computer competent individuals.

ACKNOWLEDGEMENTS

We would like to thank Thea de Poel and Jos Spetter for their helpful comments on this paper.

Women, Work and Computerization: Forming New Alliances
K. Tijdens, M. Jennings, I. Wagner, and M. Weggelaar (Editors)
Elsevier Science Publishers B.V. (North-Holland)

WOMEN, RECURRENT EDUCATION AND NEW TECHNOLOGIES

Sabine Gensior, Lydia Hohenberger and Hildegard Schicke *
Berliner Institut für Sozialforschung und Socialwissenschaftliche Praxis e.V.
Berlin, FRG

1. INTRODUCTION

In social policy debates and within women's studies, recent discussion of the issue "women and new technologies" has focussed on identifying the obstacles which hinder girls and women from gaining access to further education, school and vocational education relating both generally and specifically to programming and computer skills. Based on the idea of "women's relative distance to technology" a vast number of studies, practice reports and model programmes meanwhile deal with the specific origins and forms of this so-called distance to technology or aim at investigating the specific social learning of women and at developing practical solutions ('women learn microcomputer technics'). At present it is almost impossible to assess what research is being carried out under what premises, what has been tested in pilot programmes and which applications have been successful. The assumption of the so-called female technological distance has not been adequately researched: neither are its origins clear, nor is there sufficient analysis of the consequences that the postulate of a specifically female approach to technology could have with regard to labour, labour market, educational and further education policies.

Discussion of women's work and employment has received impetus recently from a number of sources. The most important of these relating to our topic are:

* The concept of "recurrent education" has acquired new signifance. In the educational policy debates of the 70's (OECD, Paris 1973), this concept served to emphazise the need to plan and structure the distibution of educational opportunity and of occupational work. The underlying assumption was that a background of occupational experience and a mature motivation would result in a higher return of education (Edding 1963), "Recurrent education" aimed at refreshing and expanding basic qualifications and stressed the importance of spreading phases of education through the whole life course.
* We are presently confronting the beginnings of a profound reorganization of work in industry and the service sector, based on the introduction of new technologies and materials. New conceptions of production, organization and communication are the basis for creating new structures of work organization. In the course of this, women may begin competing for jobs in new areas.

---

* "Der Senator für Jugend und Familie Berlin - Die Frauenbeauftragte -" initiated this model programme which was evaluated by the B.I.S. under the direction of Sabine Gensior and Carol Hagemann-White. The empirical work on this project was carried out by Lydia Hohenberger and Hildegard Schidke during 1987. The present paper was written in corporation of the whole team.

* Both women's studies and political concern for equal opportunity have directed attention to characteristic traits of women's biographies, the circumstances and perspectives of their lives. Labour market, social and family policies have failed to meet the orientations, the experiences and the needs of women in regard to education and to work and employment.

Education, especially the vocational further education offered to unemployed women, has begun to identify and concern itself with a new target group: women in mid-life. Recent educational concepts postulate that women's lives are characterized by discontinuity and by periods of transition. Many women reduce their labour market participation or their occupational potential in favour of homemaking or child-rearing. Later, when they seek re-entry to the labour market, their skills are obsolete and they have difficulty finding work. Moreover, they are often socially isolated. And yet, they have half of their lives ahead of them.

2. LIFELONG LEARNING AND NORMAL FEMALE BIOGRAPHY

The biographical event of "becoming a mother" has justly been characterized as a "normative expected interruption" in the vocational history of women. At the same time, both the employment participation and the demand for employment of mothers have steadily increased. This phenomenon indicates that life schemes of women no longer follow the pattern of a "normal female biography". The concept of a normal biography for women imputes that women's life concept subordinates employment to family duties.

The numerous model programmes for the reintegration of women into the labour market respond to the trend of the increasing number of working mothers. Thus these programmes fill a need for further education that was previously neglected. Company training in the Federal Republic of Germany as well as publicly funded further training are - in fact although not by intent - addressed to those who can meet the male labour market standards; this is the implicit prerequisite for the opportunity to shape a career by further training or to secure and enhance one's potential for vocational development. These standards imply the exemption from reproductive family duties as well as a continuous employment history.

2.1 Model Programmes for Women in Mid-life

A number of West German programmes are specifically aimed at women in mid-life, aiming to offer them the chance to reconsider their previous biography orientations as well as improving their chances of re-entry into the labour market. Evaluating these measures centered on women's lives and needs enables us to formulate broader policy recommendations relating to education and to the labour market.

At the same time, recognition of the situation of women in mid-life breathes new life into the old educational principle that practical experience is profoundly educational. It is confirmed that the life experience of these women has provided them with what can be described as extra-functional skills such as responsibility, flexibility, prudence, and skills in planning and in cooperation. These correspond with those generalized, social skills called for in modern high-technology jobs.

We have recently completed the evaluation research on a programme for "Further education and reintegration of women". Since 1984 this programme has provided orientation and work-field related further education to un-employed housewives who have interrupted their careers for family reasons. In 1987 half of the 40 courses offered were motivation and orientation courses. The majority of work-field related courses were in office-work. In addition there was a refresher course for nurses, preparation for work as an assistant home nurse, one preparatory course for training as a certified housekeeper and two courses for women considering starting an independent business. This model programme has made the interruption of employment in the normal female biography an issue in educational planning irrespective of labour market demand.

The model programme under study breaks with the tradition of sex-neutral further training. On the one hand, it takes into account the special situation of unemployed housewives and allows for their family responsibilities. On the other hand, the attempt is made to meet women's needs for a second stage of vocational development. These are the innovative impulses of the model programme conception.

Our results have so far shown:

* Participation women are highly motivated to seek employment and further education. - More than 80% of participating women expressed their wish to start work or a vocational further training.
* The self-reflective learning processes initiated by female instructors are based on the principles of identify learning and of a feminist political orientation. During the learning process the typical devaluation of women's family work as a mere career gap is made an explicit issue in order to stimulate women to revaluate their family phase.
* The return to work in mid-life is discussed as a possibility for a second career orientation phase: Assuming a working-life of up to 60 years of age, mid-life women have on the average another 20 years of work ahead of them.
* However, the interests of many women in a second chance in their former vocation encounters a very limited range of refresher courses inside and outside public funded measures.
* With regard to their interest in a certified vocational qualification women returning to work are confronted by only a limited offer of vacancies by the labour market administration. The only alternative in the FRG are the primary occupational training programmes of two or three years duration; not a few of these have age limits.
* The right to further education that is adult-oriented and also recognizes the numerous skills of mothers may best be realized in less formalized further education. Since those courses, however, aim at fields of work which have hardly been developed as occupational fields, their utility on the labour market remains uncertain.
* Through their participation in model programmes women seeking reentry to employment begin to reshape their life and career. However, restrictions in the labour market as well as the organization and structure of the system of occupational education in the FRG severely impair their changes of compensating for the consequences of employment interruption.

3. RECURRENT EDUCATION - WITH REGARD TO EDUCATIONAL PREFERENCES AND PRECONDITIONS OF WOMEN

In the 70's the programmatic reform concept of "recurrent education" was, for the most part, discussed sex-neutrally. Its reference frame, however, of "increasing equal opportunity" has set a number of objectives and reform proposals which today can (and ought to) be transferred to the problem of the "equality of the sexes". For example, one major claim of recurrent education, the "reallocation of education phases", implies rejection of the male norm of continuous employment.

Newly allocated education and employment phases as such will not necessarily pave the way for the equality of sexes. This - rather formal - aspect would have to be accompanied by a qualitative improvement of women's position in the labour market. Here, too, demands and objectives derived from the recurrent education discussion could be drawn upon. In the following these ideas will be exemplary introduced and applied to the situation of women and their further education needs.

(1) Reallocation of Education Phases

"Recurrent education" as an educational policy approach would dissolve the compact life stage of initial education before employment, replacing it by intervals of learning alternating with practice or employment phases. The aim is to "reallocate education phases" throughout the whole life. This emplies creating institutionalized and socially desirable interruptions of full - time employment. The dominant patriarchal labour market norm of full - time and continuous employment would "en passant" be ruptured

Applied to the situation of women "a periodical alternation of education and practice" (this is the OECD translation of the term "recurrent education") might make it less difficult to take on productive responsibilities while maintaining career orientation. This would be particularly the case if reallocation of education phases successfully avoided the present overlap of entry into (especially qualified) employment with the generative phase of women (cf. Edding 1985:170). A further reduction of conflict between career and family interests would result if the "family phase" itself could be used for further education activities (as already suggested by Myrdal/Klein or Mushkin 30 years ago).

As soon as a (family-related) career interruption is no longer equivalent to stagnation or even occupational dequalification, the burden of decision would be taken off women. The option of a family-related career interruption could, at the same time, also be an attractive option for fathers if (further) vocational qualification could be pursued during the family phase.

(2) Curriculum Development

A second central objective of the recurrent education conception is the development of an adult-oriented and practice-related curriculum that should also include dimensions of personal enrichment and political awareness as well as qualification. The demand for more practical relevance in further training was initially raised with a view to improving efficiency. The motivation to learn is higher if "learning comes close to the experience of vocational practice" (Schmitz 180:287f.). Nevertheless, there is more to it than merely attaining a more efficient reorganization of vocational training. Friedrich Edding, who initiated the development of the conception of recurrent education, also stresses its implications for personality development as well as social andpolitical consciousness).

The demand for an intregration of vocational personal and political education suggests, in our opinion, a link to women's education. There the demand that learning refer to both life and work, as we have noted, takes as central a position as the demand for reflective, emancipator learning processes.

Model programme courses, e.g. for "Women in mid-life" satisfy by their target-group definition the "old" demand from the recurrent education discussion, that further education be adult- and participant-oriented. Moreover, specific problems such as the distance to technology that women and girls owe to their socialization is both given particular didactic consideration and at the same time explicitly discussed as topic within courses.

(3) Optimizing the Career Decision

A central objective of recurrent education is to optimize the decision for a vocational training, for instance by providing a "better choice of career" by additional chances in the course of time (cf. Bodenhofer 1976:304). Such "additional chances of training" in later life phases were sought bya considerable number of women in our study. The objective of a "better choice of career" proved especially important for those women who, during an orientation course, became aware of their nonvoluntary "choice" of career and who wished to revise an earlier wrong decision through further training. As we have pointed out above, the conventional further education system does inadequate justice to such needs. Recurrent education could, however, offer a solution. This conception aspires to optimize occupational decisions by way of creating options for repeated and extended education steps open to revision. Such small steps correspond to the orientation needs of unemployed housewives and would facilitate their "return to work". After five or ten years employment interruption and the resulting distance to their occupation, and with continuing responsibilities in the family, the women in the model programmes under study often approached their reentry to employment in small steps; they appreciated the absence of pressure and the variety of personal decisions accepted in the courses.

The option of small and repeated steps that could be revised might not only support reentry to the former occupation after a long pause in employment. It could also facilitate a transfer or entry into typical fields of employment. The conditions of women's employment could thus be improved, i.e. today's women's employment could diminish in importance.

(4) Adoption of Qualifications
Along with the goal of an "optimal career decision", "adoption of qualifications" (underlined in conceptions of recurrent education) is also applicable in an especially high degree to the further education claims to respond to the increasing and rapidly chancing requirements of the labour market by alternating phases of learning (cf. Smitz 1980:287). Given that women's jobs have been especially hard-hit by the introduction of new information and communication technologies, the adoption of qualifications for women to this development should have high priority.

The recurrent education objective, which considered the adoption of qualifications as a public responsibility, was met in the 70's by the objection that such adoption was a "traditional element of the work task itself" and ought to be provided, for the most part, by on-the-job-training or by short-term external training measures (cf. Bodenhöfer 1976:306). This argument - expression of the sex-neutral discussion of the 70's - ignores the chronic underrepresentation of women in such further training measures. After all, participation in such measures presupposes a high degree of integration into the employment system, something that women who enrol in courses following the family phase, precisely want to achieve in the first place. This has been shown convincingly both by labour market research and by specific studies on women. That is why it is necessary to develop alternatives to the selective public and company-provided training in order to put an end to the marginal position of women in the labour market. This does not apply only to women after their family phase. In view of the speed of diffusion and development of new technologies, the necessity of further training is for the most part incontestable. However, the conception of recurrent education differs from the usual pattern of further education, which promotes especially employed men with middle- and higher-level positions and qualifications by its egalitarian claim to open such late learning phases to everyone, whether male or female.

4. HIERARCHIC OR FUNCTIONAL DIVISION OF LABOUR? FUTURE EFFECTS OF NEW TECHNOLOGIES ON ENTERPRISES

An organizational revolution based on new technologies is presently taking place. Experts have coined the phrase 'systemic rationalization'. This refers to different and new conceptions of organization which mark a transition from the purely technical and hierarchical division of labour traditionally predominant in large scale industrial manufacture to a more functional division of labour. The large industrial enterprises are in many respects considered to be models or trendsetters. In the most progressive cases - but as of yet only these organization forms close to a functional division of labour - have already emerged. However, it must be remarked that our discussion is restricted to a few selected developmental aspects; one cannot yet speak of trends materialized at plant level on a broader basis.

What could be the effects of such a new functional division of labour? It could lead to a greater autonomy of workers, to a different definition of workplaces, a new deployment of labour, transformed personnel structure. For instance, a more homogeneous work and personnel structure aimed at reciprocal replacement of staff could lead to other power structures and social relations than those in a hierarchical organization. Such trends have been observed in some pilot companies; in the future they could well apply to a majority of workers and, most important, not only to males.

It is known that the use of new technologies requires from workers more abstract, systematic thinking, a more intelligent mode of work or production, faster reaction capacity and on the whole a higher level of responsibility and readiness to accept this.

Education, training and further education policies are attempting to meet this by a wide range of measures. In the West German industrial sector, the general qualification requirements thus described in the abstract have already influenced the reorganization of recognized industrial training in the field of metal and electronically occupations. It took altogether 14 years before a reform of these training programmes was achieved in 1986. Because such reform processes in the FRG depend on negotiations between collective bargainers and the government, they are traditionally extremely lengthy. At present, occupations in the service sector, like office work, for example are undergoing reorganization. In our view, this reform or reorganization process should not only be regarded as an administrative act of adoption; at the same time it is an attempt to determine the contents of future-oriented training. Under the direction of the government, attempts are being made to respond to the requirements set by new technologies.

Another level of response primarily linked to technical-organization policy at plant level can be identified by asking the following questions:

* In what way will work organization be altered and shaped?
* How will the power structure change?
* What changes will take place in subcultures and mentalities at the workplace? Can these change the discrimination of women which until now has characterized personnel deployment policies?

To put it more generally: How can the modernization processes taking place be used for improving the position of women on the job? For at present it appears incontestable that reforms within companies do, at the same time, change their power structures. Consequent the question arises how women are affected in this process of change and, most important of all, how they can actively influence it?

4.1. Situation of Technology

It is today possible to list some significant general conditions for the design of technology, organization and labour for company level in the next ten years:

(1) The technological supply of the 90,s is basically known (microelectronics, laser technology, biotechnology etc.). Trends in technology application that are presently

recognizable will continue (numerical controlled machinery and plants, complete information and control systems, computer linkage of manufactures, users, sales etc.). It can be assumed that the realization of innovations takes approximately 10-20 years.

(2) The present macroeconomic framework will continue to exist for companies: increasing world-wide division of labour and growing international competition, pressure for permanent productivity increase, no return to full employment,altered relations between economy and ecology (newly developed tertiary services etc.).

(3) The question is how labour demand will change. Many are of the opinion that demand will grow for formally high and medium qualifications and diminish for the unskilled and semi-skilled force?

(4) Future problems, such as key areas of humanization of work, qualification, new stress and strain, participation, flexibility etc. can be identified and solved, provided that well-founded assumptions are the starting-point with regard to trends in the company's rationalization policy, personnel management, implementation of technology etc.

New developments in technology, especially in microelectronics, have loosened the ties between the technical structure of production processes and human labour. The tendency towards flexible reorganization of production processes seems to suggest a requalification of personnel on a broader basis. The greater distance of human labour to the technical process opens up the theoretical possibilities of alternatives, which would also mean more qualified forms of human labour application. Management, however, often claims that qualification gaps impede promising technological innovations. This presumption refers mainly to technicians, engineers and to a certain extent to skilled workers.

Another problem in labour policy is raised by telecomputing, which is technically feasible but far from being applied on a large scale. The possibility of decentralizing white-collar-jobs by means of new information and communication technologies, which would transfer them from offices and administrations into homes or satellite or neighbourhood offices, has been widely discussed in the FRG as elsewhere for the last two or three years. This has diverted attention away from a third trend, namely the development in the so-called tertiary sector, especially in the area of clerical-administrative jobs. The application of modern electronic data processing systems based on data banks and corresponding systems programmes facilitates:

* the systematic storage and examination of extensive quantities of data,
* a more or less direct access to data required by EDP users for information purposes,
* the automatic implementation of certain calculation and typing processes following a given impulse,
* whole operation procedures in direct contact with the computer which immediately processes the data input.

5. PRELIMINARY CONCLUSIONS

The trend of work organization on a technical basis will probably be within the scope of the above summarized lines of development. These suggest a chancing organizational structure on plant level. Our thesis is that the position of women in companies can be changed as well, along with this process. For these changes that the vision between heavy physical work and other work is becoming obsolete, i.e. that the historically evolved dividing lines between men's and women's work, also reflected in the protective labour legislation, may gradually become a thing of the past.

Moreover, it is probable that the future development of qualifications cannot be pressed into the simple patterns of qualified versus unskilled work. We now encounter different, new qualification components which cannot be ascribed in a simple way to one or the other sex and which do not carry over directly the patterns of operation from existing occupations and workplaces. As we are at the moment confronted with a multitude of hypotheses, or rather a mix-up of hypotheses, that all try to interpret qualification developments, this confusion is not merely a lack of knowledge and interpretation, but it also indicates a lack of action by companies, and it ought to be used for improving the position of women.

REFERENCES

Bermann, T., Not only Windmills. Female Service Workers and New Technologies, in: Women, Work and Computerization, A. Olerup, L. Schneider, E. Monod (ed.), North-Holland, Amsterdam 1985, p. 231-248

Bodenhöfer, H.-J., Ökonomische Aspekte des wiederholten Lernens. Kommentar zum Beitrag von Friedrich Edding (Economic Aspects of Recurrent Education. - Comment on Friedrich Edding's Essay), in: Zeitschrift Für Wirtschafts- und Sozialwissenschaften (19. Jg.) H.4, S.303-313

Clement, W., Edding, F., Recurrent Education und Berufliche Flexibilitätsforschung (Recurrent Education and Research on Vocational Flexibility), Schriften des Vereins für Sozialpolitik, Neue Folge, Bd. 103, Berlin 1979

Edding, F. Ökonomie des Bildungswesens (The economy of the Educational System), Freiburg 1963

Edding, F., Über die Verbindung von Berufsarbeit und Lernen in einem alternativen Studium (Linking Vocational Practice and Learning - an Alternative Model of tertiary Education), in: Krise der Arbeitsgesellschaft - Zukunft der Weiterbildung, E. Schultz (Hg.), Frankfurt/M. 1985, S. 167-176

Frauen am Computer. Was die neuen Technologien den Frauen bringen. Eine Einführung (Women and Computers. Advantages and Disadvantages of New Technologies for Women. An Introduction), E. Vogelheim (Hg.), Reinbek bei Hamburg 1984

Gensior, S., Lappe, L., Arbeitsmarkt und Frauenerwerbsarbeit (women in the Labour Market), in: Arbeitskreis Sozialwissenschaftliche Arbeitsmarktforschung (SAMF), Arbeitspapier 1983-2

Gensior, S., Moderne Frauenarbeit, Arbeitszeitflexibilisierung und elektronische Heimarbeit. Der frauenspezifische Arbeitsmarkt (Modern Female Employment, Flexibility in Working Time and Electronic Home-Work.The Female Labour Market), in: Karriere oder Kochtopf? Frauen zwischen Beruf und Familie (Career or Cooking-pot? Women between Employment and Family, Jahrbuch für Sozialökonomie und Gesellschaftstheorie, Hochschule für Wirtschaft und Politik Hamburg, Westdeutscher Verlag, Opladen 1984, S.58-75

Gensior, S., New Technologies - Possibilities for a New Valuation of Women's Work, in: Women, Work and Computerization, A. Olerup, L. Schneider , E. Monod (ed.), North-Holland, Amsterdam 1985, pp. 171-180

Gensior, S., Neue Technologien und Weiterbildung in Büroberufen (New Technologies and Further Training in Clerical Work. Report on an Research Project), Europäisches Zentrum für die Förderung der Berufsbildung - CEDEFOP - (Hg.), in französischer, italienischer und deutscher Sprache, Projektbericht von S. Gensior, B. Schöler - unter Mitarbeit von S. Knuth, Berlin 1986

Gensior, S., Better or Worse? - The Development of Work in the Coming Decade in the Utilization of New Technologies, in: k. Fuchs-Kittowski, D. Gertenbach (ed.), Proceedings of the IFIP TC9/WG9.1 Conference on "System Design for Human Development and Productivity: Participation and beyond", Akademie der Wissenschaften der DDR, Berlin/DDr 1987, pp.44-99

Gensior, S., Teilzeitarbeit und frauenspezifischer Arbeitsmarkt. - Zur 'Interessenidentität' von Frauen und betrieblicher Personalpolitik (Part-Time Employment, and Female Labour Market. Women between Economic Strains, House-Work, and Restrictive Personal Policy), in: Rechtsalltag von Frauen. U. Gerhard, J. Limbach (Hg.), Frankfurt/M. 1988, S.61-75

Goodmann, S.E., Perby, M.-L., Computerization, A. Olerup, L. Schneider, E. Monod (eds.), North-Holland, Amsterdam 1985, pp.43-51

Lappe, L., The Advance of Microelectronics in Assembly Operations. New Opportunities for Working Women, in: Women, Work and Computerization, A. Olerup, L. Schneider, E. Monod (eds.), North-Holland Amsterdam 1985, pp. 161-169

Lappe, L., Leichte Arbeit - schwerwiegende Folgen. Frauenarbeitsplätze in der Industrie (Easy Work - Far Reaching Consequences. Workplaces for Women in Industry), in: G. Elsner (Hg.), Was uns kaputt macht, Hamburg 1984, S.61-66

Lappe, L., Die Arbeitssituation erwerbstätiger Frauen. Geschlechts spezifische Arbeitsmarktsegmentation und ihre Folgen (The Working Conditions of Employed Women. Gender, Labour Market Segmentation and its Consequences), Frankfurt/M. 1981

Mushkin, S. (ed.), Recurrent Education, Washington 1974

Myrdal, A., Klein, V., Die Doppelrolle der Frau in Familie und Beruf, 3. Aufl.,Köln/Berlin 1971

OECD, Recurrent Education: A Strategy for Lifelong Learning, Paris 1973

OECD, Recurrent Education - Trends and Policies, Paris 1975

Schmitz, E., "Recurrent Education", "Lifelong Learning", "Education Permanente", in: Wörterbuch der Weiterbildung, G. Dahm, R. Gerhard, G. Graessner et al. (Hg.), München 1980, S.285-289

Wagner, I., Women in the Automated Office. Contradictory Experiences - Individual and Collective Coping Strategies, in: Women, Work and Computerization, A. Olerup, L. Schneider, E. Monod (eds.), North-Holland, Amsterdam 1985, pp.53-64

Wagner, I., Zwischen Erfolg und Selbstbeschränkung. Arbeitserfahrungen weiblicher Lehrlinge und junger Facharbeiterinnen in gewerblich-technischen Berufen aus der Perspektive der moralischen Urteils- und konsensfähigkeit (Success and Self-limitation. Analyzing the Work Experience of Female Apprenticeships and young

Female Skilled Workers from the Perspective of Moral Judgement and Consensus), in: Soziale Welt (35. Jg.) H. 1/2 1984, S.215-235

BIOGRAPHICAL NOTE

GENSIOR, Sabine, Social Scientist, research on labour market problems and policy, development of work organization and technology, and on women's labour market problems and working conditions. During 1977-1988 (1977 to 1981 and 1985 to 1988) researcher at theScience Center Berlin (WZB). Member of the executive committee and manager of Berliner Institut für Sozialforschung und sozialwissenschaftliche Praxis e.V.; Professor at Fachhochschule für Wirtschaft Berlin for 'Technological Development, Conditions of Work, and Employment Opportunities - in Special Consideration of Working Women. Member of the Executive Committee of the German Association of Industrial Sociology and Sociology of Work.
HILDEGARD SCHICKE, born in 1958. She achieved her diploma of pedagogics in 1985. Since 1981 she has worked in further education in politics. From 1987 to 1988 she has worked as a scientific research assistant in the evaluation research project 'Further education for women in mid-life' at the Berliner Institut für Sozialforschung und sozialwissenschaftliche Praxis e.V. (B.I.S.).
LYDIA HOHENBERGER, born in 1959. She achieved her diploma of political science in 1984. Since 1985 she has been employed based on a fee in various scientific research projects. From 1987 to 1988 she has worked as scientific research assistant in the evaluation research project 'Further education for women in mid-life' at the Berliner Institut für Sozialforschung und Socialwissenschaftliche Praxis e.V.
Corresopndece address: B.I.S., Berliner Institut für Sozialforschung und Socialwissenschaftliche Praxis e.V., Pfalburger Str. 72, 1000 Berlin 15, FRG

Women, Work and Computerization: Forming New Alliances
K. Tijdens, M. Jennings, I. Wagner, and M. Weggelaar (Editors)
Elsevier Science Publishers B.V. (North-Holland)

PART 4

EMPOWERING STRATEGIES FOR WOMEN

Kea Tijdens
University of Amsterdam, The Netherlands

INTRODUCTION

In order to develop strategies for the women's movement different actors have to be identified and their strategies need to be studied. Important actors in the field of women's work are employers, the state, female employees, male employees, trade unions, etc.. In this part of the book papers are included about the strategies women have (see Tuender, Manacorda, Osimo and Defever and Tijdens), the strategies men have (as Cockburn studied) or the strategies management has (as Tremblay studied). Vast issues are at stake from different actors: main interests of employers seem to concentrate around low wages of women, main interest of male employees seem to concentrate on excluding women from skilled jobs, main interests of the state concern the key role of the family within society, main interests of trade unions are to organize powerful groups of workers. The conspiracy idea of men, whatever position they have, occupy key positions in order to consolidate power needs some readjustments in the field of computerization and technological change, because there might be different and even conflicting interests. The strategies of different actors should be related to the strategies of women and women's groups.

The strategies of employers towards the internal labour market is discussed by Diane Tremblay. The internal labour market is changing due to management strategies modeling the outcome of computerization and technological change, in terms of women's work among other aspects. She distinguishes two broad categories of employers' human resources management strategies: one which aims at the flexibilization of the employment system mainly through the flexibilization of work; the other which tends to construct employment stability through the development of internal labour markets. The developments on the internal labour market seem a useful starting point to analyse the employers' strategies. The author argues that the outcome of technological changes is not yet determined, and there is room for alternatives.

Affirmative action in the computer field has not been very successful yet, apart from creating favourable conditions for training and education as Gien Tuender-de Haan argues on the basis of a small-scale study of the impact of office automation on job segregation in a clerk's office in a municipality. Only in individual cases successes have been reached: women who manage to enrich their own career-paths. This suggests that women more often have an individual strategy than a collective strategy, although affirmative action presupposes a collective strategy of women and their representatives. Tuender suggests that with more support and stimulation from the organization others could become interested as well. She brings up a very interesting point: an affirmative action strategy for user groups could be helpful to incorporate women in the process of system designing.

Two contributions about affirmative action came from Italy. An Equal Opportunity Information System is described by Paola M. Manacorda and Clelia Bertello. The system was designed for the purpose of a course to equal opportunity advisers. The system should provide methodological knowledge for creating, collecting, storing and interpreting information concerning items as recruitment, training figures, career structures, etc. The system is applicated for an IBM-compatible PC.
Bruna Osimo describes the introduction of affirmative action in a telecommunication company in Italy. A research was made in two departments of the company: a white and a blue collar area, in both half of the workforce is female and in both new technology is introduced. It appeared that women could not influence new technology. The activity of the group was agreed upon an internal agreement between the personnel management and the trade unions representatives. The aim was to focus the attention on discrimination practices towards women in the company, to find out their origins and to define a series of proposals of affirmative actions to redress unbalances on the workplace.

Marleen Defever and Kea Tijdens describe the new technology in the local branches of one of the biggest Dutch banks. Female employees, the works' council and the women committee use several strategies concerning new technology. The authors distinguish three levels on which strategies are concentrated at and items are discussed. Courses are judged as the key and one of the main items in the strategies and struggles of female workers concerns the power and knowledge of computerized systems. Other important items employment, monitoring, job rotation and positive action. The tension within the labour process between taylorism and flexibility seems to give a good perspective for strategies for improving jobs of female workers.

WORKSHOP DISCUSSIONS

On affirmative action many interesting points were discussed: The definition of affirmative action should not only include career paths for individuals, but especially job improvement for all, including the formal recognition of the diversity of women's work. A central theme in affirmative action must be the creation of opportunities for women to define their own skills. A bottom-up strategy supported by formal policies is central for affirmative action strategy. Coalitions should be made between different interest-groups, like trade unions, management, women's groups, researchers, but different roles and conflicts in the organization should be recognized. The need to bare strategy on the specific characteristics of the organization should be recognized. And automation can be seen as a back door approach for affirmative action.

In the workshop about the banking sector an interesting discussion was held about banking systems and the position of women in different countries, especially Canada, The Netherlands, UK and Malaysia. The role of technology was discussed. Looking at the impact of technology does not allow options for a process of choice, it is a deterministic view. Another point of discussion was the question what is so special abut technology and women's work, because many of the same issues arise about access to work, wage differentials and occupational segregation. Time of rapid technological change is time of destructuring of the labour market - therefore, a time of influencing changes. The internal labour market offers a framework for understanding the position of women within banking. The advantage of this is that it highlights the active role of firms, employers and management. Thus, there is another possibility of influence for women. It is not just the state on the socialization process, which need to be forced to change. Also, the role of firms in reproducing socialization of women and men needs to be examined. With both technological and organizational change, the labour market is more fluid - there are new entry ports and new job opportunities. In some countries it is women who are seen as most flexible by firms seeking flexibility in their work force; in other countries it is young or old people or different ethnic groups.

Within the strategies of women and women's groups the goals need to be described exactly and levels of influence and most appropriate forms need to be identified precisely. The role of autonomous women's groups and their relation to trade unions needs to be discussed: if trade unions are the only body recognized by employers, it is important for women to work with them. The importance of having autonomous women's groups is that they can formulate collective demands, develop self-confidence and that they are easier and more attractive for women to participate. On different levels of management can be negotiated different and it is important to remember that management is not homogeneous: conflicts within it can be exploited. Division of tasks can sometimes be negotiated with immediate colleagues, while pay rationally can be negotiated within the industry, depending on industrial relations structure within the country. Conflicts of interest can arise between different groups, for example between male and female trade unionists or between different groups of women.

Women, Work and Computerization: Forming New Alliances
K. Tijdens, M. Jennings, I. Wagner, and M. Weggelaar (Editors)
Elsevier Science Publishers B.V. (North-Holland)

SEXUAL DIVISION OF TECHNOLOGY: THE SAME AGAIN OR DIFFERENT?

Cynthia Cockburn
City University
London EC1, Great Britain

The women's movement of the latter behalf of this century has been characterized by a feature missing from earlier phases: a demand for access for women to technological knowledge, skills and jobs. Indeed this appetite on the part of women intensified in the 1980s while some other aspects of feminism were in decline. In Britain there have been many practical women's training initiatives designed to bread men's all-but-exclusive grip on the building trades, engineering and information technology. In parallel, more academic feminists in Europe and North America have attempted to theorize the relationship of man to machine and to trace its history (Schwarz-Cowan 1979; Rothschild 1983; Zimmerman 1983; Faulkner and Arnold 1985).

When approaching the issue of "women's absence from technology" it is always necessary to clear out of the way some conceptual problems. First, what precisely is meant by technology? In one interpretation women have been as involved with technology as men. We have undoubtedly invented and developed many technologies, from the primitive digging- stick to advanced culinary method. Not all technology is hardware, and the term covers more than it is popularly used to include. As Maureen McNeil has pointed out, the widespread propaganda for men as technologist should not lead us to assume that what women do and what we know is not technological (Mc Neil 1987).

Second, there is the problem of women's invisibility in a history mainly written by men. We should not assume just because Ada Lovelace is the only woman who features in a global history of the machine that women really were inactive. Autumn Stanley's encyclopedic rediscovery of women inventors is a valuable corrective in this sense (Stanley, in progress).

Third, there is the issue of whether women are really right to hanker after access to those technologies that repose in men's hands. The labour of a high proportion of the technically-qualified workforce is devoted to the military; to weapons production; nuclear power; and the manufacture of harmful products. With this perception of course feminists diverge from Marx's unquestioning assumption of 'technological progress'.

I would argue however that even after acknowledging such cautions there does remain some validity in the common sense view of many women that men have appropriated a certain technological sphere, that this is not new, and that there is a politically significant issue of power involved. It is after all difficult to dismiss as misguided the aspirations of young women to become engineers and technicians and to disregard what they say about the resistance they encounter from men.

I believe a helpful approach to the history of technology, skirting some of the pitfalls noted above, is to differentiate technologies in a hierarchy of production. In particular we should differentiate the use of tools from the making of tools, and, in turn, the making of tools from the making of tools-for-tool-fabrication. I would argue that all technology, from that of hunting or gathering to that of interplanetary travel, has to be seen as a medium of power and the higher the level of the technology in the hierarchy of production the more the power it affords to the group that has tenure of the skills involved.

Technology of course does not conceptually supplant ownership of the means of production as the ultimate source of power. Rather it significantly mediates the possibilities inherent in that ownership. Those who lay claim to the means of production, from slave-owning emperors to centrally-planned nation states, all depend for the making of wealth on a yoking in tandem of labour and machinery. They may be expected to pay well in cash or food, freedom or status, for the skills needed to effect this linkage and improve its productivity.

It is common sense to suppose that technology, as a medium of power, will be used in any system of dominance to further the interests of any dominant group. Historical evidence suggests that, though things may have been different in the neolithic, from the bronze age the human race developed, through successive modes of production, as a patriarchy. It is not surprising then that in a where men as a sex were in possession of all other vehicles of power from state organization to marriage, they were also possessed of the knowledge and skills con trolling the most powerful technologies. The 'mighty five' - the lever, the wedge, screw, wheel and inclined plane - that made it possible to move mountains and build pyramids were the technical armoury of the male sex. And the metallurgical knowledge and skills to produce those and other implements have been highly significant.

It will be clear that the skills that gave some men authority among other men of the manual working classes also enhanced the power over women of men as a sex. Not only were women firmly subordinated to men in the patriarchal family, they were also dependent on them for many important practical processes of everyday life. Certain technological skills, defined as male property, were therefore both a cause and an effect of male supremacy. As Paola Tabet puts it, 'le monopole de certaines activités clés est nécessaire aux hommes pour s'assurer le contrôle des instruments de production et, finalement, l'utilisation globale des femmes' (Tabet 1979).

It is something of this that women experience today when we sense ourselves enfeebled by our dependence on men not only to keep function ing our workplace technologies (from the industrial sewing machine to the word processor) but also to keep our dish-washer running, our car on the road, and our roof from letting in the rain. Men's exclusive grasp of certain kinds of knowledge combines with other media of power - economic, legal, social - to reproduce male supremacy. And that supremacy conspires in turn, in a dialectical process, to keep women's technologies - such as our historically-given knowledge of health, nutrition and psychology - from becoming significant or powerful.

Production of course made a quantum leap with the advent of capitalism - an entirely new set of social relations that was to find the organizational means to bring together science and technology, harnessing the two for production. The history of the early stages capitalism is too familiar to need retracing here. But what is significant for us as women and for our understanding of technology and power is the change capitalism signified for the relationship between producers and 'their' technological know-how. The craftsman had owned his own tools. This included the tools owned by those men who made the tools that other producers used. As Marx put it, "the labourer and his means of production remained closely united, like the snail with its shell" (Marx, 1887). Now the snail had to be prised out of its shell if capitalism was to fulfil its potential: in the new factories the employer owned the instruments of labour and put the worker to work on them. All the labourer possessed to sell was his or her own labour power - and the level of specialized technical know-how inherent in that labour power was crucial.

The tools too were altered. At first they were simplified and multiplied to suit the detail tasks of the new rationalized production process. Then individual tools were combined into machines and, further, into machine-like factory systems with vast new possibilities of accumulation for those that owned them. Women had long used tools. But in the period preceding the industrial revolution it had been men who on the whole made tools. Industrialization therefore offered men as a sex opportunities that were not accessible to women, for the key technologies involved were those of making and maintaining tool-making tools and ultimately machine-making machines.

Often, as David Noble has pointed out, two visions - that of the employer and that of the craft engineer - "took shape as one in a single mind: the capitalist was frequently an inventor of sorts while the inventive craftsman shared not a little of the entrepreneurial spirit of the capitalist" (Noble 1979). But the capitalist and the skilled man did not need each other equally. The capitalist was always ready to turn the skilled man's inventions against him and the latter's trade union in turn responded by demarcating and defending areas of competence. Instrumental in bringing about the subdivision of work, the deskilling of traditional craft engineers and the substitution of fixed capital for labour, was a new breed of formally-educated engineer. Electrical, chemical and other specialized engineers came into existence as the 19th century advanced and secured professional status.

Women for their part worked in repetitive, super-exploited and often heavy occupations in the new capitalist industrial system. Frequently they brought to the work skills that were not recognized or paid for. But their relationship to tools and machinery was almost always subject to a regime of knowledge and control enforced by and through, as well as for, men - from the maintenance engineer to the boss.

What we can see then is not only a situation in which women were excluded by organized technical working men as a potential source of cheap substitute labour, but a system patriarchal to the core - in the extended sense of a system of hierarchical and conflict-ridden relations among and between men.

And this brings me to the present - to electronics, the computer and information technology. Given such a history it is perhaps surprising that, in the 1970s, in the flush of sex equality legislation,one heard it so often and optimistically stated: "now, with the electronic revolution, you'll see - women will become engineers". What is not surprising is that by 1984 it was felt necessary to launch a national campaign in Britain to bring about by positive action what was clearly not occurring by force of history: "Women Into Science and Engineering", or "WISE Year". Neither is it surprising that WISE Year had no very notable success.

For both the original optimism and the more teeth-gritting determination of WISE Year were based on a series of misconceptions. First, they assumed the reason women had not been engineers in the past was such physical constraints as dirt and danger, no longer applicable in the electronic era. Second, they assumed that what needed changing was women's consciousness. "If only we could encourage more girls to take it up". "If only they could see how much more they would earn as engineers than as hairdressers". But young women were wiser than the WISE-promoters. They knew very well their presence in engineering would be unwelcome to most men and were aware of the sanctions that would be employed against them should they try to enter.

Third, the initiatives ignored the frank misogyny and the will to sexual domination that lay behind the cold-shouldering of women by trade unions, professional institutions and colleges, and workplace colleagues. Finally, even if some individual men were to respond positively, men as a sex were tied up in a fiercely competitive hierarchy of skills and occupations and in a deteriorating bargaining position with regard to the employer. Their own survival in the world of men depended on their holding onto and renewing in ever-updated forms their technical qualification. It is every bit as important to men in the 1980s as it was in the 1890s to limit competition for jobs by excluding women, young people and ethnic minorities from the white man's labour market.

What has changed however is the social conjuncture. Equality and anti- discrimination laws have made outright exclusionary practices by unions or employers illegal. Although few in number, some women are pushing forward through school exams and college courses into the technical job market. What my own research has shown (Cockburn 1983, 85 and 87) and that of other researchers too is showing (Berner 1983; Chabaud-Rychter et al 1987); Flament 1987; Rudolph 1987) is that the reproduction of male control of key technologies is today being secured predominantly by cultural means.

'Culture' may sound soft and malleable when set against law and economics but the deployment by men of particular forms of social interaction with each other and with women, and of immensely flexible and effective ideologies in regard to technology, is proving a powerful mechanism for marginalizing and alienating women, for causing them to 'fail' and to withdraw from competition with men for the more skilled, interesting and well-paid occupations. Deeply implicated in the social relations of technology is heterosexuality. Girls' upbringing leads them to experience a need to be desirable sexual and domestic partners of men. It leads them to respond to the socially-sanctioned imperative of complementarity: "what he is I must not be; what I am he is not".

The implication of such an analysis is that, while gaining technological access for women is important, there are two things more important still. One is to develop and maintain women's organizational strength through feminism. The other is to transform the social relations, the products and the processes of technology as we get to grips with them. The three things together however make a formidable and inspiring project for us as women, as we invent a future different from our past.

REFERENCES

Berner, Boel (1983), University of Lund, Sweden: Women, power and ideology in technical education and work, paper to the International Conference on the Role of Women in the History of Science, Technology and Medicine in the 19th and 20th Centuries, Hungary.

Chabaud-Rychter, Danielle et al (1987): Division Sexuelle des Techniques et Qualification, Programme Interdisciplinaire de Recherche sur la Technologie, le Travail, l'Emploi et Les Modes de Vie, Centre National de la Recherche Scientifique, Paris.

Cockburn, Cynthia (1983): Brothers: Male Dominance and Technological Change, Pluto Press.

Cockburn, Cynthia (1985): Machinery of Dominance: Women, Men and Technical Know-how, Pluto Press.

Cockburn, Cynthia (1987): Two-Track Training: Sex Inequalities and the Youth Training Scheme, Macmillan.

Faulkner, Wendy and Arnold, Erik (eds) (1985): Smothered by Invention: Technology in Women's Lives, Pluto Press.

Flament, Catherine (1987): De la Technique comme Enjeu dans les Rapports de Sexe, paper to the Table Ronde sur Rapports Sociaux de Sexe: Problématiques, Méthodologie, Champs d'Analyse, Atelier Production Reproduction, CNRS, Paris.

McNeil, Maureen (ed) (1987): Gender and Expertise, Free Association Books.

Marx, Karl (1887/1954): Capital, Vol.1, Lawrence & Wishart.

Noble, David (1979): America By Design: Science, Technology and the Rise of Corporate Capitalism, Oxford University Press.

Rothschild, Joan (ed) (1983): Machnina ex Dea: Feminist Perspectives on Technology, Pergamon Press, USA.

Rudolph, Hedwig (1987): The Cultural Construction of 'Engineering' as Male Territory in the Federal Republic of Germany, paper to the Conference on Women and Engineering, Berlin, February.

Schwarz-Cowan, Ruth (1979): From Virginia Dare to Virginia Slims: Women and Technology in American Life, Technology and Culture, no. 20, January. Stanley, Autumn (forthcoming): Mothers of Invention, USA.

Tabet, Paola (1979): Les Mains, les Outils, les Armes, L'Homme, XIX (3- 4) July/December.

Zimmerman, Jan (ed) (1983): The Technological Woman: Interfacing with Tomorrow, Praeger.

Women, Work and Computerization: Forming New Alliances
K. Tijdens, M. Jennings, I. Wagner, and M. Weggelaar (Editors)
Elsevier Science Publishers B.V. (North-Holland)

STRATEGIES OF FEMALE WORKERS TO IMPROVE THEIR POSITIONS WITHIN THE PROCESS OF AUTOMATION IN BANKING

Marleen DEFEVER and Kea TIJDENS
Amsterdam, The Netherlands

Automation in the banks means the automation of clerical work. Low skilled and routinized jobs will change and partly disappear by this type of automation. Higher skilled desk employees will see their job changing fast. Because of occupational segregation mainly female workers do the clerical jobs in this field. They try to influence their labour position, either stopping deterioration or getting change for the better. Formal and informal strategies with more or less success are to be used, either by a group or by individual female workers. Co-operation, rotation of tasks, allocation of tasks or education are be goals of their efforts.
In this paper we describe the new technology in the local branches of one of the biggest Dutch banks. This new technology aims the automation of the labour processes concerning the retail banking, one of the biggest and most labour intensive processes within the bank. Clerical jobs as well as cash and desk jobs have changed already and will change in the near future. We describe several strategies concerning new technology as used by women and by the works' council and we distinguish three levels where items are discussed. Although several important items within the strategies exist, courses are judged as the key. Introducing automated equipment means that workers have to know something either about the application programs or the packets which are in use for the computer. Knowledge about this can improve one's position. In the local branches the entrace to different types and number of courses are narrowly related to salary levels and type of jobs. Because male and female workers are not equal spread about salary levels, there are differences between male and female workers concerning vocational training. We argue that one of the main items in the strategies of workers concerns the power and knowledge of computerized systems.The tension within the labour process between taylorism and flexibility seems to give a good perspective for strategies for improving jobs of female workers.
At the end of the paper we pay attention to the question whether positive action plan in the bank might improve women workers position.

1. THE BANK

One of the biggest Dutch commercial banks is ABN with about 700 offices all over the country and more than 16.000 workers in 1986. 45% of them is female and the staff is highly qualified. The districts within the bank advise and support the branches, which carry their own administration as well as their own profit and loss accounts. Branches support offices and agencies in that region. The personnel occupation varies from three in an agency to hundreds in a branch.

Sexual segregation in the bank is wide spread and it does not occur at random: women are over-represented in executive and supportive jobs, men are over-represented in managerial and commercial functions. As a result, men and women are not equally distributed among departments and locations: women are under-represented in the headquarters, in the computer departments, in local office management, in commercial departments and in credit- departments. On the other hand, they are over-represented in local offices and agencies, in counter and clerical departments and in the clerical centres. Female and male workers in the lower salary groups do about the same amount of VDU work. Men in higher salary groups do considerably less VDU work. Only women fulfil data-entry tasks and mainly women are employed in departments or offices where employment has diminished (Tijdens, 1986).

2. PAYMENT TRAFFIC

In the 1960s and 1970s retail banking has been followed by an explosion of offices, bank accounts and clerical work. Nowadays retail banking, especially payment traffic, is the most labour intensive production processes within the bank. Payment traffic includes giro money, clearing and cash transactions in the offices. It has been one of the earliest objects of automation. Daily mutations in giro accounts and daily debits were also automated. Recent automation in banking generally aims at ripping all transactions for booking as nearby the input source as possible. Bank management strives for having data typed at the counters, in the offices or by the clients in their homes. Those data are filled in in the central payments files directly, reducing female data entry jobs. This technology presupposes a data transfer network between the hundreds of offices and the computer centre. Within several years the bank will have changed from batch processes into real time processing for most of the services. The network between local branches and computer centres is a necessary infrastructure for the further automation of the payments traffic. For example, since a year cash dispensers are in use at some local branches and quite a lot of them will be in use within a few years.

Front-office and back-office automation has to be distinguished. Terminals on the counters of the offices are called front-office automation. They are used on behalf of:

* direct authorization in case of cash transactions;
* direct administrative processing of various counter transactions, like cash withdrawals and payments as well as foreign currency transactions;
* information retrieval based at client files;
* (sometimes) direct delivery of printed daily copies, containing an up-to-date bank balance.

In the field of back-office automation, workers in the clerical departments of the offices re-type hand-written payments orders. Nearly all of the workers in this field are female. Formerly, these activities were carried out in the clerical centres, which are disappearing. Millions of such payments orders are processed daily in the Dutch banks, but a decreasing part of them needs to be typed in.

3. AUTOMATION WITHIN THE BANK

In the first instance, ABN concentrated at back-office automation. In 1978, the development of the SON project started. SON is a re-typing system, meant for the data-entry of mutations in clients' deposits. The project is the continuation of an already existing system, that however only had 73 connections throughout the countrie, a number that was judged too small. SON was introduced in the local offices from 1982 on. After

some years, 200 offices were integrated in the network. The applications to follow can be found in the front-office field. Recently, the ABN management presented plans for the so-called Open Bank Project, aiming at the installation of counter terminals in the offices on behalf of authorization and data-entry of cash transactions as well as on behalf of client file data retrieval.
Four tasks (counter employee, cashier, authorization clerk and clerical employee) are unified in one person by the introduction of counter terminals: the employee behind the counter terminal. The clerical tasks and the authorization tasks decrease within these four tasks. At the counter only the transfers can be done for which applications are available in the computer system. Clients with other questions are referred to the client-adviser, the so-called division between desk and commercial activities. Experiences in another Dutch bank showed that the degree to which tasks are divided depends on the amount of tasks that were already divided before the introduction of counter terminals (Appeldorn, 1984). In small offices, where everyone does all executive tasks, the division of labour did not change after the introduction of the counter terminal. On the contrary, in bigger offices where tasks were already divided, the division increased after the introduction. Now, counter employees are expected to remit clients to other employees for advice, while formerly they gave the advice themselves.

The administration of the cashier's desk has become redundant, because of front-office automation. Tasks at the counter will diminish as well as clerical tasks done by counter employees. Cash dispensers also will decrease tasks at the counter. In the ABN the number of jobs in the offices will be reduced during five years at a pace of 250 yearly because of this Open Bank Project. Counterwork is done by young workers, approximately 70% of them are female. But, employment losses are less than was calculated in the planning phase. In fact, employment losses in the back-office is just as large as it is in the front-office, while it was expected to minimize back office work completely. More clerical work was left than expected by the system designers. One of the reasons might be that their analysis of the labour process was not adequate enough.

4. TYPOLOGY OF STRATEGIES

The female workers, who are touched by the front office automation, are positioned at the point of intersection of two types of strategies. The first type is the strategy of trade-unionists within the bank and of the members of the works' councils. In the bank four trade unions (a left-wing, two right-wing and a Christian) develop their activities. The aim of their strategy is to influence the automation and its consequences. Improvement of the position of female workers is not a specific goal of one of them. But since mainly women work in the front office, the strategies of these groups are relevant to the female workers at stake. The second type is the strategy of female workers, who are organized within their own committee. Their strategies concern explicitly the improvement of the position of female workers. Both types of strategies concern three levels:

* the shop floor, where the workers fight to influence the local management;
* the firm, where negotiations between management and work's council takes place as well as negotiations between management and shop stewards' committee;
* the industry branch level, where collective bargaining takes place between branch organizations and trade unions. On this level female committees of several banks have their own regular meetings and a women network for the banks exists.

Strategies of workers concerning automation are concentrated on specific items, on one or more levels. The main items in these strategies are described below.

5. PRESERVING EMPLOYMENT, A STRATEGY IN THE FIRM

A so-called Social Plan about automation was the result of negotiations between the bank and four trade unions in November 1986. In this Plan movements within the internal labour market are regulated: workers, who will loose their job because of automation, will get a proper other job within the bank. They have priority on the internal labour market. The impact of the Plan was that guarantied employment for all workers at the bank. Another part of the Plan was a salary guaranty for workers who would loose their job. The salary scale of the former jobs will be maintained up to a maximum of three years. The works' council is obliged to look after the execution of the plan. The subscibers of the Plan will discuss the execution yearly. The works' council is obliged to look after the execution of the plan. The signers of the Plan will discuss the execution yearly.

Although there is an formal employment guarantee in the Plan up to 1991, about a thousand jobs will disappear between now and 1991 by natural turnover. Most of the jobs will be clerical jobs, that is to say female jobs.

6. PRIVACY AND MONITORING, A STRATEGY IN THE FIRM

In the beginning the strategy of the works' council concerned technical aspects of automation. An important issue was monitoring labour productivity. The results should be used for the judgement about the workers. The works' council was terrible against this, but management was hard and firm. They declared that they would continue to register measurements of individual workers, but they promised to use the figures "carefully" and "nuanced". In fact the works' council had no real influence on the technical aspects of automation.

7. TRAINING, A STRATEGY IN THE LOCAL BRANCHES

The possibilities for training are not equal among different groups of clerical workers. In 1987 in a research 80 clerical workers at ABN were interviewed. 76% of them said that the entrance to training was very bad (Dijk, 1987). 44% said that the workers of the front-office had more possibilities than the workers of the back-office. 23% said that a selection had taken place on forehand: the workers with the highest qualifications got the best possibilities and the longest training courses.

In fact, the policy of the bank is to ignore the problems of the low qualified workers in the back-office, most of them women. They don't get any courses, they are not selected and if they are not brutal enough, they won't get any training with the new automated equipment. Their chances for other jobs within the bank are low and become even lower during the implementation process. On the other hand the higher qualified counter employees have better chances to go to courses. In the before mentionned research the clerical workers supposed that workers in offices who were ahead with the implementation, would have better possibilities to go to courses then the workers of the offices with later implementation. Counter employees with the highest qualifications get the longest courses, although there seems to be no visible reasons why those workers should get more than others. Office-managers make the decisions who can and who cannot follow the introductory courses as well as the continuation courses.

During the preparation and the introduction of the new computer systems male and female workers struggle about the new jobs as well as the entrance to training courses. They question which jobs will be the best in future, which training will be needed and what possibilities they have in getting those new jobs. Patriarchy seems to be a leading thought by some of the office-managers while making choices about new jobs and access to courses. In some cases female workers are excluded from training, especially from the continuation courses.

Training in the case of the Open Bank Project is a basic course of five full-time days, which are spread over five weeks. At the moment the courses have started already. Items in the course are handling the terminals as well as some commercial training. Although the character of the course is limited, management does not allow all workers within the field of the cash and the desk activities to participate. Far out most of the workers want to follow the course, because they consider it as an important instrument for their actual work. Workers who do participate in the courses, feel it as a confirmation of their performance, because management invests in them. Those who are not allowed to participate in the courses are supposed to learn the basic elements of their colleagues, but in fact they will have to learn it themselves.

In future, more applications will be implemented on the network and every time a course will be organized. Now, about 70% of the workers within the cash/desk area are female. The basic course is, as a consequence, mainly given to female workers. The question is whether women will have and will take their chances in the future, for the further courses and for other more specialized courses. A broader range of items within the courses is probably a boundary condition for enlargement of integration of tasks and for job rotation.

8. JOB ROTATION, A STRATEGY IN THE BRANCHES

Several organizational choices are left, independent of the specific technology. In the situation before front-office automation the division of labour concerns routine clerical jobs for the administration of payment traffic on one side and commercial jobs concerning advising and selling services on the other side. The depth of the division was dependent of the number of workers within the branch. In the case of the introduction of front-office automation, the organizational choices were more or less division of labour. Two items were very important for these organizational choices: the possibilities for workers to take the course and the question whether or not there should be job rotation.

In a pilot branch of the Open Bank Project, management introduced a scheme for job rotation between the cash and desk area and the clerical area. This project that was led by a female local manager, was very successful. Job rotation will be maintained in this branch. As well as in other branches 70% of the workers in this branch were female. The experiences in the pilot branch showed that job rotation improves mutual communication and a greater variety of tasks. Especially women workers attach great interest to a well performing organization and appreciate if they are successful in this sense. For the local management job rotation proved to be good to improve the team spirit and the team flexibility. For the workers job rotation proved to be a way to update their knowledge and skills.
Bank management in general has two contradictory points of view:

* on the one hand they are interested in more flexibility of the workers, that is to say that they want more workers in the cash and desk area to do commercial tasks as well, as was seen in the above example;
* on the other hand they intend to keep the existing division of labour, as they did in emphasizing the division of labour as a consequence of the limiting participation in the courses.

This field of tension is a favourable condition for shop floor politics in the branches. The aims might be a less limited participation in the courses as well as job rotation in the team. Another goal might be that job segregation should be broken down, for example by requiring a better entrance for female workers to courses, more female trainees and a positive action plan.

9. POSITIVE ACTION, A STRATEGY IN THE FIRM AND IN THE BRANCH OF INDUSTRY.

The aims of the Committee of women workers, as mentioned before, concern improvement of the position of female workers. The committee advises the works' council on this item and organizes meetings for female workers. In 1987 they organized a meeting for a hundred women on the item of positive action. In that meeting management promised that:

* they would seriously look after child-care, but they intend to limit participation just for higher qualified workers (child-care facilities in the Netherlands are very poor). As an example they mentioned a few very high qualified women who worked at the computer centre;
* they would organize periodical meetings between management and some female workers as well as women of the committee and an external expert; a plan for the improvement of the position of female workers within the bank.

Although since then some meetings have been organized, it is still not clear whether these promises will be executed.
The Committee of women workers has calculated that the career ratio of female workers is half of that of male workers. Although men and women are nearly in equal numbers working in the bank, 70% of the women workers do work in five jobs and within those jobs they work at the lowest levels. The Committee strives for improvement of career ratios via the internal labour market as well as equal treatment within the bank. Necessary are:

* measurements for stimulating careers of female workers; these measurements should aim to take away those structural and individual barriers that are relevant for women;
* measurements to break through the sexual division of work;
* measurements anticipating on future developments concerning automation and its effects on female workers on behalf of the above points.

These measurements should be quantified in a model for Positive Action. Because of strategic reasons the Committee emphasizes the careers of women, who work already on a certain level within the bank. The Committee emphasizes as well that management might have interests in the Plans. Arguments are that potential capacities within the bank will be used and that the legal obligation for equal pay and treatment is made operational in a handy way.

On the level of the branch of banking industry a Committee exists, called VIVO, with representatives of women working in several banks. This committee is without any status and only meant for the informal contacts. Information exchange is the main item on the meetings, which are held every 3 months. Another item that is discussed is the successes and the failures of the Committees within each bank.
In 1986 VIVO made a proposal to the employers to organize an orientation day for female workers inside and outside banking industry to stimulate women to take starters jobs in the information departments in the banks. They would get internal courses in the bank. After initially enthusiasm the employers gave a negative reaction on the plan in the end. Although there is still a shortness of labour in the information departments in banking industry, shortness is said to be due to high qualified specialists. There was no need for female starters anymore.

Apart from the Committee there exists a 'Women Network in Banking Industry'. This Network is mainly meant for women, who work in qualified and mostly male occupied jobs. They try to influence their careers, their strategy is creating an informal circuit and exchanging information.

10. CONCLUSIONS

Workers' influence on new technology is in this banking case concentrated on a few items: employment, monitoring, training, job rotation and positive action. On several levels in the organization there are disputes and conflicts on these items. We distinguished three levels on which workers make strategies: in the offices, in the banks and in the branch of industry. It is difficult to say whether the described strategies in the Dutch banking sector will be successful for female workers in the long run, partly because of the problem of coherence and unambiguity of these strategies, partly because of the different actors in the strategies on the different levels. But the tension within the labour process between taylorism and flexibility seems to give a good perspective for strategies for improving jobs of female workers.

ACKNOWLEDGEMENTS

The authors would like to thank Diane Whitehouse of the London School of Economics for her comments on this paper and Iske ter Haar of the Dutch FNV, who chaired the workshop on the conference.

REFERENCES

Appeldorn, H.M.G.: De balieterminal (The counterterminal), scriptie Sociale Academie Karthuizer, Amsterdam, Februari 1984

Asch, A. van en K.G. Tijdens: 'Netwerken bij de vier grote banken (Networks within the four biggest Dutch banks)' in Informatie, jrg. 26 nr. 3, p. 178-188, March 1984

Dijk, J.A.G.M. van: Onderzoek Open Bank Project ABN (Survey Open Bank Project ABN), Utrecht, November 1987

Tijdens, K.: Automatisering, interne arbeidsmarkt en segmentering, een studie naar de positie van vrouwelijke medewerkers bij de vier grote banken (Automation, internal labour market and segmentation, a study after the position of female workers within the four biggest Dutch banks), Amsterdam, June 1986

BIOGRAPHICAL NOTE

Marleen Defever works at a large bank and is an active member of the women's group. She also studie politics at the Free University. Kea Tijdens worked at another large bank and has been an active member of a women's group within the bank as well. Now she works as a researcher at the University of Amsterdam.

Correspondence address: Kea Tijdens, University of Amsterdam, Herengracht 196, 1016 BS Amsterdam, The Netherlands.

Women, Work and Computerization: Forming New Alliances
K. Tijdens, M. Jennings, I. Wagner, and M. Weggelaar (Editors)
Elsevier Science Publishers B.V. (North-Holland)

TECHNOLOGICAL CHANGE, INTERNAL LABOUR MARKETS AND WOMEN'S JOBS: AN ANALYSIS BASED ON THE CASE OF THE BANKING SECTOR

Diane G. TREMBLAY
Télé-Université, Université du Québec,Canada,

The theoretical basis of this paper is that there is no technological determinism and that firms' human resources management strategies play a very active role in modeling the outcome of computerization and technological change, in terms of women's work among other aspects. In this perspective, the context of technological change can be seen as an opportunity to be seized, since it generally destructures existing internal labour markets (ILM) and makes room for organizational and other changes. This paper defines two broad categories of employers' human resources management strategies on the basis of research in the financial sector: one which aims at the flexibilization of the employment system mainly through the flexibilization of work (precarious forms of work, forced mobility, etc.); the other which tends to construct employment stability through the development of internal labour markets. We develop on the ILM which seems a useful starting point to analyse the employers' strategies and eventually to construct alternatives. The outcome of undergoing technological changes is not yet determined, and there is room for alternatives; the importance of human resource management decisions is however determinant as concerns the place of women in tomorrow's employment systems.

INTRODUCTION

On the basis of a research on the banking sector, our paper will shed light on the evolution of internal labour markets and women's place in the job ladders or hierarchies, in the context of technological change.
First, we explain the perspective which has oriented our research on women and technological change. We then present the concept of internal labour markets and explain why it seems useful to analyse women's position in job systems, and eventually to construct strategies of access to new jobs arising from the shifts in job content and structure in relation with technological change. Thirdly, we present the place occupied by women in the financial/banking sector in Canada and Québec, thus establishing the existence of sexual segregation in many bank jobs, as is often the case in white collar occupations.
We then go on to analyse employers' strategies regarding employment of women and internal labour markets, which leads us to distinguish two categories of manpower management strategies. The first, observed mainly in Canadian banks, is based on the

flexibility of the employment system. The second is one aimed at workers' stabilization in the firm, which usually means development of internal labour markets. This second strategy is visible in Canada as well as in France (which we both studied), but mainly as regards male occupations.
However, given current changes underway in the financial sector, that is technological changes as well as deregulation, the employment systems of banks are undergoing some important changes. In this context, women's access to internal labour markets may change, as well as their job content, as we will indicate in this paper.

WOMEN AND TECHNOLOGY

Whereas much of research done in the 70's and even some done in the 80's tends to be dominated by technological determinism, our more recent research has brought us to adopt a more open vision of technological change, considering that the impact of this change is not easily isolated from other elements, such as organizational change for example (Tremblay, 1986a). In any case, although hard to isolate, the outcome of technological change must be viewed in this perspective as not being one or unique. Our analysis has brought us to consider that human resources management strategies and organizational strategies of firms are closely related to technological change and contribute to model the outcomes in an important fashion. Given this fact, it is important to analyse these management strategies if we want to try to change situations observed for women.
To resume, let us say that technology (given hardware and system design) delimits a *range of possibilities*, amongst which human resources management strategies and decisions will make a choice.

As concerns more precisely women's work, we consider that technological change may pose some new issues, but it mainly poses the *same old problems* of elimination of gender segregation, or more concretely of access to jobs, to good jobs, as well as access to wages equivalent to those of men's jobs. In this perspective, women's alternative strategies would aim at two main goals in the context of technological change: desegregation of the labour market and reevaluation of women's work in terms of wages and social status.

It is these basis of our analysis that brought us to put the accent on the concept of internal labour market in our research as well as in this paper. It is an interesting concept in this context because it puts the accent on firms' practices to explain the differentiation of situations of various groups (women or occupational categories) in the firms as well as in the labour market in general. It sheds light on the *active role of firms in creating and reproducing differences* based on gender.

INTERNAL LABOUR MARKETS

Let us now present the concept of internal labour markets as it was developed primarily by Doeringer and Piore. Although the concept goes back to earlier years (Dunlop), Doeringer and Piore (1971) were the ones who popularized the analysis in terms of internal labour markets (ILM). ILM can be defined as "an administrative unit... within which the pricing and allocation of labour is governed by a set of administrative rules and procedures", to be distinguished from the external labour market of conventional economic theory "where pricing, allocating and training decisions are controlled directly by economic variables."

This co-ordination through a system of rules is the main difference with the traditional vision of the labour market, governed by prices. In this perspective, an internal labour market can be considered as an "institution" which regulates the system of human resources management through a series of rules (explicit or implicit), often related to collective bargaining issues but also often based on habit and customary procedures.

As for the origins of ILM, various authors have attributed their appearance to different factors. Three main explanations are given: a tendency towards a certain stabilization of the firm's activity, the need for specific qualifications, often related to the specificity of the technology used, and finally the desire to reinforce the social cohesion of the firm's labour force.

The first explanation relates to the variability of economic activity and more specifically to the firm's strategy with regards to this. If a firm dominates a market or can at least reject fluctuations upon other firms, it will tend to create islands of stability within its structure. Also, the increase of certain types of activity such as professional/management activities related to the core of the firm's activity also tends to favour the development of such islands of stability or ILM.

The second factor, that of specificity of qualifications required, is very much related to technological advancement of the firm: if one considers that technology at least partly defines job contents (although not completely, management and work organization being very important in our view), then one must admit that a specific technology would require specific qualifications, and therefore would tend to favour internal training (on the job), an element which largely contributes to the establishment of ILM.

Finally, the reinforcement of the social cohesion of the firm is often also related to situations of change, either change in technological systems, or in work organization or manpower management. In this perspective, the existence of ILM is seen as a way for the firm to ensure the profitability of its "investments" in the training of its labour force. A strong integration of workers in the firm can increase productivity, ease reorganizations, by diminishing the threat of job loss or precarious forms of employment because of the existence of the ILM; finally it can help prevent strikes or labour disputes due to technological or organizational changes.

In the context of technological or organizational changes, ILM can be seen as a precious tool of human resources management. In our view, the main interest of the concept of ILM is precisely the fact that it sheds light on the fact that firms do intervene very actively in terms of manpower management, more particularly as concerns paths of mobility or career lines within the firm. It is precisely this aspect of management, through rules, through defined paths of mobility or career lines, and therefore through the establishment and modeling of internal labour markets that this theoretical model of ILM appears useful for analysis of women's place in specific jobs, specific sectors, and the possibilities of changing these specific places.

This concept puts into light the fact that labour market instability can be created and that firms can to a certain extent contribute in modeling the characteristics of "labour supply" (workers). This perspective is all the more interesting since it can take the "blame" of given situations (unemployment, job instability, etc.) *off* of the individuals concerned by showing that "labour demand" (firms' strategies or practices) can also be part of the explanation. This does not necessarily mean that workers are totally dominated by firms' strategies, but that there might be need for compromises, and that some (women's) situations might not be totally chosen.

Before we analyse employees' strategies in terms of ILM, we will therefore have a look on women's place in the banking/financial sector.

WOMEN'S PLACE IN THE BANKING/FINANCIAL SECTOR

In the banking sector, or the financial sector in general, women are the majority of workers. In Québec, in 1986, they represented 60,3 % of workers in the Financial Institutions sector (Banks, Insurance and Real Estate), the percentage being 60,8% for the whole of Canada. As can be seen in the following table, based on Canadian Census data, the percentage of women in the Financial sector has continually increased over the last thirty years. Let us add that from another perspective, this represents about 7 % of all female employment in Québec, against 8% for Canada.

Percentage of women in Finance, Insurance and Real estate

	Québec	Canada
1961	41,3 %	45,7 %
1971	47,4	51,4
1981	59,8	61,0
1986	60,3	60,8

The Financial institutions are the third sector of employment for Canadian women: Community, Business and Personal services come first with some 45% of women's employment (1981 Canadian Census data), followed by Trade occupations (19%).

Of all women in the financial institutions sector in Québec, 22% are book keepers, or assimilated occupations, 19% are tellers, 21% are secretaries (or stenographers), 9% have other office jobs, 9% are specialized in sales jobs and 10% have "other" jobs, while only close to 9% occupy administrative or management jobs (intermediate or top level).

Even with these figures, the picture is somewhat more positive than it really is in the banking sector only, and in bank agencies more particularly. Office jobs represent roughly three quarters (3/4) of jobs in bank agencies, which is what we studied more closely. Amongst these office jobs, that of teller is still the most common, although in the banking institution we studied (1400 agencies in Québec), the percentage of tellers has diminished from 44% in the seventies to 40% in the mid-eighties. The other categories have evolved as is shown in the following table, the figures being relatively representative of the situation in other institutions.

Table 1. Evolution of job categories in institution A

Occupational categories	1975	1980	1985
Junior Clerk	12%	4%	5%
Teller	44	40	40
Senior Clerk	11	11	16
Secretary	5	3	3
Agent (financial counsel-current operations)	3	4	11
Accountant	8	6	1
Director	10	10	8
Other	7	18	18

The major transformation of job structures concerns professional and technical jobs, which are on the rise, while office jobs, mainly junior clerks and tellers, are losing ground. The situation in Canada therefore resembles closely that which we observed earlier in France (Tremblay, 1986). As for the reduction in director and accountant jobs, the first is quite generally observed in most Canadian banks, where high level management jobs have been the object of a "slimming" strategy: in the case of the accountant jobs, the prototype of one of the rare categories in which women had managed to make some room for themselves, it has diminished in this institution mainly because many jobs were in fact not "real" accountant jobs and have been reclassified as "senior clerk" jobs (which explains the increase in this category).

MEN'S JOBS, WOMEN'S JOBS ......

In banking agencies, women are the majority of workers. There is a clear predominance of women particularly in the most "visible" jobs of teller, cashier or clerk. Men are generally in the higher positions such as director or assistant director of an agency, although one of the important changes in job structures is that more men are appearing in the "new" jobs, such as financial counseling, private banking, professional jobs in general. The following table gives an indication of the male/female distribution in the specific institution which we studied, which is however close to that of other institutions for which we did not always get data, but had interviews of a more qualitative nature to verify the representativity of our data.

TABLE 2. DISTRIBUTION BY SEX, 1980-1985. INSTITUTION A

	1980		1985	
job titles	men	women	men	women
Junior Clerk	6,5	93,5	7,8	92,2 %
Teller	4,7	95,3	3,9	96,1 %
Senior Clerk	10,2	90,8	6,3	93,7 %
Agent (counsel)	44,7	55,2	39,2	60,8 %
Agent (current optns)	33,3	63,3	29,4	70,6 %
Asst Director-Consel	83,0	16,9	82,4	17,6 %
Asst Director-Cur.Op.	93,0	7,0	75,9	24,1 %
Director	82,3	17,7	83,4	16,5 %

What is important here is of course not the specific percentages; we mainly wanted to attract attention to the fact that women and men do simply not occupy the same jobs, in the banking/financial sector as in many other sectors of economic activity. It is taking into account this fact that many realized that the goal of "Equal pay for equal work" had to be transformed into "Equal pay for work of *equal value*" in order to attain/approach equality between men and women. The problem of equivalence between men and women's jobs however still remains to be solved in many areas of employment (see Dussault, 1987), and the entry into ILM traditionally reserved for men often seems an "easier" (relatively easier) route to better paying jobs...

Another privileged route is to enter new jobs resulting from the evolution of technological change and, more and more, from the deregulation of the financial sector (CEC, 1986). Regarding these new jobs, these are mainly related to financial counseling, and will probably evolve towards insurance, trust and other financial activities which will become "legal" for banks with deregulation. (CEC, 1986; Québec, 1987).

As can be seen in Table 2, the "new" categories of employment (Agent-Counsel or Current Operations) are the stage of an "invasion" of men into office jobs. Whereas they were almost totally absent from non-management jobs previously (5 to 10% at the most), they now represent between 30% and 40% of the new intermediary jobs (agent...). Men are therefore entering the banking sector internal labour markets through new "ports of entry", that is the professional and technical categories of employment, precisely those which are developing the most and will continue to develop in the future, given the context of deregulation and technological change (the latter contributing to the reduction of basic banking tasks, and therefore of teller, junior clerk and secretary jobs). It must however be noted that the outcome of the restructuring of ILM and the appearance of new entry port jobs in the banking sector is not yet evident. Indeed, the data collected in Institution A and the interviews in other Canadian banking institutions tend to indicate that women's place in these categories of employment may also be increasing. It still remains to be seen however if the present employees (women) can manage to gain access to the new mobility paths emerging with the appearance of new professional categories of employment, or if female office employees will remain restricted to their entry port jobs as was traditionally the case. In this perspective, let us now turn to the analysis of manpower management strategies, which undoubtedly play a determinant role in this issue.

MANPOWER MANAGEMENT STRATEGIES AND INTERNAL LABOUR MARKETS (ILM)

My interviews and research in the banking sector in France and in Canada (Tremblay, 1986a, 1987) indicate clearly that there exists a very developed ILM in this sector. Banks have always had very clear lines of promotion (from entry port jobs to high management was not impossible), strong rules and conventions governing the employment system, very developed systems of job training, and a very systematic organization of the allocation of work. It must however be noted that the ILM or lines of promotion/career are clearly (almost exclusively) more open to men than to women, in Canada as well as in France.

In the context of technological change and deregulation, the employment system or ILM is however being destructured and remodeled. Criteria, rules and various characteristics of the ILM are touched, some being destroyed and replaced by new rules or criteria, new ones appearing, particularly for the new jobs.

All employment parameters are not destroyed instantly, but many traditional characteristics of the employment system are modified. In the banking sector, reorganization is largely based on more professional (counseling) knowledge, stronger commercial (sales) capacities, and some more technical qualifications related to the use of micro-computers in particular.

In our interviews, it appeared that the acquisition of these new qualities for in-house personnel represents a more violent process in North America (Canada) than in Europe (France) because the latter values more "socialized" adjustments to change, whether they be organizational or technological in nature. On the contrary, anglo-saxon systems do not manifest such a tendency or preoccupation for the "social impact" of technological or other changes.

Given these different attitudes toward change, firms still do have some choices to make in terms of manpower management. Various strategies are open to firms, and these strategies will have an impact on female labour, the women present in the firm's personnel as well as those who could eventually enter the labour market.

Our analysis of employers' strategies in the banking sector has led us to define two main categories of manpower management. The first can be called a strategy of "flexibilization" (Michon, 1987) of the employment system. It rests mainly on the use of the external labour market for access to jobs, but also on the development of particular or non typical (normal) forms of employment (short term contracts, occasional work, part time, etc.), as well as on wage adjustments, such as the reduction or elimination of cost-of-living adjustments (COLA clauses). The banking sector usually uses the two first methods of flexibilization, wage adjustments in COLA clauses being more common in the manufacturing sectors (automobile, steel, etc.).

The second strategy is one which aims at workers' stabilization in the internal labour market, their motivation and involvement in the "firm's project". This usually means more occupational training, development of a strong ILM, sometimes a more "democratic" work organization, etc.

Let us add that it is not at all impossible for a firm to use both strategies, either in different establishments or in one particular plant. It is even common to use different strategies for different occupational groups, and apparently even more so for different sex categories (Maruani, 1988). In the banking sector, it is particularly evident, since the ILM is very developed, but rather closed to women, who are generally restricted to "entry port jobs" (Tremblay, 1986,1987).

Both these strategies are of course very different in terms of methods, and even more, of impact on the labour force. The first is a strategy of pure rationalization, of cost minimization through direct or indirect wage compressions, or through close adjustment of working hours to the volume of production (part time, occasional work, etc.) in order to ensure higher productivity of work.

The second is a more qualitative strategy, the involvement of workers in the firm's project, their identification to the firm's "culture" and their motivation to work being considered determinant for productivity. The development of clear ILM paths is also often associated with this type of strategy.

The research done in France and in Canada shows there are important differences between the two countries' attitude towards the use of these strategies. The second strategy is more often used in French banks than in Canadian banks, but the first strategy (flexibilization) is more common in both countries for female labour. However, research also indicates that important changes are underway in both countries and, as we indicated earlier, the outcome of the changes in job content and ILM due to technological changes and deregulation is quite unclear at the moment. There is still place for the development of a strategy for women to ensure their place in the employment system... But let us have a glimpse at where things stand at the moment.

WOMEN'S PLACE IN THE EMPLOYMENT SYSTEM: TODAY AND TOMORROW ....

In Canada and in Québec, as we indicated earlier, women form the majority of workers, and particularly of non-management workers, in the banking sector. As concerns forms of employment, occasional work is not uncommon, for example sometimes when it is associated with technological changes; on a short term basis, employees will be hired to do the job "as it used to be done" while the regular employees learn the new way to do things. Also, part-time work is developing very strongly in the banking/financial sector. A survey done at the Canadian level indicates that in financial institutions, 57% of part-time employees work part-time on a "permanent" basis, while some 26% work on a "temporary" basis and 16% on a "seasonal" basis (Canada, 1986: 21). Part-time employment represents 10% of total employment in the Finance sector, having increased from 5% in 1975, at a rhythm five times superior to the increase of total employment from 1975 to 1983. Amongst part-time workers, office workers represent 92%, while only 6% are "supervisors" (intermediate management), and 1% are management as well as professional and technical employees, who also work part-time (Canada, 1986: 30).

Obviously, women are more concerned by part-time employment, and our interviews reveal that this situation automatically excludes these persons from career paths or ILM, if this was not already organized by the employment system and rules, which often, restricts women's access to entry port jobs.

On the contrary, in France, part-time work is very marginal, but it is developing for female bank employees (3-5% of the personnel), as our interviews indicated (Tremblay, 1986a). In this perspective, it can be said that French banks have preserved the "normal" form of employment (i.e. full time, indeterminate contract). We also observed a certain interest for employees' participation and involvement in the "firm's project", modern management forms and occupational training. This interest is slowly developing in Canadian banks, but American corporations in general (Canada + USA) more often opt for the first strategy identified earlier, and do not seem very intent on modifying their attitude.

How do these factors impact on women's situation in the banking/financial sector, and in the Canadian labour force in general? How does this influence their position regarding the evolution of ILM in this sector, particularly as concerns the new jobs, professional and technical jobs, which are developing the most? In the banking/financial sector, it is evident that the need for office employees, and particularly for the least qualified ones (tellers, jr clerk) has diminished massively, and will continue to do so, while professional and technical employees (financial counseling and computer specialists...) are in stronger demand. Obviously, it is almost impossible, even with the best internal training plan, to turn bottom level office workers into high tech specialists. However, commercial and sales activities related to the development of new financial products, as well as basic financial counseling (computer aided) CAN be open to women actually in the banking sector. It is precisely here that a firm's strategy towards this group of employees will be determinant in the future of this category of workers.

From the point of view of management two options are open for these new professional jobs. The bank can either fire some women (or let attrition work) and hire professional employees at the beginning of their career on the external labour market, or it can give professional training to the women in office jobs, open up mobility paths for them and therefore reconstruct the ILM, transforming "dead end" entry port jobs into the first step to the new restructured ILM. This second strategy would permit female employees to enter ascendant career lines through the new ILM.

As concerns the more technical jobs, more frequently found in head office than in bank agencies in any case, the strategy is more clearly one of hiring new employees on the external labour market. This of course leaves a possibility for women's participation to increase, but this aspect depends more on the educational system which must attract women in these fields, firms being responsible, ex post, for the design of an "employment equality" strategy aimed at getting some women into the technical internal labour markets.

AND WOMEN'S STRATEGIES

The issue of these undergoing changes is still unclear in Canada, as it is in France, but a few experiences of programs aiming at equality of employment in France and in Québec do confirm that women's place in the labour market, and particularly in banks' internal labour markets, does depend strongly on the firms' strategies in terms of human resources management. (Tremblay, 1987a).

In our view, the context of technological change can be seen as an opportunity to be seized by women. As we mentioned earlier, it often destructures existing internal labour markets, it can change rules and procedures to a certain extent and it may make room for organizational and other forms of change. But this of course remains a *potential*, a *possibility*.

In order for women to be able to seize this opportunity, they must know and understand the employers' or firms' strategies to which they are confronted. This appears to be a prerequisite to eventually be able to criticize these practices along the lines presented earlier and present alternative human resource management strategies, for example to propose substituting "stabilizing" strategies in the place of (women's) work flexibilization strategies, or equipment flexibilization instead of or combined with work flexibilization.

It is in this perspective that the concept of internal labour markets appears to us very useful in this context of technological change, as it puts the accent on the active role of firms in (re-)structuration or (re-)modeling of employment systems, and therefore in attribution of places and establishment of wages for women in a given employment system.

REFERENCES

Canada. (1986) A Survey of Part-time Employment in Federally Regulated Industries. Volume 1. Study done for Labour Canada by Hay, Management counseling. Ottawa: Travail Canada. 52 p.

Conseil économique du Canada (1986). Concurrence et stabilité, L'encadrement du système financier. (Competition and stability in the financial sector. An Economic Council of Canada Report. Ottawa: Supply and Services Dept. 53 p.

Doeringer, P. and Piore, M.J. (1971). Internal labor markets and manpower analysis. Lexington: D.C. Heath and co.

Dussault, G. (1987). "A travail équivalent, salaire égal": la portée de la revendication. ("Equal wage for work of equal value": the importance of this demand) Montréal: Institut de recherche appliquée sur le travail. Bulletin no. 27. 80 p.

Maruani, M. (1988) Le clavier enchaîné; la construction de la différence. (The construction of gender difference: the example of a newspaper firm) in Interventions économiques, no. 20 Montréal: Ed. Saint Martin.

Michon, F. (1988). Flexibilité et segmentation. (Flexibility and segmentation). Interventions économiques, no. 19. Montréal: Ed. Saint Martin.

Québec (1987). La réforme des institutions financières au Québec. (The reform of financial institutions in Québec). Report to the Minister of Finances. Québec: Government of Québec. 128 p.

Tremblay, D. (1986). Informatisation et remodelage des filières d'emploi dans la banque. (Computerization and restructuration of mobility lines in banking). in Les Informa-G-iciens. Conference papers. Namur: Presses Universitaires de Namur. 177-187.

Tremblay, D. (1986a). L'articulation travail-technique et les mutations de l'activité bancaire. (The articulation between work and technology and the transformations in banking) Paris: research report, 200 copies distribution. 200 p.

Tremblay, D. (1987). L'articulation travail-technique et les stratégies d'entreprises. (The articulation between work and technology and firms' strategies) in Tremblay, D. (ed) Diffusion des nouvelles technologies: stratégies d'entreprises et évaluation sociale. Montréal: Interventions Economiques/Ed. Saint Martin. 304 p. pp. 67-84.

Tremblay, D. (1987a). Où en sont les PAE? (How far have we come with Equal Opportunity Programs?) La Gazette des femmes. nov.-déc. 1987. pp. 7-8.

BIOGRAPHICAL NOTE

Diane tremblay is professor in labour economics and sociology of work at Télé-Université, Université du Québec, 4835 ave. Christophe-Colomb, Montréal, Québec, Canada, H2J 4C2

Women, Work and Computerization: Forming New Alliances
K. Tijdens, M. Jennings, I. Wagner, and M. Weggelaar (Editors)
Elsevier Science Publishers B.V. (North-Holland)

CONSONANCE OR CLASH

A preliminary report on computerization and affirmative action in the municipality of Rotterdam.

Gien Tuender-de Haan
University of Amsterdam
Department of Andragology
Amsterdam, The Netherlands

In this paper the results of a small-scale study of the impact of automation in a municipal office are presented. The focus is on the possibilities of preventing job-segregation by applying a policy of affirmative action.
On the level of job-structuring the pursuit of affirmative action proves to have met with little success as yet. For training and education favourable conditions have been created, notably for part-timers. The organizational culture has not incorporated the affirmative action goals very deeply. On the individual level however some inspiring examples were found of women creating their own career-path in automation. This suggests that with more support and stimulation from the organization other women could become interested as well.

1. INTRODUCTION

This paper presents the results of a small-scale study of the impact of office automation on job-segregation in a municipal office. Since the municipal organization concerned has adopted an affirmative action-program[1], the main question is: *Which practical possibilities have been found to carry through the policy of affirmative action within the process of computerization?*

The study took place in the town clerk's office of the municipality of Rotterdam.[2] This organization was chosen because Rotterdam is one of the few cities which have formulated a policy of affirmative action regarding its female employees as well as a policy of automation. Sources of information are:

1. reports and leaflets on both policies;
2. interviews with eleven female employees and with six personnel- and automation-officers.

The study as reported here is a descriptive analysis of the results of affirmative action in the computerization process. Its aim is to look for possible improvements, to be tried out in future processes of automation.

---

1 Affirmative action is a policy aimed at the improvement of the traditionally weak labour market position of women, ethnic minorities or young people. It can be pursued on the organisational level or be regulated by rule of law.

2 A group of five students took part in the research: Mirjam Brouwer, Henny de Bruin-den Haselaar, Joke Schouten, Elvyre Smits, Heleen ten Woude.

Municipal organizations offer an interesting field for the investigation of the interplay of affirmative action and computerization. Many municipal authorities have adopted a program of affirmative action in some form or other. Most are just starting to develop a more systematic policy of automation. How to combine both policies is at present an urgent problem.

First a short overview of the theoretical backgrounds of the research will be presented. Next the results are reported. Finally some concluding remarks will be made, leading up to the suggestion of improvements.

2. THEORETICAL VIEWPOINTS

2.1 Office automation and women's work

Without intervention, the impact of office automation gives rise to a bleak view of the future quality and quantity of women's work. Office work will become more structured and standardized. Jobs will be polarized into routine functions on part-time or flexible contracts on the one hand and a few higher, more interesting and more secure positions on the other. Since most office work is done by women, their jobs will be the first to suffer. (i.e. Weggelaar en de Boer, 1984; Tijdens 1986).
This development will be in a direct line with existing segregation of functions between women and men in the office. Yet examples of the opposite possibility do occur. A case study by Craig and associates (1985) in Great Britain shows that technological change in various branches of industry can lead to an extension of tasks and responsibilities in women's jobs. Comparable results are found in the graphics sector in the Netherlands, though here only in the exceptional case of one all female department where women decided for integration of jobs in an automation process. (Bunck et al., 1985). In research on office work usually the secretarial function is noted as open to job-enrichment by computerization. (Wagner, 1985).
None of the above instances was inspired by motives of affirmative action. They suggest however that job structure can be open to influence by affirmative action programs.

2.2 Affirmative action and segregation of jobs

In the Netherlands at present research on affirmative action mostly concentrates on the organizational level. (i.e. de Jong et al, 1986). The present study is no exception to the rule. Nonetheless aspects like governmental policy (Triest, 1986), general cultural climate and preferences and expectations of individual women (Sociale Atlas van de Vrouw, 1982) should be kept in mind.
The available documentation on practical experiments in affirmative action shows an emphasis on the goal of gaining access for women to more varied functions in the organization on all levels. Only sporadically changes in functional structure have been proposed and tried out. An example is the creation of career-paths from secretarial functions to the managerial level. (Ter Berg and Wegner, 1987)
Segregation between women's and men's work on the labour market has been interpreted within different theoretical frameworks. Craig's investigation of segmentation in industry, mentioned above, has shown that it is not the exclusion of women from high-skilled or high-productivity jobs which accounts for the unfavorable labour market-position of women. In different enterprises within the same branch new technical jobs may be either open to women or exclusively filled by men. If women enter into the technical jobs however, this does not end segmentation. Those jobs are then defined as 'women's jobs', with low status and renumeration. Where the same jobs are fulfilled by men, they are considered 'technical jobs' with higher status and pay. A striking dif-

ference is found between enterprises with a fixed system of functions and salaries, where women cannot easily obtain new technical tasks, and those with more fluid job structures where women can enter into technical jobs, without the appropriate extra payment however. The authors therefore consider feminization of jobs as the central mechanism in segmentation. The study indicates that jobs take on the characteristics of secondary jobs if they are filled by 'secondary-type' workers.
Although unionization is the most important factor associated with more favourable employment conditions, it has not usually sufficed by itself to raise women's pay much above the level of non-unionized male labour. Women are prepared to take low-paid and unrewarding jobs because of existing constraints on their choice of jobs and the importance of their earnings to family living standards. (Craig et al., 1985).
In the maintenance of segregation in industry custom and accepted practice seem to play at least as central a part as qualification or productivity. A recent Dutch investigation of career opportunities for women in three large service-organizations shows how customs and expectations intertwine: women are employed in simple operating or clerical jobs because they are expected to be more tolerant of simple, repetitive work lacking career perspectives. On the other hand women will entertain less ambitions with respect to their jobs when opportunities are absent. (Weggelaar et al., 1986).

2.3 Persistency of segregation as an ideological problem

To interpret the persistency of customs and practices in social life in general theories of ideology have been most enlightening, since they take into account cultural and structural aspects simultaneously. Ideology can be understood as the representations people form about themselves and the world, which organize and legitimate the social order. It defines reality, determines what is just or unjust, what is possible or impossible. (Gramsci, 1971). The daily structuring of social life is founded upon ideology and in its turn confirms the self-evidence of existing representations and practices. (Hall et al., 1978). In this vein what is 'fit work for women' can be interpreted as part of an ideology of femininity. (de Bruijn, 1981; van Arnhem, 1986).
A central part of any ideology is the conception people form about themselves. A changing ideology should not only cause a shift in people's perception of the world, but also address them in a different way, articulate new forms of subjectivity. (Mouffe, 1981).

2.4 Computerization as a challenge

The difficulty of changing existing customs and structures lies in their self-evidence. Thus the slow pace of results of affirmative action might be explained.
In processes of computerization however, accepted ways of organizing work are already under scrutiny. This situation may open up more opportunities for affirmative action than are usually present.
Moreover, if desegregation of jobs during computerization succeeds, in a municipal office chances to include an appropriate reward structure may be greater than elsewhere. Government bureaucracies are generally characterized by a relatively clear structure of functions and renumerations.

2.5 Questions to be studied

From these considerations the following specification of the research problem can be derived:
In carrying through a policy of affirmative action within computerization of a municipal office, have any practical possibilities for change been found:

1. on the structural level, i.e. job content, required qualifications, facilitation of career development for women,
2. on the level of organizational culture, i.e. dominant views on femininity, justifications of existing segregation,
3. on the individual level, i.e. women's individual perspectives, their ambitions and expectations?

3. RESULTS: PRACTICE IS TOUGHER THAN POLICY

3.1 Affirmative action in the Rotterdam municipal organization.

Affirmative action regarding women, ethnic minorities and young people by the municipality as employer - of 29.000 employees - is part of a policy of affirmative action for the population of Rotterdam in general.
The general policy is co-ordinated by the Emancipation Bureau established by the Municipal Council in 1982. The municipal services are responsible for initiating and executing emancipatory measures in their own fields. The bureau, supported by a Council Committee, is to supervise these activities and to develop an integrated policy. It is established on a temporary basis until 1989 and has a staff of four. The services have their own sub-committees.
One task of the project is to advise on affirmative action in the internal personnel management of the services. In the municipal service studied here, the town clerk's office, measures of the affirmative action program have been agreed upon. Higher functions will in case of vacancy be reserved for women, functions on the middle level for ethnic minorities and lower-level functions for young people. Personnel management is responsible for the execution.
It is not possible to ascertain how fruitful affirmative action in the town clerk's office has been in practice. A strong autonomy of the fifteen departments composing the service causes a lack of overview centrally and impedes the execution of a uniform policy. According to the central personnel management officer there are marked differences between departments:
'Some departments need a lot of consultation, others enthusiastically exceed what is asked of them.'
Some of the measures taken are exemplary, such as the ruling that part-time employees have the same rights as full-timers to reimbursement of training- and education-costs. A system of career development has however not been realized; necessary conditions like registration of future vacancies and of interests and capabilities of women are not met.
Efforts to stimulate women to take an active part in decision-making have met with success. A structure of central and departmental women's committees has developed and women are preparing to participate more in the official representative system for employees. Not all women are interested in this development: *'I'm not a feminist!'* is one of the comments we heard. Among men also there are signs that acceptance of the policy is not unequivocal. The opinion of a personnel manager illustrates this:
'The policy strengthens the position of women as long as it stays within the range of what is reasonable and decent. Too much pressure could be counter-productive. The trouble with people promoting affirmative action is that they have a limited view of the organization as a whole.'
The central personnel manager is already quite satisfied with existing opportunities for women:
'Men and women have the same career opportunities, provided some conditions are met. A part-time job should not be under 20 hours a week, for instance.'

In four departments experiments have started recently, establishing target figures for the number of women to be employed. If results are encouraging this will be extended to a general rule. Sanctions, either positive or negative, are not included in the experiment.

3.2 Computerization and women's jobs

After a decade of rather un-co-ordinated decision-making and the purchase of different brands of computers a more systematic automation policy has been adopted recently. Plans and budgets are determined by a central automation committee. Departments have sub-committees on automation, representing management and employees.
An automation expert describes the course of the procedure in practice:
'We prepare the proposals for the automation committee. Usually they are accepted with at most some marginal alterations. Expertise on the subject is generally lacking, even in the committees.'

In the committees predominantly men participate. One department has two female representatives. They are hampered however by a lack of expertise, according to the department's personnel manager:
'Two of the representatives on the committee are women, but if there is some problem with the apparatus they are not the ones to enlighten you. If there really is some difficulty it is always the men from the committee who solve it. A reason we have women on the committee is that our sub-department consists only of women.'
The general opinion is that employees do have a say on computerization. As the central personnel manager observes:
'Wishes and requirements of employees are taken into account. They are the ones who have to use the apparatus.
Men and women have the same say in the automation process. This concerns questions like: What are the advantages of automation above the existing situation? What are the consequences for my job? Will it mean a change in tasks? The personnel officer should anticipate these questions and give the necessary information beforehand.'
Not all women are completely satisfied with the influence they have. One of them says:
'If they are satisfied, I'm satisfied. Only if I would have complaints in my work I would appeal to them, but this has not happened yet.
They listen to you, but whether you have any real influence? They have a fixed pattern, you cannot change that. They want it this way, that's how you do it. I'm satisfied. If something is not right they will find out for themselves. Then they'll say: yes, you were right after all.'
It is not always thought necessary to include considerations of affirmative action in the automation process. One manager formulates this clearly:
'In our department the process of automation is still in the planning stage. It is much too early to say anything definite about pursuing a policy of affirmative action. I don't expect specific measures in behalf of women will be necessary. There won't be any important changes in the division of tasks.'

Introduction and training have been organized more systematically as well. Everyone seems to agree that women get the same training opportunities as men do. If there are differences they lie in women's computer-anxiety or in their lack of mathematical education.
A problem for some women who have tried in vain to register for a training course is that it is not clear to them who decides on their application.
Most courses are given by outside personnel. The automation officer is considering a change in this respect:
'It might be more efficient to have one or two permanent instructors instead of hiring people incidental. If we had more people on our staff I might be able to anticipate things

instead of just answering questions. In fact we should have an information centre, with facilities for training as well as for daily counseling.'
It is in stimulating women to take training courses that the affirmative action policy seems to have its strongest effects. One of the staff members of the Emancipation Bureau recounts:
'Last year we organized a one-day seminar on computerization to inform women about future developments. Instead of the expected number of 90 participants 320 women took part, one out of two of the total. As soon as anything on the subject is organized specifically for women they come running. Afterwards about 70 or 80 registered for introduction courses.'

As to changes in the quantity of women's jobs, no figures are available. All departments have reduced their staff. Since this is done without forced dismissals it has not given rise to strong feelings. The possibility that in natural turnover women may be over-represented is not noticed. Neither is the fact that temporary contracts are usually filled by women commented upon.

3.3 Changes in job content and structure

Changes occurring in the content and responsibilities of women's jobs are not easy to pin down. In one department the automation-plan includes a leveling of the organizational structure. More responsibilities will be assigned to lower level functions. A woman in this department notes the following changes:
'The work is the same, only faster. And you can work more independent, execute things on your own.
Also, if somebody new arrives in the office I do the introduction until they can enter a training course. Usually they say: You can explain it clearly and simply, you do it if you have the time.
If your tasks are extended, they can put you in a higher salary-group.'
Others feel that task extension has taken place without a raise of salary:
'The job is underpaid, with the introduction of word processing some tasks have become more complicated. The work with a computer is more responsible, it takes more brainwork. This should be reflected in a higher salary - but we have raised the matter so often already!'
Most of the women we talked with find that their work has become more challenging. Yet for typists in a typing pool some loss of tasks and responsibilities is noted by a personnel manager:
'Particularly the typists are anxious about becoming redundant. Staff members now type their own reports as well as agendas and notes of meetings. What the typists miss is the production of a beautiful report, which gave the satisfaction of work well done and the added advantage of contacts with the staff member who wrote it.'

In most departments changes that do occur are not planned. Rather they are considered as the outcome of a process of growth - suggesting a dominant view of automation as a natural phenomenon:
'Jobs and tasks have changed. Employees have been informed before-hand, but really changes came more or less naturally.'
'It has grown slowly, people were not conscious of it beforehand.'

The lack of structure in the changes of job content does not restrain some women from extending their own jobs or seizing career opportunities arising from automation. Maybe it is even the absence of planning which creates those individual possibilities.

In selecting whom to invite for interviews we made sure two of the women would be able to tell about their individual initiatives. One woman mentions her career advancement as well as her motivation to work in computer programming:
'I have always been interested in automation. I bought a little PC at once. I took an introductory course, then a PC-user course. At the seminar on automation last year I made known that I would be interested in automation, but there were no further courses. Then the Bureau of Training and Education was offered one place on the Vrouwenvakschool (Women's School for Informatics) in Amsterdam and they invited me to go. I had to go to Amsterdam every day for six months, it cost f. 10.000 plus travel-expenses, but they let me go.
Now I'm back here as a trainee. I have been trained for application-programmer. If a system does not function a 100% the irritation influences the atmosphere at work. So I think that if I can make my programs more user-friendly people will be more enthusiastic about computers.'
The second woman we interviewed had to take more hurdles:
'At first no one wanted to do word processing. I was very interested and when I was completely qualified, then the others started as well. Then someone had to do the system-management. A position was created with a higher salary and I was allowed to fill it. This did cause some jealousy in the office.
The work was more interesting, more contacts with people. I trained colleagues and that task was to be extended. But then I became pregnant and I reduced my work to two days a week. They had to find me another job, I work now at the day-care centre for children. The trouble is, you are constantly working in the same field, at home as well as here: taking care of children.
I want to go back into automation. Now I'm doing a programming course, after that I can go on for systems-analyst.
The job of system manager I built up myself. Later they rated the job.'
Although only a small number of women profit from opportunities in automation, their experiences suggest that a more stimulating career-policy could be of interest to more women.

4. CONCLUDING REMARKS AND PROPOSALS

The pursuit of affirmative action in the computerization process has not been very successful yet on the structural level. Apart from creating favourable conditions for training and education, notably for part-timers, results so far have been reached only in individual cases.
In particular changes in job-structure in connection with automation are treated as a more or less natural phenomenon - 'it kind of grows.' This view precludes the possibility of consciously rearranging the emerging new tasks into more interesting and rewarding jobs for women in the lower echelons.

Neither are cultural changes very marked. Management seems to consider affirmative action as an extra complication in the automation process, which should be dealt with after everything else has been arranged. If women do not profit from the new technical developments, reasons are predominantly sought in their lack of technical interest and schooling, or their 'personal' choice to work part-time.
Women mainly experience the impact of automation positively. This implies however, that they perceive computerization as a process directed by experts on which they have only marginal influence.

It is on the individual level that the most inspiring examples are found: women who manage to enrich their own jobs or arrange their own career-paths. This suggest that with more support and stimulation from the organization others could become interested as well.

As a result of the study some suggestions for improvement will be offered to the management and to groups of female employees to be commented upon. We consider to make the following suggestions:

- start a small-scale experiment in job enrichment in a sub-department about to start a new phase of computerization;
- include affirmative action as a subject in courses for managers and automation-officers;
- establish an information centre for automation to provide employees with the necessary expertise, making sure the centre is sufficiently attractive to women.

ACKNOWLEDGMENTS

I am grateful to Sarah Goodman for her encouragement in acting as a referee to this contribution.

REFERENCES

Arnhem, Corina van, 1986: Beter nu de vrouwen geweerd dan straks zelf het huishouden geleerd. Enkele opmerkingen over het belang van betaald werk voor vrouwen. SocFem Teksten nr. 9, p. 92 -103. [Better to avert women now than to learn housekeeping presently. Some remarks on the importance of paid work for women]

Berg, Judith ter, en Karen Wegner (eds.), 1987: Loopbaanbeleid stapvoets. IWA rapportenreeks nr. 10, Amsterdam. [Career development step by step.]

Bunck, T, et al. (eds.), 1985: De verdwenen kopij. Druk en papier FNV, Amsterdam. [The lost copy.]

Bruijn, Jeanne de, 1981: Vrouwen en de arbeidsmarkt. Kongresbundel Zomeruniversiteit Vrouwenstudies, Amsterdam, p. 285 - 292. [Women and the labour market.]

Craig, C., E. Garnsey and J. Rubery, 1985: Labour Market Segmentation and Women's Employment: a Case-study from the United Kingdom. International Labour Review, Vol. 124, nr. 3, p. 267 - 280.

Gramsci, Antonio, 1971: Selection from the Prison Notebooks. (trad. Q. Hoare, G. Nowell Smith), Lawrence and Wishart, London.

Hall, Stuart et al. 1978: Policing the Crisis; Mugging, the State and Law and Order. London.

Jong, A.M. de, 1983: Gelijke behandeling en het personeelsbeleid. De positie van de vrouw in de arbeidsorganisatie. Kluwer, Deventer. [Equal treatment and personnel policy. The position of women in the organisation.]

Mouffe, C., 1981: Hegemony and the Integral State in Gramsci; Towards a New Concept of Politics. In: G. Bridges et al.: Silver Linings. Lawrence and Wishart, London, p. 167 - 187.

Sociale Atlas van de Vrouw, 1983, (ed. Corrinne Oudijk). Sociaal en Cultureel Planbureau, Sociale en Culturele Studies nr. 3. Staatsuitgeverij, 's Gravenhage. [Social Atlas of Women.]

Tijdens, K.: Automatisering, interne arbeidsmarkt en segmentering, een studie naar de positie van vrouwelijke medewerkers bij de vier grote banken, Amsterdam, 1986 [Automation, internal labour market and segmentation, a study after the position of female workers within the four biggest Dutch banks]

Triest, M., 1986: Positieve actie in vijf landen: Canada, Australie, Noorwegen, Nieuw-Zeeland, Frankrijk. Emancipatieraad, 's Gravenhage. [Affirmative action in five countries.]

Wagner, I., 1985: Women in the Automated Office. Contradictory Experiences - Individual and Collective Coping Strategies. In: A. Olerup, L. Schneider and E. Monod: Women, Work and Computerisation, Opportunities and Disadvantages. Elsevier, Amsterdam, New York and Oxford.

Weggelaar, M. en K. de Boer, 1984: Micro-elektronica en vrouwenarbeid, Deel 2: Gevolgen van de invoering van micro-elektronica voor de arbeid van vrouwen. Verslag van vier case-studies. Ministerie van Sociale Zaken en Werkgelegenheid, 's Gravenhage. [Micro-electronics and women's labour, Part 2: Effects of the introduction of micro-electronics on women's labour.]

Weggelaar, M., W. Trommel en F. Molenaar, 1986: Doorstroming van vrouwen in het bedrijfsleven; feiten en visies. Ministerie van Sociale Zaken en Werkgelegenheid, 's Gravenhage. [Women's career-patterns in trade and industry; facts and visions.]

BIOGRAPHICAL NOTE

Gien Tuender-de Haan majored in Social Psychology. At present she is a staff-member of the Institute of Andragology, University of Amsterdam. Her research interests have been Youth Unemployment, Affirmative Action and Office Automation.
Correspondence address: University of Amsterdam, Department of Andragology, Grote Bickersstraat 72, 1013 KS Amsterdam, The Netherlands

Women, Work and Computerization: Forming New Alliances
K. Tijdens, M. Jennings, I. Wagner, and M. Weggelaar (Editors)
Elsevier Science Publishers B.V. (North-Holland)

USING COMPUTERS IN THE INTERESTS OF WOMEN: AN EQUAL OPPORTUNITY INFORMATION SYSTEM

Paola M. MANACORDA,
Miriam BERGAMASCHI,
Clelia BERTELLO
Milano Italy

This paper describes an Equal Opportunity Information System designed within a vocational training course for Equal Opportunities Advisers, most of them were women. Instead of arranging a traditional EDP course, we tried to involve participants in designing a complete Information System in the interests of women, from their points of view and with their participation. A set of statistical and non-statistical indicators and measures had been designed and an application had been made to a real organization, using PCs and a statistical package. The designing phase required an in- depth discussion about the women's interests and their achievement.

1. INTRODUCTION

In recent years computers have been used for many different purposes: finance, business, health, education and so on. Several of these sectors, such as health and education, are populated by women, both as workers and as users; thus one would expect that women's interests will be taken into account in using computers in these fields. In particular, one would expect that the interests of women will be taken into account in designing Information Systems in sectors in which both manpower and users are prevailingly women. For instance, in such Information Systems one should find that the presence of women and their double role (as workers and as users) in a specific service sector is put forward, assessed, or even stressed in order to transform that service sector in the interest of women.

We can easily verify that this does not occur. Information Systems in health, education, as well as in other fields populated by women are usually "neutral" with regard to the women's conditions and problems. This is due to three main reasons.

First of all, women, as a special group, are still reflecting about their values and expectations with regard to work, power, science and technology. Different women, with different backgrounds and attitudes, have different needs and objectives. Even the meaning of "equality" and "difference" is under scrutiny in the women's movement, and this makes it difficult to find a totally common definition of "women's interests".

Secondly, the historical exclusion of women from science and technology makes it difficult to transform women's values and expections (even when clearly identified) into scientific concepts and technical tools. One can easily recognize, for instance, how statistical tools can be shaped in different ways in order to assert or to deny specific

women's situations. As far as other scientific and technical concepts are concerned, things are not so easy. This important subject is at present being debated in the women's movement.
Thirdly, the persisting and ever- increasing division of labour in the Information Technology sector results in a lack of complete control by women - as a social group- over their product. In fact, despite the fact that an increasing number of women are involved in implementing Information Systems as programmers or systems designers, nevertheless their values, processing, interpretation and use of information.

2. TRAINING

This being the situation, even the EDP vocational training usually offered to women does not eliminate their feeling of "impotence" in designing Information Systems, although it may enhance the personal achievement of any single woman. Furthermore, the EDP vocational training that women receive usually stresses the technical and operational know-how, rather than the capability of designing technical systems, thus reinforcing the gender division of work in the EDP sector. On the contrary, we think that the best way to train women to EDP is to involve them in a complete project, from system design to its implementation, in a project whose aim should be the interests of women, as they are indisputably recognized by the group of women themselves.

The opportunity to try to achieve this ambitious goal was offered to us by a vocational training course held in Milan during 1987. The course was aimed at training thirty "Equal Opportunity Advisers". Although this career is not yet established by the law in our country, nevertheless the Lombardia Regional Authority decided to give thirty people the opportunity to prepare themselves for this career. People involved were twentynine women and one man; they were all employed in different companies, among which were banks, industries, local authorities, service sectors. Most of them had a secondary school diploma, while few of them had a degree, all in non- technical matters. None of them had any previous skill in EDP. Since the students came from different sectors, the project had to have a general, non sector-specific character. The course, which lasted for one year (600 hours) was sponsored by the European Social Fund.
Once recognized by their employers as Equal Opportunity Advisers, the task of the people trained should be that of:

- analyzing and assessing the women's situation in their companies, in terms of "equal opportunity" on the job;
- designing and implementing affirmative actions in order to promote real equal opportunities for women.

The syllabus consisted of law, economics, labour economics, labour sociology, statistics and EDP, each of them for 100 hours.
At the start of the course, the EDP section was seen as a traditional vocational training. It was expected that, at the end of the course, participants would be able to create a simple data base, to process information through the use of one of the most common software packages, to use a word processor to prepare reports. We, as responsible for this section of the course, thought that the opportunity could be taken to achieve a broader and more ambitious goal, that is to involve women not only in technical training but also to design an information system in the interests of women, with their participation and from their point of view.
Thus we and the subgroup of ten participants, among which was the only man present in the course, decided to design an "Equal Opportunity Information System" or, in other words, an EQUAL OPPORTUNITY BALANCE (E.O.B.). The other 20 participants had a more traditional training, consisting in a capability of using the LOTUS 1- 2- 3 package for building a simple data base and for statistical processing.

3. GOALS

Our goals were as follows:

1) to overcome the initial fear of computers experienced by most of the participants (including the only man)
2) to give the participants the technical know-how useful to understand the main features of a traditional company Information System. We realized, in fact, that not all the future Equal Opportunity Advisers could have the opportunity to build their own Information System. In many cases, they would have to cope with an existing one, built with a "neutral" approach by traditional EDP people (often men).
3) to give the participants the methodological know-how useful to create, collect, store, process, retrieve, interpret and use information inside and outside the company, as far as possible in the interests of women. That is, collecting, processing and evaluating information to assess the equal/unequal opportunities for women in a specific sector or company.

According to us, the future Equal Opportunity Adviser should become able - as a minimum- to use a traditional Information System in the interests of women; as a maximum, to design his/her own Information System, or at least to integrate or complete the traditional one with information useful to assess the women's situation. We started by defining the "phenomena" that an Equal Opportunity Adviser should, in our mind, take into account and keep under scrutiny in any company or organization. We defined the following:

- recruitment
- training
- occupation
- career
- salary
- productivity
- flexibility
- unemployment and retirement

For each of these items we discussed in the group the measures and facts more suitable to describe the phenomenon in terms of equal/unequal opportunities for women. This discussion, perhaps the most interesting feature in our project, led to a number of not so insignifant choices and assumptions, since they showed how the values and points of view of women can be taken into account. For instance when we wanted to analyze the "recruitment" phenomena, we thought that it was not enough to simply count how many men and women were hired by the company in a specific period. Since we assumed that, in a general situation of unemployment, women are less likely to obtain a job than men with similar characteristics, we tried to verify this "unequal opportunity hypothesis", by calculating (separately for women and men) the ratio between persons actively applying for a job to the company and persons obtaining a job, and then comparing the two ratios. Furthermore, according to the "unequal opportunity hypothesis" that the recruitment tests are built taking into account the historically and culturally consolidated male capabilities and skills (such as logical and physical skills), we thought that it would be important to evaluate the imbedded discriminatory character of those tests. Similarly, the analysis of the careers requires not only the calculation of the percentage of women and men staying in the highest (and, viceversa, in the lowest) hierarchical levels, but also an accurate calculation of the time spent by women and men to achieve a specific hierarchical level, other conditions than sex being equal. In several cases the participants realized that even the most complete but traditional company Information System does not allow them to assess the women's condition, both for lack of information and for its unsuitable processing and representation. For instance, the composition of family in terms not only of children but also of elderly and infirm persons, is not known to the company (perhaps this is a good thing) and the relative information must be collected directly by asking women. Other information, although

relevant to women's conditions, is not suitable for statistical analysis and for regular collecting and processing. Because of this it must be approached through "ad hoc" enquiries and surveys. therefore we took particular care to distinguish the different kinds of difficulty that can arise during the designing of an Information System for the Equal Opportunity assessment. They are:
THEORETICAL problems, inherent to the lack of agreement and consensus among women about a specific phenomenon. As we mentioned above, the meaning itself of "parity" and "difference" is one of the main problems emerging from the debate. For instance, there was not complete agreement about how to consider the prevailing logical character of the recruitment tests. Should we always consider it as a discriminatory character against women ?.
CONCEPTUAL problems, due to a lack of conceptualization of certain phenomena, even in the management sciences. When we want to verify or exclude the hypothesis that women are less productive than men, we should rely on a generally agreed concept of productivity, and this is not the case.
SOURCE problems. These are related to the fact that many sources, inside and outside the company, are and presumably will continue to be inaccessible to the Equal Opportunity Adviser. As is well known, for instance, the actual salaries obtained by the senior managers are very different from those officially shown, due to the presence of individual incentives. All the economical and financial aspects of a company have usually a limited circulation, although some of them could be reported in the official balance sheet. Furthermore, all the information about people outside the company, that is people not hired or retired, as well as information about the destiny of ex-employees, are difficult to collect from the inside of the company, and thus require ad hoc surveys.
LEGAL problems. In Italy we do not have, as in other countries, a Data Protection Law. However, an important comma in the fundamental law regulating the rights of the workers, does not allow the employers or any other people for any purpose, to collect any information on personal characteristics of the worker. The only personal characteristics that can be collected are those strictly related to the need of job assignment, that is the education, the skills and attitudes. Therefore it is strictly forbidden to collect and store information about, for instance, the marital status of the worker, his/her family composition, and even the education if not useful for the job assignment. However, in certain circumstances the collection of some personal information is permitted with the employee's consent.

4. EQUAL OPPORTUNITY BALANCE

Once we had defined the phenomena that we thought to be relevant to an "Equal Opportunity Balance" and the related measures, we were able to design a set of statistical and non-statistical indicators. We assume here the broader meaning of the term "indicator", that is not only a synthetic measure of a phenomenon such as an average or a ratio, but also a correlation table, a contingency table and so on. For each indicator, we filled a frame containing the following items:

- Name of the indicator
- Its description
- Its meaning (which hypothesis it allows us to verify)
- Information needed to build it
- Source of the previous information
- Range
- Terms of comparison (other companies in the same sector, regional or national similar indicators, etc.)
- Related Indicators and measures, if any
- Methodological notes.

In fig 1 one can find an example of the description of indicators.

Once having completed the methodological part of the project we tried to apply our Equal Opportunity Balance to a real case. For this purpose we choose two small departments of the Municipality of Milan, (the Accounting and the EDP Departments) from which we obtained the information needed, that is the data base containing the individual data of the workers (205 people), obviously in an anonymous form. The attempt to build a real Equal Opportunity Balance in a specific situation was very useful, since it made clear to the future Equal Opportunity Advisers the problems and difficulties they would have to cop with. Beside the difficulties mentioned above, we realized that even an important organization such as the Municipality of Milan, has a very unsatisfactory manpower Information System. Data related to different aspects of a worker, such as his/her salary, career, absences, are stored in different files with different structures. Record linkage of information referred to these different aspects of a worker is on going in the EDP Department of the Municipality but it would have been be problematic to achieve it inside our course. Data about the number of children were collected or not, according to the fact that the birth had occurred before or after a specific law, so hindering a significant analysis of the situation of women with and without children. Therefore we were compelled to apply only that part of the Information System for which we had reliable information, that is: occupation, salaries, careers and other elements. We found, for instance, that the average salary of women is less than that of men at the same hierarchical level, despite the fact that in our country any salary discrimination according to the sex of the worker is strictly forbidden. The lower hierarchical level of women with respect to men (most women are concentrated in the levels 4, 5 and 6, while most men are concentrated in the levels 6 and 7), as well as the smaller number of overtime hours, are responsible for this situation. The comparison of the two different departments showed that women are prevailing over men in the administrative department, while men are prevailing in the technical EDP department. Nevertheless in the former the women's situation, in terms of career and salary is better than in the latter. This is due probably to the longer length of service of women in the former department with respect to the latter.

5. THE ACHIEVEMENTS

The processing of data was accomplished on two IBM- compatible PCs, working under MS-DOS. The software package used for word-processing and statistical calculations was LOTUS 1-2-3. The group obtained not only matrixes and other statistical tables but also graphic representations such as histograms and pies. At the end of the project the achievements of the group can be summarized as follows:

- great involvement in the project phase, with a sense of intellectual creativity which has been highly appreciated
- overcoming of "computer fear"; some of the women became enthusiastic, and offered a great support to the more anxious colleagues. Other women were less enthusiastic but succeeded in considering the computer as a useful tool to achieve a complex goal
- an EDP technical know-how still limited, but capable of improvement through continuous use in the future career as Equal Opportunity Adviser
- an important methodological know-how concerning the models, techniques, tools and cautions needed to collect, process and use information in specific situations.

The capability to take into account the interests of women, the cooperative attitude among the group, the willingness to improve their ability and skill were, according to us, the main conditions for that result. Therefore we are grateful to the group that allowed us to take such an opportunity, in the interests of all of us as women and researchers.

BIOGRAPHICAL NOTE

Paola M. Manacorda, a graduate in Mathematics, is a free-lance systems analyst and Information System designer, mainly in the social field. She is author of several essays and books of social impact of EDP. In the last years she has been involved in women's studies, especially about women and technology.
Correspondence address: Via Santa Croce 3 - 20122 MILANO (ITALY)

Miriam Bergamaschi, a graduated in Economics, has been responsible of the Research Centre of the CGIL-FIOM Union in Milan. As an expert of affirmative actions, she is currently at the PARI& DISPARI Agency, an institution established by a group of women in order to design and implement affirmative actions in Italy.
Correspondence address: PARI & DISPARI, via Pacini 22 MILANO (ITALY)

Clelia Bertello, a graduated in Mathematics, has been a system analyst in a software company in Milan. Since several years she is a teacher of Information Science in the secondary school.

Women, Work and Computerization: Forming New Alliances
K. Tijdens, M. Jennings, I. Wagner, and M. Weggelaar (Editors)
Elsevier Science Publishers B.V. (North-Holland)

AN EXPERIENCE OF AFFIRMATIVE ACTIONS AT ITALTEL

Bruna Osimo
Italtel, Italy*

1. THE CONTEXT

In spring 1986, a work-group[1] was set up to carry out a research for the introduction of affirmative actions at Italtel. The activity of the group was agreed upon an internal agreement between the personnel management and the trade unions representatives. The aim was to focus the attention on discrimination practices towards women in the company, to find out their origins and to define a series of proposals of affirmative actions to redress unbalances on the workplace.[2]
Since 1980 Italtel[3], a state controlled company has been facing a technological transformation from electromechanical to electronic products. Over the last few years, the company has reduced the number of its employees by more than a third and the ratio between the number of women and men is still in favour of the latter. At the same time men are much more numerous among white collars, women, on the contrary, among blue collars.
Italtel's offices and factories are spread all over the national territory. The research is being carried out over seven different areas (production, research, service and sales)[4] situated in Milan or on its outskirts. The work force varies according to its hierarchical

* The author takes upon herself the responsability of the paper's content.

1 The work group was composed of R. Barbieri, M. Bergamaschi, F. De Cindio, F. De Filippi, F. Lomuscio, S. Morganti, B. Osimo, R. Pavan, W. Pezzi and C. Simone.

2 The research is under way. Till now (June 1988) four areas out of the seven chosen have been taken into consideration. The work group is also carrying out a monitoring activity on the way the affirmative actions proposals are being realized providing, when it is necessary, corrective interventions.
Each area is investigated as following: - statistical analysis; - analysis of the organization; - questionnaire supply aimed at finding out workers' perception on the relationship between work and technology.
Moreover each area may require investigations focused on peculiar aspects (for instance in area A the recruitment's procedures and the access of the workforce to reconversion training course have been reconstructed because they were thought to be useful for the analysis).

3 Italtel is Italy's leading manufacturer in the telecommunications industry, its large product range is today predominantly electronic (digital switching system, transmission, power supply system and so on). Its main domestic customer is SIP (the Italian telephone services supplier). Italtel sells its products to other firms in the domestic market and abroad. Its production plants are located in Milan, L'Aquilla, Palermo, Santa Maria Capua Vetere. Product research and development laboratories are on the premises of all plants; Italtel offices and branches are in Italy's main cities as well as abroad. Its total personnel numbers roughly 17.000 employees.

4 "Area" shall mean a microcosm chosen as representative sample. Each area has got a homogeneous workforce that belongs to the same function. Another characteristic is bound to the work place (for instance in Area B the white collars attend to administrative jobs and their offices are all located at the second floor of a building of the firm in Milan).

level, to its professional positions ("at the top", "at the bottom") to the weight of the area in the organization (with consequences on the professional chances of its personnel). The members of the work group were appointed by the two parties involved in the agreement. Italtel chose the members (two men and a woman) among its employees while the trade unions decided, by mutual consent with Italtel, to appoint either some trade unionists or university-researchers (all of them being women). None of the members work full time on the research.

Two more remarkable points :

- the chief executive officer of the company is a women (and that fostered the decision to invest time and resources in a programme of affirmative actions - no other firms did it in that period (1986)
- in the Milanese trade unions there is a "Co-ordination" of women trade unionists. It represents a "landmark" for the debate and the reflections on women's work. The co-ordination has given its spurring contribution to the development of Italtel's A.A. programme

2. ELEMENTS WHICH EMERGED FROM THE ANALYSIS

These observations referred to two areas (examined till March 1988) : the former called "testing area" (Area A); the latter "suppliers accounting area" (Area B).

Area A belongs to the productive process, its work force consists of 43 women out of 99 blue collars.

Area B is administration and its task is to settle the suppliers invoices. The workforce is made up of white collars and the women are 20 out of 43.

Since the early eighties the previously mentioned technological reconversion from electromechanical to electronics has re-shaped the production processes and changed the products. Area A is an example of this change. Its activity consist of testing parts of the electronic switching systems. There are different kinds of tests, more or less complex that are applied after every stage of assembly.

In this area the assembly of the components on the board (the stage that precedes the testing) is completely manual (even if in other sectors of the same company this process is being automatized).

The test is considered to be the most skillful task of the whole process.

The workers belonging to the area were hired when the production was still electromechanical but later followed training courses for the change over to the electronic technology. It's interesting to note that the women (81% have an average long service of 15 years compared to 27% of men) have faced during their working experience not only the technological change but the environmental change as well. At the end of the sixties (when all of them had been just hired and were very young), the work rules were so strict that they could be fined because they were accused of having combed the hair during working hours, and a comb found in their overall's pocket could be a sufficient proof of their breach. Nowadays it is much better even if they sometimes seem doubtful about the fact they are still working in the same company. The burden of the bygones is still clear in their memory and makes them to show their distrust in the present and in the future.

Area B, on the contrary, already existed before the technical change: the terminals are tied to the computerization process regarding the suppliers' invoices entries: in this second case the new technology exists together with the traditional way of working (still more widespread in the office).

Half of the workers have been working in the same area for 14 years: the others for four or five years when similar activities were spread throughout the company and which have now been all transferred to area B.

The work group "affirmative actions" has verified the presence of two common elements in both areas:

- women are only users of computers; they do not participate in systems development and do not choose the solutions for their implementation.
- women are concentrated in the lowest levels of the professions. As far as technical positions are concerned, for example, in area A, women's presence is tied up with the complexity of the tasks; the more the tasks are complex (with regard to technical knowledge) the less are carried out by women.

The work group has surveyed the women of the two area in order to investigate either their attitude towards the job or towards new technology.
It seems to be interesting to focus the attention on the following question: "which technology do you use in your job?" Women's answers have been very different such as: the experience, the testing program, the typewriter, the calculator, the computer and so on. All the questions concerning technology seem to have obtained doubtful answers. The "machine", in no case, has been perceived in itself as threatening.

Some women of area A consider the machine helpful because it avoids heavy tasks. Some others remark its introduction can mean working in shifts. Moreover, the technical knowledge is considered to be valuable by the women of area A. It explains the major chances of men to move up the ranks. It is not considered to be so relevant by the workers of area B.
Some answers concerning the relationship between work and family can not be easily explained. For example the question: "Does your work affect your family life ?", has received a positive answer on the part of area B workers; a negative one by those of area A.
It may be true that female blue collars' job is usually considered to be less important than their husbands'. As a consequence there is no great rivalry in the family. It might be diametrically opposed for the women of area B.
Some other researches regarding female blue collars in Italtel have pointed out how much importance women attach to their effort to make the best of their role as mother and worker as well. It must be added that during the meeting between some members of the work group and workers of the area, it was only the men who seemed to be interested in mentioning the roles as they are at present while the women, on the contrary, did not say a word about their family life.
Two more interesting elements:

- no systemic discriminations between women and men have been found out in area A with regard to the personnel policies and practices. The work group did not find workers with the same requirements treated differently. Each element of differentiation seems to have its peculiar reason (i.e. the fact that men succeed in finishing the whole training courses for the reconversion while women do not, can be surely attributed to their different previous educational qualifications). Nevertheless all these elements put together give as a result a working situation where women are concentrated in unskilled jobs with no professional chances at all.
- In area B, taking apart the people with a diploma (8 men and 3 women, ten of them with the same diploma women occupy lower contractual level than men and, furthermore, a passage to an up level always requires a longer period of time for a woman than for a man.

3. RELATIONSHIP BETWEEN THE WORK GROUP, THE MANAGEMENT AND THE TRADE UNIONS INVOLVED IN THE RESEARCH

It seems to me quite significant to mention something on this point, keeping in mind that my point of view comes from my own working experience in the company.

In my opinion trade unions and management are joined, as far as this research is concerned, by a mutual attitude of "let's wait" even through not knowing what they are waiting for.
This experience of affirmative action has underlined some limits concerning the representation, and more in general the meditation.
A bunch of women of area A, with no contractual power, has compelled either the company or the trade union to change the traditional way of bargaining and to find out together a suitable solution for the women. It must be remembered that the management and the trade union have something in common in their historical attitude towards women: both, even if in different ways, tend to preclude the access to the top to women and to consider them a weak factor in the organization. On one side companies have always thought that women are to easily diverted from their job by family considerations; on the other trade unions are always afraid of being neglected by women as far as their involvement in strikes are concerned.

4. THE WORK GROUP

The last part of the paper deals with the way of working chose by the group till now in order to add some elements of reflection on this peculiar experience. As previously said, none of the members work full time on the research. At the beginning the members spent a lot of time and dedicated many efforts to the definition of the methods of conducting the analysis. The attention was formerly focused on the opportunity of investigating women's attitude towards their job and secondly which means could be used for. At the same time the methods of analysing the work organization, of collecting and interpreting the data at the table were almost taken for granted and were not discussed deeply. After having defined the way of working, the group was divided in small groups each of them with a specific task. It must be kept in mind that the the work group hadn't yet settled their dispute about the survey in area A. The relationship between the group and the people examined might be influenced by this.

For example our first proposals of affirmative actions in area A (testing area) introduced organizational changes and training courses suitable to women in order to increase their presence in skillful occupations. When the women offered resistance to the proposals, the work group that took for granted women would have accepted them, realized the usefulness of getting a deeper knowledge of area A female workers' demands.

It is also appeared clearly that nothing in the analysis should have taken for granted. Each element needs to be deeply studied. The work group has verified, once again a well-known fact: women do not decide the procedures only use them. To understand women's work also means studying how procedures are applied. To neglect this point, as far as our experience is concerned, may bring only to study men's work because it is the men who have either defined the procedures or implemented them in the organization.

BIOGRAPHICAL NOTE

Bruna Osimo, degree in philosophy. Training activity addressed to different kind of people (for age, cultural level, job , nationality). Starting from 1980 activity of education, vocational guidance and member of a research group aimed at analyzing the infosystem guidance in Milan. Since 1984 in Italtel she has undertaken training activity, job analysis, job evaluation. Since 1986 member and representative of Italtel's Affirmative Actions work group
Correspondence address: Italtel - via A. Tocqueville, 13 - Milano - Italy

Women, Work and Computerization: Forming New Alliances
K. Tijdens, M. Jennings, I. Wagner, and M. Weggelaar (Editors)
Elsevier Science Publishers B.V. (North-Holland)

PART 5

WOMEN, WORK AND COMPUTERIZATION IN THE THIRD WORLD

Ria Hermanussen
The Netherlands

INTRODUCTION

The idea that third world countries take a backward position with regard to the use of new technologies is too simple to be true any longer. The expansion of computers and automated equipment takes place on a world-wide level and has its effects on the work of women in the third world as well as it has in the first world. This was one of the main messages of the four women from Mexico, the Philippines and Singapore who attended the third IFIP-conference on 'Women, work and Computerization'. It was the first time that women from developing countries took part in the international debate about 'Women, Work and Computerization'. The two former conferences in Riva del Sole (1984) and Dublin (1986), were mainly confined to the way information-technology transforms the work and the position on the labour-market of women in Europe and the United States. This was not on purpose. Also then it was realized that the use and the effects of new technologies don't keep within the bounds of the industrialized countries. However, a real international evaluation, was hindered by lack of information and of third world participants.[1]
In order to broaden the international character of the conference, the organizing committee of the third conference on Women, Work and Computerization, has actively sought the participation and contribution of women from third world countries. Although the amount of reactions was not yet very numerous, we might say that the contributions of the four participants from the Philippines, Singapore and Mexico gave the mainly western participants, a good first impression of the process of technological innovation in some developing countries.
Because of the arrears of the third world participants in the international debate about technological innovation and women's work, a special plenary session was devoted to technological innovation in developing countries. Beside this plenary session, the four women from the third world took part in the workshops and presented a paper or commented on other papers from the point of view of developments in their own countries.
The social impact of information technology is connected with the socio-economic situation in a country. This is true of third world countries as well. Singapore for

---

1 Ellen de Ru, Loes Steensma en Kea Tijdens, Women, Work and Computerization, report of a conference, Dublin 27-29, 1986

example, is a small island state, in a phase of transition from a third world country to a developed country . A fast adaptation of (western) information technology is stimulated here as one of the main instruments to increase business-efficiency and labour-productivity.[1] Because of the small population of the country and the low rate of unemployment, computerization and new technologies are not so much seen as a threat for employment-security as for example in the Philippines or in Mexico. In these countries, a majority of the people lives below subsistence-level ,has now work at all or works in the informal sector with low job- and income-security.

The speeches and papers of the four participants of the third world,clearly reflect these differences between the socio- economic context in which technological innovation takes place. While Rosalinda Pineda-Ofreneo from the Philippines and Pilar Vasquez from Mexico focus on the dangers of automation and computer technology for the work of low-skilled women in the industrial and administrative sector, Lim Gek Kheng for Singapore mainly deals with the question why women are less successful then men to make use of new opportunities created by the information technology.
Although jobs in the computer industry in Singapore are open to men and women and even 58% of the programmers and 52% of the analysts/designers are females, Lim finds that a relatively small number of women is represented in the data processing management. In her paper Lim points at several cultural mechanisms in society which should be kept responsible for the sex-segregated structure of computer-professions in this country.[2]
Pineda-Ofreneo, a researcher from the Philippines, foresees in the short term more dangers then new opportunities for the employment of women. She illustrates her anxiety with examples of the working conditions of female workers in the electronics industry. In the past ten years, the electronics industry in the Philippines experienced spectacular growth, but deep crisis as well. This industry, however has no life of her own and depends completely on imported technology and raw materials. In the past the main threat to the employment of the numerous girls and women were the periodically downfalls in the world market of electronic products. In the future Pineda-Ofreneo foresees that an even more disturbing development is the introduction of automated equipment. This enables one worker to produce ten times as much a she does manually, but may also enable multinational corporations to reallocate manufacturing to their home bases.[3]
Also Pilar Vasquez, a trade-unionist who studied the introduction of digital telephone systems, is rather pessimistic about the employment prospects of the female telephone operators at the Mexican PTT. Without training and retraining, Vasquez foresees that the new opportunities created by automation, will not be accessible for the operators. The workers, together with the trade-union, should develop a more active bargaining policy with the management to create alternative employment and training- or retraining facilities. This is the only guarantee that mass-lay-offs can be anticipated.[4]

Beside differences, the papers and speeches of the four representatives of developing nations showed some parallels as well. To some degree, they all established the fact that although the use of computers and computer controlled machinery is expanding very fast, the social debate and research about the impact on society and more specific on the position of women is just beginning to start. Until recently, the debate about the need for technological innovation was limited to only a small group of policy-makers and experts. Especially national governments of third world countries give strong

---

1` Lim Gek Kheng, Women's participation in the computer industry Singapore

2 Ibid.

3 Rosalinda Pineda-Ofreneo. Women, Work and Micro-electronics: focus on the Philipines; see also: Maartje van Putten, Flexibilization in an international perspective

4 Pilar Vasquez. Automation at the Mexican PTT: does it displace the operator?

encouragement to a quick adaptation of new technologies. In their ambition to modernize the low- or under developed economies, government-officials hope to create a new stage in the industrial and economic development of their countries and to free themselves from the image they have in the world economy as suppliers of simple and cheap labour and products. Furthermore many countries consider a quick technological change of their production structure as a necessary condition to get hold on a part of the world market. An indiscriminate adoption of new technologies by third world countries, however, is not without risks. In her paper, the Mexican researcher Soledad Robina warns that information technology is not neutral. New technologies are mostly imported from abroad and the information is stored and processed in large data bases belonging to transnational corporations. According to Robina, third world nations, therefore should formulate their own need in terms of social and economic development, which means more adequate planning en cooperation.[1]

In the debate about the necessity of automation and information technology, the role and contribution of the trade-unions is still rather limited. Apart from legal restrictions reducing trade-union freedom, many countries don't have a long and out standing trade-union tradition. As Pilar Vasquez states, this hinders the development of workers-strategies with regard to complex social-economic issues such as the expansion of new technology.[2] What's more is that the impact of automation on the employment situation is still a rather new phenomenon. Especially in the Philippines and Mexico, the trade-unions didn't recognize in an early stage the full and possible impact of technological innovation. Whereas the greater part of the working population is employed in local firms or in the agriculture, the impact of new technologies is not yet felt by the main part of the workers. For this reason, Pineda-Ofreneo as well as Vasquez, call women whose work will be affected by automation to organize themselves and stimulate trade-unions to develop an adequate collective bargaining policy with regard to the negative results of new technologies on the work of women.

1 Soledad Robina, Notes on the use of new technologies in Latin America
2 Pilar Vasquez. Automation at the Mexican PTT: does it displace the operator?

Women, Work and Computerization: Forming New Alliances
K. Tijdens, M. Jennings, I. Wagner, and M. Weggelaar (Editors)
Elsevier Science Publishers B.V. (North-Holland)

AUTOMATION AT THE MEXICAN PTT: DOES IT DISPLACE THE TELEPHONE OPERATOR?

Pilar Vasquez
Journalist, Mexico

In this paper I am going to present a general view about the automation process at the state owned company 'Telefonos de Mexico' (TELMEX). More in particularly I will discuss the automation process at the Long Distance Traffic Department and the effects of this for the jobs of the telephone-operators. In 1987 TELMEX employed 11.839 operators. All of them are women.

TELMEX is a public company which means that the majority of the shares (51%) belong to the state. As far as the number of employees is concerned, TELMEX belongs to the ten biggest companies of the country and monopolizes the traditional telephone services.

The modernization and the automation of the telephone stations at TELMEX became in an accelerated state, since the disastrous earthquake of September 1985 ruined 27 of the PTT-buildings in Mexico-city. This however doesn't mean that before this time no automated equipment was used by the Mexican PTT.

The automation process at TELMEX dates from 1967, when TELMEX started to introduce analog-systems in the towns for the long distance (and partly the local) telephone traffic. The automation of the telephone service lasted 14 years and right now in all towns some form of analog automatic system has been implemented. Except the towns,this automatic service covers also 32.9% of the villages and the rural areas.

This means that in almost every part of the country no direct interference of the telephone operator is needed anymore for making a telephone call. Also a greater part of the international and intercontinental calls is connected automatically, without the assistance of the operator.

In the meanwhile the role of the telephone-operator has undergone dramatically changes. While she formerly had to ask people with whom they wanted to be connected, she now stays at the backstage. She only takes care of the technical side of the connection and disappears at the moment the technical connection is made.

As stated before, the introduction of the automated systems took place slowly and gradually and still today much of the telephone-traffic is arranged manually.

Recently, however, some digital telephone systems have been introduced, of which the effects for the employment of the telephone-operators will be more drastic. These digitalized systems make phone-connections possible without any interference of the operator.

The efforts of the TELMEX-management to introduce digital equipment have to be understood in the context of the severe crisis the company experiences at the moment. Not only because of the damages caused by the earthquake, but also because of some corruptive officials, bad management and reduced profitability since the

economic crisis, the financial and organizational condition of TELMEX is rather poor. These internal problems are responsible for a high rate of inefficiency of the TELMEX-organization which has a severe negative impact on the public opinion about the TELMEX-services.
In order to resolve the problems of stagnated growth and inefficiency, the TELMEX-management elaborated a program to improve and diversify the supply of the telephone-services.
One of the essential elements of this program is the introduction of digital telephone-systems.

Up till now, however, the employment effects of automation for the operators are still minimal.
Because of the expansion of the telephone-traffic in Mexico and because of the strength of the trade-unions, the number of operators even expanded with 21.4% between 1980 and 1985.
This, however, did not according to the management result in an increase in productivity. On the contrary. The productivity decreases with approximately 25%, which led the management to the conclusion that it would be sufficient to employ 8.300 operators instead of the actual 11.839.
A problem with these figures is that the trade-union nor the operators have their own productivity-graphics, which makes it difficult to judge if the company-information is reliable.
Furthermore, the company representatives consider that because of the introduction of automated and digitalized systems, the Long Distance Traffic Department can almost do without operators. They expect that by 1990 the operators will operate only 6.4% of the Long Distance Traffic. This prediction is based on the information about the price difference between the automatic service and the service of the operators. In towns this difference is almost 200%.

Because of the expected decrease in employment, TELMEX contemplates alternatives like the relocation of workers to other zones and departments. Also the company is studying on starting new services or improving the old services, which means that more attention is paid to quality control, the development of new telephone-sets, better repair services etc. TELMEX is as well proposing the relocation of her activities to new areas like telemarketing, tourism, the development of data-banks and TELMEX self-promotion.

It is to be noticed that the automation of the Long Distance Service affects the work of the operators in the sense that it will slowly substituting the operators. But up till now the automation has not notified the working-situation, nor the organization of TELMEX very far-reaching. In fact, at TELMEX there are still three systems in operation: the digitalized system, the automatic system and the manual system; even as I have stated above the last two systems are decreasing slowly.

In order make clear the reaction and the answer of the operators on their employment-threatening, I first want to discuss the labour conditions at TELMEX and the position of the trade-union.
Until 1979, the TELMEX-workers worked under very unfavorable conditions; under a close and arbitrary vigilance which meant that the operators could not even leave their commuters for the satisfaction of their personal needs. There were hardly any arrangements between the workers and the management about the length of the rest-periods, about overwork, or about the maximum work-load etc. The trade-union in that time was more on the side of the management and had little attention for the poor working conditions of the female operators. The influence of the operators on the policy and the negotiating-issues of the trade-union were very limited.

In 1976, the telephone operators initiated a movement to democratize their union in order to exert more influence on the negotiations between the management and the trade-union-officials, who are by tradition mostly men. This resulted three years later in the first Collective Bargaining Agreement by which the management agreed to accept fixed working-conditions. This agreement was preceded by a strike, showing the company's resistance to subscribe an agreement related to issues as fixed working-hours, fixed breaks and paying or compensating overwork.

In 1980, the Collective Bargaining Agreement had to be renewed and this time the union succeeded in getting accepted an automation-clause. This clause was related to personnel-readjustments in case automation would lead to a surplus of workers. This meant that forced retrenchment would only be accomplished if there were no possibilities to relocate the affected workers in other working places of the national territory where the company is operating. If the relocation application would be higher then the needs, the exceeded workers were to be relocated in other places with possibilities.

The company and the union agreed on the point to give the workers a 30 days lapse of time to adapt to the new working-situation and eventually to be retrained. After that lapse of time the relocation would be considered as definitive.

In case a personnel reduction was proved to be inevitable, the reduction would take place in the concerned department only and should affect the non-syndicalized in the first place. If the syndicalized workers have to be retrenched the company has to take into account the seniority of the worker. The retrenched workers receive an indemnification of the company of five months payment as well as 20 days payment per services year. Compared to what is arranged in the law, this indemnification-agreement of TELMEX is far better.

The above clause shows that the union contemplates, at least since 1980, the possibilities of personnel readjustment and eventually lay-offs. Via this clause the union succeeded in getting guarantees about relocations and as a last possibility a good indemnification.

According to my point of view, in spite of the automation clause, the operators are not comforted in their feelings of uncertainty about the future of their work. A displacement of their jobs because of the automated service is a ghost that occasionally passes through the department, causing fear for employment loss.

The most important reason for this uncertainty is the lack of daily and precise information about the automation process. The operators feel that they are not well-informed and that they receive only pieces of the information they need for analyzing and commenting their situation.

To become better informed, a commission of operators visited in 1988 the telecommunication company ATT in the United States. From this visit they learned that is urgent to develop new activities for alternative employment. Almost half of the formerly employed operators at ATT lost their jobs. Also the union is becoming more active in learning from experiences of PTT-workers in other countries and some union representatives paid a visit to some European telephone companies.

Likewise, in 1987, the union organized a Latin American meeting to discuss the social impact of new technologies.

All these efforts, however, haven't succeeded in accomplishing strategies or alternatives to defend the present rate of employment at TELMEX. Neither did this international information exchange lead to a deeper understanding and knowledge about the position of the workers. This has several reasons.

Firstly, the union doesn't have enough and relevant information about the management plans with regard to the use of new technologies. The management is not very cooperative and open about her strategies.

Secondly, the union-official haven't found the proper mechanism to strengthen her negotiating-power with regard to the digitalization process, nor has she found a way to generalize the discussion among the operators. To my point of view the participation of the operators is necessary and useful, because they have the experience and their creativeness can help to formulate alternatives. In order to give proper answers to the changes caused by automation a frame-work of analysis in need, which means that the operators have to be trained in interpreting the information and their knowledge about the changes taking places. The union has not succeeded in developing such a systematical approach to start the automation discussion, nor with the management-representatives nor with the operators. This means that rumours are the main information channel at the moment, with all risks connected to it.

One may not conclude however that everybody is staying at the fence. Right now a special commission is installed, which consists of female union members and company-representatives to study the operators' situation and their employment-alternatives. The most important agreements accomplished by this commission are:
- the approbation of training-programs;
- more attention for the deficiencies of the old manual equipment which hinders the operators to carry out their work efficiently.

Also the commission reached an agreement about becoming informed about modernization- and digitalization projects. One year before implementing new automation plans, the management has to inform the commission in order to allow them to analyze the plans and to have enough time for formulating alternatives.

Lastly, I want to express my opinion about the affects of automation on the jobs of the operators. Even though the use of automated equipment shows a tendency to displace the workers, it does not mean that they will be disappeared in a short time.
The next coming years the operators will continue to be in charge of a gradually decreasing Long Distance Traffic Department.
This means that the present situation represents to the operators a new opportunity as well as a tremendous risk.
The opportunity for the operators consists in having the opportunity to improve their working situation through negotiating alternative employment which might be in the future more interesting then the work they are doing now. This is low-skilled work, routinely and not very exciting.
The fundamental risk of the actual situation is that the workers cannot come to an agreement with the management and that they will be laid off in the not far future.
This would be a disastrous situation for the TELMEX-women. Because of the economic situation in Mexico and the high unemployment rate, the chances to find other jobs outside TELMEX are very limited. Furthermore are the chances limited because most of the operators have only primary school and the qualification they learned at TELMEX are not useful outside the company. So if they will succeed in finding other jobs, which will be certainly not the case for all of the more then ten thousand women of TELMEX, than it will be very low paid jobs. This is not a very satisfying prospect if one realizes that the wages of the operators are three till four times higher then the legal minimum-wage in Mexico.
At the moment both scenarios, the optimistic and the pessimistic one, are possible and the only thing we can do is to fight for the best.

Correspondence address: Pilar Vasquez, Eugenia 626 ED 1-702, Col. del Valle, Mexico 12, DF, C.P. 30100, Mexico

Women, Work and Computerization: Forming New Alliances
K. Tijdens, M. Jennings, I. Wagner, and M. Weggelaar (Editors)
Elsevier Science Publishers B.V. (North-Holland)

NOTES ON THE USE OF NEW TECHNOLOGIES IN LATIN AMERICA

Soledad Robina
Mexico

1. RAPID DEVELOPMENT

While the debate about the impact of new information and communication technologies has continued to be a main topic in national and international forums, technology in increasingly has continued to expand, its use becoming increasingly wide-spread in every type of institution, corporation, organization and even at home. The irruption of informatics into modern life has affected almost every field of concern.

There remains virtually no branch of science or aspect of human activity that has yet to be touched by the affects of the use of computers and their expansion. It is impossible to even think about the concept of development without taking informatics into account. However, the expansion of the use of informatics is not a neutral process; it can either further development or it can lay the bases for a state of even more profound dependence.

To the extent that informatics is intimately related to the area of information in general, this latter field has also experienced the impacts of a technology which has revolutionized the manner in which data are processed, recovered and exchanged.

The enormous amount of information generated in the world each day surpasses geographic and social boundaries. Little by little, computer technology has produced more efficient and more economic models for data storage. The development of information banks[1] has played a fundamental part in this process of automation and processing of information.

A recent turning point in technological innovation has been marked by the "connection" of data bases to personal computers. This change may be basically due to two facts: first, the tremendous surge which has ocurred in the microcomputer market (the number of small machines in existence in the United States alone has jumped from 5 million in 1983 to 30 million in 1987) and second, to the expansion of "intelligent modems", which has lowered costs and facilitated the connection of personal computers to telephone lines, allowing data to be sent directly from computer to computer.

In recent years information and communications technologies have followed an explosive pattern of growth, while undergoing, at the same time, spectacular changes. Generally speaking, the most relevant changes have included the lowering of costs, together with the reduction in their physical size. Nevertheless, another significant change which has affected this field has been a modification of the relationship between the user and the machine. The development of microcircuits has reduced to an amazing degree the complexity of the user-computer interphase. The complexity of this relationship has

---

1 We include in the concept of Information Banks, both bibliographic and factual data bases.

been increasingly transferred to the machine, with the result that the operation of the majority of computers today no longer requires highly trained operators.

2. THE MICROCOMPUTER INVASION

Latin America has also experienced a rapid development in the areas of information and communications technologies however, the introduction of these so-called new technologies, microcomputers among them, has not been carried out in the same manner in each of the countries in this region.

One of the most notorious differences to be observed in this respect has to do with the role played by the state in the area of informatics. Brazil, for example, since 1971, has considered technological development as a strategic part of its economic development and has tried to protect its national microcomputer industry by enacting a law governing the area of informatics. Chile, on the other hand, has relinquished its right to oversee the behavior of the information industry and has decided to limit the development of information technology to that technology supplied by the makers of data processing equipment and to the perceptions that the private and public sectors may have as regards their needs in this area.

Venezuela has not yet passed a law concerning informatics and which might promote and form a national industry capable of satisfying internal demand. Mexico has applied a policy which emphasizes development of an electronics- informatics industry whose main objective is the fabrication of computers and peripheral equipment.

Argentina attempted to create and apply a policy relating to the field of informatics, but these efforts have not gone farther than the formulation of theoretical bases for initiating the process, while any actual progress in the field of informatics has been left to the isolated efforts of different economic agents.

And finally, as our last example, we will mention the case of Cuba. Cuba has been the first Latin American country to promote the development of informatics on a national level and in an independent and autonomous manner in relation to the computer technology used in the rest of the continent.

The fact is that microcomputers, with or without policies governing their production, sales, distribution and use, have invaded Latin American markets. Their most common applications have been: word processing, access to information banks and numerical processing, mostly statistics.

It has been estimated that around 70% of the 2,500 information banks of public access existing in the world are located in the United States. This fact can be explained mainly by the greater degree of technological advancement attained by that country, its more unified market, lower communication costs and, perhaps, the lesser degree of cultural resistance that exists there as regard paying for information.

3. THE SITUATION IN MEXICO: COMPUTERS AND INFORMATION BANKS

The number of computers in Mexico has grown increasingly since 1964, when the most important transnational corporations began intensive commercial activity in the field. Armed with their experience in the United Stated and in other industrialized countries, computer suppliers launched their conquest of the Mexican market - marketing a product that had very little local demand - by selling solutions to problems that they themselves defined. Today, five of these corporations (IBM, Hewlett Packard, UNYSIS, NEC, NCR), control most of the national market. The so-called microcomputers appeared on the national market in the early seventies, gaining acceptance by the mid-seventies. In 1980 the number of small machines was estimated at 20,000 units. Today this estimate reaches approximately 90,000 units.[1](

According to UNIDO (United Nations Industrial Development Organization), 64% of local computers are used for processing administrative and commercial data; 18% correspond to data communication and 3% to industrial utilization. There are also computer systems that are used in public medicine, mail service, electricity transmissions, traffic control and in institutes of higher education. The State is, however, the main client of the informatic industry and the principal user of the national admits, at present, the country still lacks a global informatic policy which embraces the different aspects of informatic technology.

During the last few years, the development of informatics has spawned interest in the creation of information banks, a process which has, in turn been recognized as conducive to the implementation of a global development policy. This global development program undoubtedly includes as a central requirement a process of decentralization aimed at promoting the general welfare of the population and stimulating a more democratic participation of citizens in the decision-making process.

Those government policies related to the promotion of education, scientific research, technological innovation and the development of know-how constitute a privileged area for the achievement of this objective. It is foreseen that information banks will play a relevant role in developing these policies.

Until now, a great effort has been made in Mexico to create information banks. Some operate quite successfully, while others have failed. There are also some interesting cases of institutions that are now participating in the collection of national data in order to integrate them into international data bases.

There is an obvious need for more coordinated action among the different institutions creating information banks. Those institutions which have, to date, created information banks have done so in a disjointed manner, following diverse strategies, and as a result, instead of improving their usefulness by means of the interaction between their systems, gaps have been created which could have been avoided by planned action. Some of these problems have been pointed out by the National Program for the Development of Science and Technology.

Information banks, which have been a product of recent technological transformation, constitute an indispensable tool for the diffusion of knowledge. Moreover, although the use of such high-powered tools can give decisive effects on different aspects of human activity, work related to the creation and use of information banks, such as the development of directories, the design of specific characteristics, the planning of their location and coverage, and work related to their up-dating and broadening of their accessibility is very scarce and is carried out mainly in industrialized countries.

On the other hand, studies related to the characteristics of the use of information banks and, more basically, the relationship that exists between the obtention of data and the

---

1 It is impossible to know the exact number of microcomputers in existence. In spite of the application of annual import rates in 1981, a large number of machines enter Mexico illegally, across Mexico's northern border.

real application of such information in terms of the generation of practical results in different areas, have received little attention in empirical terms.

A recent study carried out by the Latin American Institute for Transnational Studies (ILET)[1] showed that although information banks constitute modern instruments which aid in the production and diffusion of information, as well as in the promotion of national development, a problem of underutilization of information banks by potential users still plagues the information industry. Underutilization also means a lack of profitability. Information banks represent a powerful tool which has not yet been adequately used. Their effectiveness, therefore, may be questioned.

The above-mentioned study cites figures that are worth considering: only 46% of university students in their last year of studies knew what an information bank is; 12% had actually consulted data bases; none had consulted an information bank more than five times. Professors and researchers were not in the habit of consulting information banks regularly. In the area of social sciences, for example, only 20% of those interviewed had consulted an information bank. The survey was carried out in two universities which have access terminals connected with national and international data bases. Only very few professors were aware of the existence of these services.

Among the principal causes of this underutilization, the following may be mentioned: the elevated cost of consulting the data bases; unqualified operators with no specialized knowledge of the area to be consulted; difficulty in obtaining the primary sources, distrust of automated data bases, and ignorance about these new information and communication technologies.

National and international information banks continue to grow in terms of numbers and topics covered, but emphasis on their sale and distribution has left the user, and the potential user in the background.

During this decade efforts have multiplied on the part of public and private institutions in many countries of Latin America, - including Mexico - aimed at developing automated information systems. Universities and researchcenters have played a central role in this process. Close to 30% of all of Mexico's information banks to which the general public has access are to be found in the Universidad Nacional Autonoma de Mexico. Several of these banks are completely unknown by both students and professors alike, in terms of their existence and range of fields covered by them.

And here is where a dialogue among deaf takes place. In Latin America today national institutions are creating a growing number of data bases. Everyone seems to want to create data banks, but apparently few wish to consult them. What should be done?

Awareness of the existence of systems already in operation is almost null, and users are few. Information needs and expectations on the part of potential users do not always correspond with the information that is available. The creation of and the broadening of access to national information banks is undoubtedly a necessary task, but it is also essential to make the public aware of their existence and their potential uses, as well as to accept the fact that, before their use may become wide-spread, a certain change in mentality must take place in those logically resistant to any innovation or change.

The above-mentioned data clearly show that the introduction of new technologies in most Latin American countries has not obeyed the real imperatives of national

---

1 Robina, Soledad, "Bancos de Información: ¿Qué piensan sus usarios?", ILET, México, 1986.

development, but rather those rules of marketing that have best served the interest of the multinational corporations.

The last several years have witnessed a process of increasing activity in terms of the international market for information banks.[1] Developing countries, including Mexico, can not afford to stay on the sidelines in terms of the development of an automated information industry. The great challenge facing these countries is the creation, organization and diffusion of national information banks oriented towards fulfilling the nation's own needs and interests. However, it is also imperative that a policy governing the use of national and international resources be formulated and applied.

4. FROM DATA BASES TO COMMUNICATION NETWORKS

The use of computers as tools for communication and for creation of networks is not a recent phenomenon. Rapid development in this direction is expected in the near future, due to the easy access which recent growth in the microcomputer industry has provided.

There has been a general lack of knowledge in the non specialized media regarding the utilization of microcomputers as tools for communication. The lack of practical experience in this area has hindered awareness of the real advantages derived from their use.

One of the factors that has contributed to this lack of knowledge has been the widespread tendency to equate the transmission of information with communication. The consultation of data bases is just one small aspect of the use of communications and networks.
The need for establishing policies related to the use of and access to information generated in industrialized countries has been pointed out on numerous occasions, as well as the need for developing national information banks. Today, now that technology so permits, emphasis should also be placed on interconnecting banks created within different developing countries, with an eye toward creating horizontal communication mechanisms.

In this sense, communication between microcomputers is not just one more function of the machine. Such communication permits the development of networks and these, in turn, allow for the sharing of resources, access to different concentrations of data and the expansion of the universe of knowledge. Personal computers acquire, then, new value, becoming in this manner a basic tool for communication, and not just for the mere processing of data.

In Latin America the development of communications networks, together with the infrastructure necessary for their development, will have several important effects, among which are included the advancement of regional cooperation in different research projects, the pooling of national and international experiences, the coordination of efforts and the rationalization of resources.

Within this context, the ILET, from its headquarters in Chile and Mexico's office participation, carried out an experimental project, establishing a microcomputer communications network between non governmental organizations in five Latin American countries.

The basic objectives of this project may be summarized as follows([2]

1 In the last decade, the world market of commercial data bases has grown 500%.

2 Contact-O ¿Cómo diseñar redes de comunicación? ILET, Marzo 1987.

The study and acquisition of practical experience in the design of a network between five research institutes in different parts of Latin America (Buenos Aires, Lima, Rio de Janeiro, Mexico City and Santiago de Chile); the identification of those technical legal and economic conditions needed for the development of networks in Latin America.

In order to set up the network, electronic mail boxes were rented from the Electronic Information Exchange System (EIES) at the New Jersey Technological Institute. It was through their computer that the exchange of messages and the participation in electronic conferences, both public and private, were carried out.

The most important function of the network was probably the diffusion of knowledge about new technologies among developing countries and the acquisition of experience in their use.

Given the diversity of centers participating in the project and their multiple areas of interest, the network's primary objective was not the coordination of the work being carried out by each of their members, but rather to offer, by means of access to the network, new horizons for their work and research. The project has corroborated the idea that the establishment of microcomputer networks has become possible in developing countries. Latin America possesses a top-level and relatively low-cost international communications infrastructure, but lacks the experience necessary for taking better advantage of the services offered.

Mexico is one of the most advanced countries in Latin America in terms of telecommunications. It has a national packet-switching network with its own electronic mail service. The increasing importance of the information accumulated in the country's different data bases mandates a more efficient communication between the different institutions developing these systems.

On the other hand, communications in Latin America generally use international networks operated from the United States (Tymnet or Telenet) for inter-country communication. In some cases, foreign computer services on these international networks are being uses for national communications. For example, the electronic mail system Telemail, offered by Telenet, is being used to send messages between different parts of Mexico and even the commercial database Dialog is being used for the exchange of messages within Mexico. At the same time, the nation's own infrastructure is not being fully used, due to a generalized lack of awareness as to the possibilities being offered for national use. The conclusion is that the services offered by international communications systems are more complete that those offered on a national level([1]

A lot of resources, intellectual and financial, are being invested in this field and the situation is changing. At the same time, sophisticated communication software packages are beginning to appear in the market. The appropriate combination of both new hardware and software developments now allows the design of new kinds of

---

1 To address some of the issues mentioned above, the ILET is carrying out research activities along two lines: the utilization of data bases in order to promote the use and development of national information resources, and the utilization of the national packetswitching network to demonstrate the feasibility of networking, its accessibility to non-specialized people, and its use in research and cooperation between the North and South and on a horizontal, South-South, basis.To address some of the issues mentioned above, the ILET is carrying out research activities along two lines: the utilization of data bases in order to promote the use and development of national information resources, and the utilization of the national packetswitching network to demonstrate the feasibility of networking, its accessibility to non-specialized people, and its use in research and cooperation between the North and South and on a horizontal, South-South, basis.

communication tools, where users in developing countries can radically alter the way they work.

On the other hand, growing numbers of people have the opportunity of accessing computers in Latin America. This new phenomenon, added to the offer of higher quality telecommunication infrastructure, is now creating a scenario that opens up the possibility for an explosive expansion of networking.

Finally, we have to point out the inherent risks involved in the indiscriminate adoption and use of informatics: the undermining of national sovereignty, the indiscriminate transfer of technology, the violation of personal privacy, and difficulties related to the right to access to information are but a few of the most frequently debated topics.

Informatics will not solve the political, economic and social problems that plague the more debt-ridden countries of the world, where millions of people lead a marginal existence. However, informatics may be seen as an indispensable tool for improving planning systems, for raising levels of education, and for spreading scientific and technological know-how. The democratization of communications will not be achieved by means of an irrational incorporation of new technologies.

As we have mentioned, the information is stored in large data bases belonging mostly to transnational corporations and the rights to access to these data bases by developing countries are not conferred under conditions equal to those governing access by industrialized countries.

Today transborder data flows basically obey the following pattern: data flow from the United States to other industrialized countries and from the United States to Third World countries. The United States receives flows of primary, non-processed data from both developed and developing countries. The United States then processes these data and offers, together with technology, a flow of specialized information products to the developed countries, obtaining huge revenues from the sale of these goods and services.

Third World nations, therefore, obtain information and information recovery services from industrialized countries and do not send to these countries flows of processed information. This produces a complete imbalance in terms of the directions of the data flows.

Adequate planning and cooperation are indispensable for advancement in this area. In order to take advantage of our resources and rationally utilize microcomputers coordinated action is essential. The reproduction of patterns developed for industrialized nations is not what is needed, but rather the satisfactory fulfillment of each country's specific information and communications needs.

A policy pertaining to the development of national information and communications should necessarily be based on global considerations regarding a country's needs in terms of education, innovation, research, and above all, the diffusion of knowledge, all within a framework promoting development while at the same time strengthening national sovereignty, reducing the country's technological dependence and generating more equitable conditions for the North-South relationship. The challenge has been voiced and developing countries must not remain silent.

Women, Work and Computerization: Forming New Alliances
K. Tijdens, M. Jennings, I. Wagner, and M. Weggelaar (Editors)
Elsevier Science Publishers B.V. (North-Holland)

WOMEN'S PARTICIPATION IN THE COMPUTER INDUSTRY IN SINGAPORE

Lim Gek Kheng
The Centre for Computer Studies
Ngee Ann Polytechnic
535, Clementi Road, Singapore 2159.

Singapore is a newly-industrializing country in the phase of transition from a Third World country to a developed country. The Singapore government has recognized the importance of information technology in increasing business efficiency and labour productivity and embarked on a national computerization programme in 1980. A pool of trained and experienced computer professionals is needed to drive this national computerization effort. Four centres were established in early 1980s to train computer software professionals. More than 50% of graduates from these centres are females. One probable reason is that more boys are channelled to the increasing number of places on engineering courses in the tertiary institutions.

Apparently there is no sexual discrimination in recruiting computer professionals. On the whole women are well represented in the computer software industry. However in terms of advancement to management ranks, men seem to fare better than women. The dilemma faced by women in Singapore is trying to balance between their career development and domestic commitments. According to 1986 Labour Force Survey, over 70% of the economically inactive persons are women who are married and above the age of 30. The Singapore government is encouraging women to stay in the workforce. The target is to achieve 50% female labour force participation rate by 1995.

The computer industry in Singapore is growing rapidly. As a shortage of trained and experienced computer professionals can be a constraint to the growth of the industry, it would be a waste if women who constitute over 50% of the trained professionals have to leave their jobs because of domestic commitments. As information technology is a new grown industry in Singapore, no studies are known to have made on women in this field. To gain a better understanding of the position of women in computing in Singapore more research is needed.

1. BACKGROUND

Singapore, a small island state with area of 621.7 sq.km., is a newly industrializing country (NIC) in the phase of transition from a Third World country to a developed country. Since 1980, Singapore's economy has been undergoing a structural change - moving from a low- wage manufacturing base to a high-wage capital-intensive and skill-intensive service economy. Recognizing the importance of information technology in increasing business efficiency and labour productivity, the Singapore government em-

barked on a national computerization pro gramme in 1980. The National Computer Board (NCB) was established to spearhead and co-ordinate the national computerization efforts. Survey result has shown that the number of firms using computers has increased from fewer than 2000 in 1982 to about 8000 in 1986 (NCB annual report 1986/87)). More and more computer systems have also been installed in government ministries and offices. Recognizing the need for an integrated information technology (IT) strategy,the National Information Technology Plan was launched in December 1986. The Plan has the dual objectives of developing a strong export-oriented IT industry and exploiting IT to improve productivity and competition in every sector of the economy.
To drive the national computerization effort, a pool of trained and experienced computer professionals is needed. Starting with fewer than 1000 computer professionals in 1980, the pool has expanded to a size of 5500 in 1986. The NCB's target is to have 10.000 computer software professionals by 1992. What is women's participation in this national programme? As information technology is a new growth industry in Singapore, no studies are known to have been made on women in this field. This paper is just an attempt to examine the educational and employment opportunities of women in the computer industry in Singapore and the potential difficulties faced by them based on the limited information available.

2. THE EDUCATIONAL SCENE

Generally it has been the government policy to provide equal educational opportunities for both men and women. Four tertiary centres were set up in the early 1980s to train over 600 computer professionals each year. They are:

* The Centre for Computer Studies
* The Japan-Singapore Institute of Software Technology
* The Institute of Systems Science
* The Department of Information Systems and Computer Science,
* National University of Singapore.

Table 1a-1d shows the number of students, analyzed by sex, graduating from these four institutions. It is interesting to note that women constitute more than 50% of these graduates, in contrast to the male dominance in similar courses in the more developed countries (NCC - Women in Computing). The high proportion of females in computer studies is an exception compared to other technical courses in Singapore. Technical courses have always been male dominated (see Table 2). The low representation of females in technical courses is expected to continue since an education policy was introduced in 1985 affecting students in secondary 1 and 2 (i.e. students aged 13 to 14). This policy allows only boys to do technical subjects in secondary 1 and 2 whereas girls have to take Home Economics. Schoolgirls can only take technical studies when they reach Secondary 3 and 4. The rationale for the change in policy was to ensure that "girls should be girls" (Straits Times 9/9/84). In the past few years with a tremendous increase in intake of students into the engineering courses in the University and the two polytechnics in Singapore (see Table 3),more boys are channelled to the engineering discipline. As acceptance on courses is based on merit i.e. academic results, better qualified girls who do not apply to study engineering are accepted for less technical- oriented courses like computer studies and business studies. This may account for the high proportion of girls studying in the four centres which were established to train computer professionals.
Before the first batch of graduates from the four centres entered the employment market in computer industry, the source of Singapore's supply of computer analysts and programmers had been mainly the science graduates (mathematics, physics and chemistry) who are predominantly females and a handful of overseas graduates in computer science.

Table 1a: Number of graduates from the centre for computer studies.
(Source: The centre for Computer Studies)

Year of graduation	male no.	%	female no.	%	total no.
Course: Higher national diploma in computer studies					
1984	48	45.7%	57	54.3%	105
1985	19	18.4%	84	81.6%	103
1986	79	40.9%	114	59.1%	193
1987	98	49.2%	101	50.8%	199
Total	244	40.7%	356	59.3%	600

Table 1b: Number of graduates from Japan-Singapore Institute of Software Technology.
(Source: Japan-Singapore Institute of Software Technology.)

Year of graduation	male no.	%	female no.	%	total no.
Course: Analyst/Programmer Diploma					
1984	26	57.8%	19	42.2%	45
1985	35	36.1%	62	63.9%	97
1986	35	35.4%	64	64.6%	99
1987	41	41.8%	57	58.2%	98
Course: Certificate in systems analysts					
1983	17	63.0%	10	37.0%	27
1984	26	52.0%	24	48.0%	50
1985	16	42.1%	22	57.9%	38
1986	29	42.0-%	40	58.0%	69
1987	23	31.1%	51	68.9%	74
Course: Diploma in systems analysis					
1984	10	66.7%	5	33.3%	15
1985	14	73.7%	5	26.3%	19
1986	15	41.7%	21	58.3%	36
1987	12	33.3%	24	66.7%	36
Total	299	42.5%	404	57.5%	703

Table 1c Number of graduates from Institute of Systems Science
(Source: Institute of Systems Science)

Year of graduation	male no.	%	female no.	%	total no.
Course: Postgraduate Diploma in systems analysts					
1982	12	37.5%	20	62.5%	32
1983	16	38.1%	26	61.9%	42
1984	22	51.2%	21	48.8%	43
1985	24	53.3%	21	46.7%	45
1986	18	36.7%	31	63.3%	49
1987	24	49.0%	25	51.0%	49
Total	116	44.6%	144	55.4%	260

Table 1d Number of graduates from the Department of Information Systems & Computer Science, National University of Singapore (Source: Department of Information Systems & Computer Science, National University of Singapore)

Year of graduation	male no.	%	female no.	%	total no.
1983	23	25.3%	68	74.7%	91
1984	19	17.8%	88	82.2%	107
1985	67	51.1%	48	48.9%	131
1986	88	45.8%	104	54.2%	192
1987	74	35.9%	132	64.1%	206
Total	271	37.3%	456	62.7%	727

Table 2 Student enrolment in engineering courses in 1986/87
(Source: Annual reports of National University of Singapore, Nanyang Technological Institute, Singapore Polytechnic, Ngee Ann Polytechnic, 1986/87)

Institutions/courses	No. of students Total	Female	Female %
National University of Singapore (electrical, mechanical, civil and chemical engineering)	2027	251	12.4%
Nanyang Technological Institute (electrical, mechanical and civil engineering)	2202	302	13.7%
Singapore Polytechnic (mechanical & manufacturing, electrical, electronics &, communication engineering)	8631	1217	14.1%
Ngee Ann Polytechnic (mechanical electrical &electronic engineering)	5757	1027	17.8%

Table 3 Growth in student population in engineering courses
(Source: Annual reports of National University of Singapore, Nanyang Technological Institute, Singapore Polytechnic, Ngee Ann Polytechnic, 1986/87)

Institutions/courses	1982/83	1986/87	Increase over 1982
National University of Singapore (electrical, mechanical, civil and chemical engineering)	2120	2027	-4.39%
Nanyang Technological Institute (electrical, mechanical and civil engineering)	585	2202	276.41%
Singapore Polytechnic (mechanical & manufacturing, electrical, electronics &, communication engineering)	5614	8631	53.74%
Ngee Ann Polytechnic (mechanical, electrical & electronic engineering)	2367	5757	143.22%

Table 4 Labour force participation rate in Singapore by age group and sex as at June 1986
(Source: Singapore Labour Force Survey 1986)

Age group	male%	fem.%	total%	Age group	male%	fem.%	total%
All persons aged 15 and over	79.4	45.6	62.3	40-44	98.0	44.6	70.3
15-19	29.3	28.8	29.1	45-49	96.6	36.6	65.6
20-24	88.5	80.8	84.3	50-54	90.5	28.0	58.9
25-29	97.2	69.8	83.5	55-59	68.7	17.4	43.7
30-34	98.3	52.3	74.9	60-64	47.3	10.3	28.7
35-39	98.6	46.8	72.2	65 and over	23.3	5.9	13.6

It appears that more females than males are educated in Singapore for employment in the computer industry. In reality they represent only the "soft sector" of the industry: the computer software industry. Computer professionals in the "hard sector" -in the areas of telecommunications, networking, computer engineering- are usually recruited from graduates from male-dominated electronics engineering discipline. The tendency for males to choose the more technical subjects are reflected by the higher proportion of boys in the Computer Architecture option in The Centre For Computer Studies. In 1986 when the option studies were first introduced, there were only 11 girls out of a total of 48 students in the Computer Architecture option. In 1987 only 17 out of 146 girls opted for it.

3. THE EMPLOYMENT SCENE

Jobs in the computer industry are opened to new entrants, both males and females, provided that they possess the aptitude and relevant training. Apparently there is no sexual discrimination on the part of the employers in recruiting computer professionals. According to the 1987 IT Manpower Survey conducted by NCB, women make up 41% of the workforce working in the computer industry. 58% of the applications/analyst programmers and 52% of the systems analysts/designers are females. However where jobs require frequent traveling, like marketing representatives, women participation rate is about 30%.

In terms of advancement, men seem to fare better than women, although there are no statistics on the number of women at senior levels in computer organizations in Singapore. However a check in February 1988 with the Data Processing Managers' Association whose members are DP professionals in management position showed that out of 67 members, only 6 are women. This figure suggests the existence of discriminatory practice in promoting computer professionals to management ranks. However some successful female data processing managers are not aware of any discrimination. Low representation of women at senior levels is common in all professions. Many reasons could be raised for the relatively small number of women in data processing management: women are not career conscious when they get married, they are less able to accept responsibility because of family commitments; they lack emotional stability, etc. These are merely speculations; research needs to be conducted to identify the actual reasons.

4. THE DILEMMA FACED BY WOMEN

All married working women face the similar problems in balancing between their career development and domestic commitments. Single women are not spare these worries either. They have to decide on whether to sacrifice the satisfaction of their personal growth need in career development in order to fulfil their roles as wives and mothers.

Women constitute a third of the labour force in Singapore. Female labour force participation rate (FLFPR) rose from 36.4% in 1976 to 45.6% in 1986. This is mainly due to an increase in better educated young entrants into the labour market. FLFPR is the highest in the age group 20-24 (Table 4) but declines continuously thereafter. According to the 1986 Labour Force Survey, women comprised 73% of the total economically inactive persons and these women were mostly housewives, married and above the age of 30. This means that many married women in Singapore leave their jobs and do not return to the work force. These women were mostly housewives, married and above the age of 30. This means that many married women in Singapore leave their jobs and do not return to the work force. These women are more likely from the lower income group who cannot afford to get help in their domestic duties. There would be more women in the economically inactive group if professional working women had not been able to be relieved from their domestic responsibility by employing maids from overseas such as the Philippines and Sri Lanka or by obtaining assistance from the in-laws.

The Economic Committee, appointed by the government in 1985 to review the progress of Singapore economy and to identify new direction for its future growth, proposed that the target of 50% FLFPR be achieved by 1995. It is felt that women should not stop working just because of child-care and household commitments unless it is their personal choice. Three specific measures were recommended to encourage women to remain in the workforce (The Economic Committee Report, 1986, page 107):

1) encourage part-time employment and job sharing
2) provide more child-care facilities at affordable prices
3) encourage women with family commitments to work at home instead of quitting their jobs

Child-care facilities, initiated by the government, has increased from 40 in 1984 to 91 in 1986 and 20-25 new centres are expected to be established each year (Strait Times, 28/3/86). Having child-care facilities is not a solution for the married women who prefers to play a more active role in bringing up their children. Part-time employment and working at home appear to be better alternatives. Women can look after their children and at the same time provide financial support to the family and keep in touch with their jobs. The concepts of working part-time and working at home are not new to women belonging to the lower-income group in Singapore. Many women in this category are contributing to their families financially by working part-time or at home e.g. working as waitresses in restaurants or working at home as child minders. However to professional women, these are new concepts. The professional women are those who are better educated and who struggle to get out of the kitchen and into a job. Their motivation is to be financially independent and to achieve personal growth by building up a career. Giving up a full-time job may not appeal to some women. It would be interesting to find out the views on working at home and part- time jobs from the professional women themselves.

5. CONCLUSION

The computer industry in Singapore is growing rapidly. Sales in software and professional services increased from S$199 million in 1986 (NCB Annual Report 1986/87). As shortage of trained and experienced computer professionals can be a constraint to the growth of the industry, it is vital that women who constitute over 50% of the trained professionals should be encouraged to remain in the workforce as long as they want. The main problem faced by women computer professionals is not so much sexual discrimination in education and recruitment, but is more likely the expectation of them to fulfil the roles of wife, mother and wage-earner.

Having realized the need to encourage women to stay in the work force, both the government and employers should seek to adopt various schemes - working at home, working part-time on site, working flexible hours - to tap on this valuable reservoir of ability. Theoretically speaking, women working in the computer software industry are most suited to work either at home or part-time because of the nature of their work. Analysts and programmers are usually self-motivated and require less supervision and their tasks can be more readily performed out of the office, provided they are able to plug into a computer at home.

Societal values and attitudes need to be changed too. Women are still expected to cope with household responsibilities even though they go out of the house to work. Child-care is mainly the mother's responsibility, e.g. a working mother has to take leave to take care of a sick child. The compulsory Home Economics course for girls in schools tends to reinforce the idea that women are responsible for household tasks. Unless the conflict between career and household responsibilities is resolved, women participation in the workforce will be limited. It would be a waste if women are given the educational and employment opportunities but have to leave their jobs because of domestic commitments.

To gain a better understanding of the position of women in computing in Singapore, more research is needed on the following areas:

a) women's participation in the computer industry in terms of the types of jobs and positions they hold;
b) employers' attitude towards women's career development in the computer industry;
c) women's attitudes towards their own roles at home and at work.

BIBLIOGRAPHY

Ministry of Labour, Report of the Labour Force Survey of Singapore (Singapore, 1986).

Ministry of Trade and Industry, Report of the Economic Committee (Singapore 1986).

Nanyang Technological Institute, Annual Report 1986/87 and Annual Report 1982/83.

National Computer Board, Annual Report 1986/87.

National University of Singapore, Annual Report 1986/87 and Annual Report 1982/83.

Ngee Ann Polytechnic, Annual Report 1986/87 and Annual Report 1982/83.

Simons, G.L., Women in Computing (NCC, England, 1981).

Singapore Polytechnic, Annual Report 1986/87 and Annual Report 1982/83

Wee, Vivienne, Women in Science and Technology: The Situation in Singapore (paper presented at the Asian Workshop on Popularization of Science and Technology to Enhance Women's Participation, January, 1987, Dhaka, Bangladesh)

Women, Work and Computerization: Forming New Alliances
K. Tijdens, M. Jennings, I. Wagner, and M. Weggelaar (Editors)
Elsevier Science Publishers B.V. (North-Holland)

WOMEN, WORK AND MICRO-ELECTRONICS: FOCUS ON THE PHILIPPINES*

Rosalinda Pineda-Ofreneo
International Studies Institute of the Philippines
University of the Philippines Law Complex,
Diliman, Quezon City, The Philippines

1. INTRODUCTION

The impact of microelectronics on women and work in the Philippines has been unevenly researched, with a lot of studies focussing on the semiconductor industry in contrast with rather scant material dealing with the introduction of computers and other forms of automation in banking, telecommunication and other sectors of economy. This short paper, although reflecting this grievous imbalance, nevertheless attempts to summarize what has been revealed and to draw some implications on the level both of policy options and women workers' response through the trade unions and other organizations.

The preoccupation with semiconductor manufacturing is quite understandable, because it has been a controversial topic for the last 15 years or more, exemplifying in the most graphic terms the transnational exploitation of women in so-called "low-wage" countries. Young girls barely out of their teens enter the industrial work force through the electronics factories, receive a mere pittance compared to their counterparts in the more advanced industrialized states, work under strict labour control with strong patriarchal overtones, and endure periodic and sometimes permanent retrenchment when the global demand for their products disappears or takes a nose-drive.

The Philippines in the mid-70s became an attractive site for the relocation of semiconductor assembly work because wages, "due to a combination of unionization and a growing labour shortage", were already rising in Taiwan, Hongkong, South Korea and Singapore, where the initial transfer occurred in the 1960s. In fact, UNCTAD reported in 1975 that among workers in various offshore locations the Filipino worker was the lowest paid, getting 20 cents an hour compared to 37 cents for the Malaysian, 40 cents for the Korean, 60 cents for the Hongkongian and 57 cents for the Singaporean.[1] In terms of hourly wages in 1981, the Philippine rate (at 30 US cents) was still one of the lowest, equivalent to only four percent of US, six percent of Australian, eight percent of Japanese, 33 percent of Singaporean, 30 percent of Taiwanese, and 44 percent of South Korean levels.[2]

Intel spearheaded the establishment of transnational subsidiaries in 1974, followed by National Semiconductor (US), Advanced Micro De vices (US), Telefunken Semiconductor (West German), Timex (US), Signetics (US), etc. In early 1978, there

---

* This is largely based on "women and the Electronics in the Philippines", a longer paper presented by the author at the ASEAN Experts' Seminar on Women and Technological Innovation held 20-22 April 1987 at the Universiti Pertanian Malaysia under the sponsorship of UNESCO.

1 Paglaban, E., Philippines: Workers in the Export Industry, Pacific Research IX, 3 and 4 (March-June 1978) p. 15.

2 Plight of Asian Workers in Electronics (Hongkong: CCA-URM, 1982) p. 22.

were already 17 semiconductor firms in the country, employing 20,000 workers, 90 percent women.[1] The number of transnational subsidiaries increased from a mere 18 percent of the total number of semiconductor firms in 1987 to 65 per cent in 1980.[2] The Filipino firms, even if not directly controlled by foreign interests lording it over the global marketplace, nevertheless remained tied to and dependent on the latter as suppliers.

Within a decade, the semiconductor industry experienced spectacular growth. From a few hundred workers in the beginning, the industry's work force swelled to 8000 in 1975, and to almost 40,000 in 1984.[3] Export performance rose from $44.6 million in 1975 to $1.3 billion in 1984. Then boom turned into bust. In mid-1985, as the industry reeled from the oversupply of microchips in the global market, employment registered a 23.8 percent decrease (from 39,818 in 1984 to 30,343).[4] Massive lay-offs continued and the number of workers employed by the industry went down to a shocking 21,275 in December 1986.[5]

It is obvious that the semiconductor firms in the Philippines do not have a life of their own. Their umbilical cords remain attached to the foreign companies they supply with their products and from which they import the necessary raw materials, primarily the silicon wafers. They take care of only the middle phase in a long and complicated production process; they have no knowledge or control of the beginning and end of this process. Without the foreign market and the foreign technology, they have nothing. They sway with the wind from the North, cutting production and laying off workers when recession strikes and the demand falls in the global marketplace.

But recession is not the only problem. An even more disturbing development is the introduction of automated processes "enabling one worker to produce 10 times as much as she does manually".[6]Even the assembly phase of semiconductor manufacturing may already be carried out in the home bases of the transnationals, with the use of highly automated factories such as the one built by Motorola Inc. in Arizona.[7] As automation becomes cheaper, the time will come when it will cost less than maintaining an operator. This means that jobs not only in the developed but also in the developing countries will be endanger ed. The relative competitiveness of cheap Asian labour will no longer be a decisive factor, thereby negating the main reason why certain phases of industrial production are relocated in the Third World. The whole rational for the new international division of labour as it has been traditionally understood collapses and economies like the Philippines' which have been tailored precisely to function within this scheme will be left holding the proverbial empty bag.

2. TRANSNATIONAL EMPLOYMENT: GOOD OR BAD?

The impact of transnational an related employment on women is well summed up in a UNIDO study:

> Because of the importance to them of low production costs, some transnationals tend to utilize unskilled and less skilled labour who are generally not

1 Paglaban, E., op.cit., p. 16.

2 ITC (Industry Tripartite Council), Newsletter on the semiconductor industry, I, 1 (August, 1985), p. 2.

3 Ibid.

4 Ibid.

5 Sectoral paper on the semiconductor industry, National Consultive Workshop for Women and Young Workers sponsored by the Bureau of Women and Minors, Ministry of Labour and Employment, August 1986, p. 2.

6 Grossman, R., Woman's Place in the Integrated Circuit, Special joint issue of the Southeast Asian Chronicle no. 66 (January-February 1979) and Pacific Research, Vol. 9, Nos. 5-6 (July-October 1978), p. 9.

7 Sharp, J., US electronics firms turn to local workforce, Bulletin today (March 11, 1984).

> utilized and therefore willing to work for low wages in often unsatisfactory working environments. It is in this context that the generation of jobs for women by transnationals should be regarded. While providing job opportunities for women, transnationals tend to have an adverse affect on women's socio-economic status. In some instances the workers are paid a subsistence wage with little hope of a raise; sometimes institutional pressures are exerted to prevent the organization of unions so as not to discourage foreign investment. The skills taught inmost of these industries ar not transferable, thus increasing the insecurity of the workers in a situation where employment is highly dependent of fluctuations in the world market .[1]

There are other costs. Electronics, according to the US Occupational Safety and Health Administration (OSHA), is one of the "high health risk industries using the greatest number of hazardous substances".[2] Migraine, headache, chronic conjunctivitis, astigmatism, and eye-strain are common among workers. Impaired vision is the cumulative effect of years of intensive assembly work. Exposure to heat and cold, to dangerous chemicals like toxic fumes and carcinogens, lead to nausea, lung trouble, acid burns, skin rashes , various allergic reactions and even serious ailments that may directly or indirectly cause death.

In general, working conditions in the semiconductor firms are not adequately monitored, and health and safety standards are not substantially enforces because of lack of resources or jurisdiction on the part of agencies attached to the Department of Labour and Employment and the Department of Health. Workers medical files are not in order, making it difficult for them to claim compensation benefits. They are victimized by or remain ignorant of the hazards they are exposed to because of management neglect.

The stress is both physical and mental as shifts are rotated to include night work and as quotas are periodically raised to increase the output per worker. An additional source of stress is sexual harassment from the male supervisors, department heads and managers who pressure the women to go out with them or be their mistresses in ex change for favourable appraisals for promotion and salary increase.

If working conditions, despite the veneer f cleanliness, leave much t be desired, living conditions are likewise far from comfortable, especially for those who have to support not only themselves but also their families by their wages. Belt-tightening as a result of limited and declining purchasing power in the midst of rapid inflation is the lot not only of women workers in the electronics industry but also most workers irrespective of sex. But even more disadvantaged are those who receive even less than the average wage during the deliberately pro longed apprenticeship and probationary period in semiconductor firms.

From the foregoing, it is obvious that women workers in the semi conductor industry have to contend with many disadvantages, which in certain respects seem to outweigh the advantages. Yet, the question also presents itself: What is better, transnational employment or no employment at all? Despite its negative aspects, some quarters maintain that the very inclusion of young women in the active labour force is progressive because it directly involves them in social production. The wages in semiconductor factories may be low by the standards of developed countries but may even be better than average for Third World states. Having a job by itself may be partly "liberating" for female workers in the "feminist sense of weakening female subordination and the sexual division of labour in indigenous Third World societies." As one author says:

---

1 United Nations Industrial Development Organization (UNIDO), Women and Industrialization in Developing Countries (New York, 1981), p. 43.

2 Lim, Linda Y.C., Multinational Export Factories and Women Workers in the Third World: A Review of Theory and Evidence, Third World Studies Dependency Papers Series, No. 51 (November 1983), p. 6.

it does improve their lives - increasing their freedom, independence, worth, status, contribution to family support, access to experience outside the family, to small luxuries and the companionship of other women - all of which are highly valued by the women themselves. It also enables young women to delay marriage and child-bearing, and to choose their own marriage partners. As more women continue working after marriage, the traditional sexual division of labour within the family changes. Working wives and husbands increasingly share the powers and responsibilities of housework, child care and family financial decision-making.[1]

3. LABOR CONTROL AND WORKERS' RIGHTS.

Since export orientation is premised on low wages, high labour productivity, prevention of work stoppages which may disrupt production, and the maintenance of a stable and conducive business environment, unions must be discouraged, strikes must be prohibited or at least strictly regulated, and privileges must be given to the electronics enterprises.

Management tries to control their workers in many ways. Quotas force them to produce at maximum speed under threat of lay-off. Loyalty to the company is assiduously cultivated by managers projecting a father image, a strong authority figure in many Asian cultures. Values like obedience to paternal fiat, hard work, honesty, discipline and self-denial are exploited so that women "may accept their subordinate position at the bottom of the factory hierarchy, as they do in traditional patriarchal societies".[2] Diversionary tactics, too, are employed, such as sponsorship of athletics, raffles, parties in five- star hotels, various cultural and recreational programs, and even beauty contests. Prizes, bonuses and other tangible forms of recognition are awarded to the most productive workers. Western-style consumerism with emphasis on tight jeans, cosmetics, disco, etc. give the women a veneer of urban chic and a false sense of superiority in being part of an "elite" work force. Says one critic: By "stressing Western versions of feminine passivity, the companies have been able to prevent the workers from realizing their potential for independence"[3], and for asserting their rights. The high value given to sex appeal, having a boyfriend or catching a husband has a disorienting affect and results in destructive competition for potential mates among the disproportionately few males within the workplace and the immediate environment.

If such techniques do not work to discourage union formation, management resorts to plain union-busting. Organizers are dismissed or suspended or a rival company union is formed to sabotage the real one. Workers who join unions are threatened with lay-off. Union leaders are bribed, harassed or even imprisoned to turn their backs on the workers' interests. Some workers are asked to spy on others for quick and easy detection - and ultimately prevention - of union organizing.

Among the unionized firms, there are concrete bases for conflict between the union members and the non-union ones (such as the casuals) who do not receive benefits, between various categories of workers (daily paid and monthly paid), between rival factions wishing to retain or capture the union leadership, and between male union officials and the predominantly female rank-and-file. The practice of having males in key positions in some unions in industries numerically dominated by women affirms the national reality of male initiative and female "followership" if not passivity in unionism.

---

1 UNIDO, Women in the Redeployment of Manufacturing Industry to Developing Countries, UNIDO Working Papers on Structural Changes, No. 18 (July 1980), p. 17.

2 Ibid.

3 Grossman, op.cit., See also: Sinay Aguilar, M.V., Partakers in International Subcontracting: Women Workers in Asia, paper presented in partial fulfilment of the degree of Master of Arts in Development Studies, Institute of Social Studies (The Hague, October 1985).

However, it is not correct to say that women workers cannot transcend their own fears and preconceived role limitations. There have lately been many cases of militant actions spearheaded by female trade unionists in the Bataan Export Processing Zone and elsewhere, showing that gender subordination is not an immutable reality in the labour movement. Some notably militant unions in electronics, garments and food processing are already headed by women. In some strike situations, married female workers already go to the extent of separating from their husbands just to remain in their unions struggle.

4. SOME OPTIONS

Both host country and local worker are not compensated commensurately by the electronics industry, specifically by its semiconductor subsector. Given this reality, should the option be outright removal of transnational-dominated assembly work? This would only deprive women of employment in a situation where they really have no alternative opportunities and worse, where massive lay-offs have become commonplace. Such employment, despite or because of its negative aspects, would lead to the development of progressive consciousness among women workers and force them to organize in a meaningful way even if this would seem to be a protracted struggle. This progressive consciousness can be on three levels: the national level, when they see themselves as Filipinas united with other sectors of society in a broad movement against foreign domination; the class level, when they act as workers leading and sustaining their own unions in striving for their just due; and the feminist level, when they are able to remove the cultural blinders and stand for equality between the sexes both at home and in the workplace.

Depending on their strength, women workers can pressure their own governments to "remove restrictions on labour organization, enact more favourable labour laws, e.g. with respect to wages and hours of work, impose regulations on health and safety, set up institutional arrangements to reduce high turnover among women workers, e.g. through better wages and working conditions, and public child-care facilities so that married women and mothers can continue working".[1]

On the union level, the following issues relating to automation could be taken up in collective bargaining negotiations: "information and consultation rights prior to the introduction of new technology; right to retention of pay rate of the newly-structured job if it requires less skill than it formerly did; employer-financed training and retraining; provision of job rotation and job enrichment program; severance pay provisions; restriction on time-and-motion monitoring on production quota setting; etc.".[2]

Over the long pull, however, the semiconductor industry will have to be critically reexamined, especially in the light of the great technological divide between the industrialized and the developing countries now made even more glaring by the fast-paced and mind-boggling advances in the field of micro-electronics. On this issue, no less than the experts meeting during the UNIDO International Forum on Technological Advances and Development (Tbilisi, USSR, 12-16 April 1983) warned that "unless the developing countries take appropriate measures to build up an endogenous capacity, there may be a never- ending cycle of imported technologies and dependence on the developed countries, the products and systems as presently developed may not necessarily be relevant to the needs of the developing countries."

In other words, there should be real transfer of technology so that countries like the Philippines can make their own computers instead of being consigned to just

---

1 Lim, Linda, op.cit.

2 Casel, J.Q., and Palafox, Juan A.F., The Impact of New Technology on Employment in Selected Banking, Electronics and Telecommunications Establishments in Metro Manila, Institute of Industrial Relations, University of the Philippines, Diliman, Quezon City, 1985.

assembling the chips that go into computers. This is the only way women workers in microelectronics can be secure in their jobs. But it is a long way, and the women workers themselves must help develop the political will to move forward.

Women, Work and Computerization: Forming New Alliances
K. Tijdens, M. Jennings, I. Wagner, and M. Weggelaar (Editors)
Elsevier Science Publishers B.V. (North-Holland)

APPENDIX 1

Report of the conference Women - work and computerization - Opportunities, organized by IFIP, Working Group 9.1.: Computers And Work, in DUBLIN, IRELAND, August 27 - 29, 1986

Ellen de Ru and Loes Steensma, Netherlands PTT (Social Science Research)
Kea Tijdens, University of Amsterdam (Department of Social Science Informatics)

1. INTRODUCTION

The International Federation of Information Processing (IFIP) organized a conference on Women, Work and Computerization in Dublin, on August 27, 28 and 29, 1986. The delegation from the Netherlands consisted of two female researchers from the PTT and one from the University of Amsterdam. They wrote this report.
The IFIP is an international organization of computer experts. The Netherlands is represented in the federation by the Dutch Informatics Society. Within the IFIP there are several working groups which study different issues. Working group 9.1. deals with computers and work.
Two years ago a number of women from this group organized the first conference on Women, Work and Computerization. It was held at Riva del Sole, Italy, and the report of the proceedings was published in 1985.[1] The conference proved such a success that it was decided to organize another one. It took place in Dublin in August 1986.[2] The Dublin conference was attended by some 60 women plus a few men, moistly from Europe. There was a particularly large group of delegates from Sweden. The delegates represented a variety of sectors. Some were employees of companies in the private sector, while others worked for government departments, universities, trade unions and international research institutes. Their professional background also varied: system analysts, instructors, social scientists, policy advisers, users, trade unionists and politicians.
The program consisted of papers given in plenary session, while additionally there were six discussion groups which met frequently but briefly over the three days and submitted reports of their talks at the end of the conference.
The conference was concluded with a "Speakers' Corner" at which delegates had an opportunity to put forward their own views. Beside evaluative comments and suggestions for a follow-up conference, two motions were put to delegates during this session. One called for the barring of representatives from South Africa from the IFIP's 10th World Congress. The other expressed condemnation of the extensive use made of computers for military purposes. Both motions were adopted with large majorities.

2. DISCUSSION OF THE PAPERS

The conference looked primarily at the question of the stance women need to adopt towards new technologies in order to improve the quality of their work (from the point

---

1 A. Olerup, L. Schneider and E. Monod (eds): Women, work and computerization: Opportunities and disadvantages, North-Holland Publishing Company, Amsterdam, 1985

2 The organizing committee consisted of Boel Carlsson (Sweden), Ingela Larsson (Sweden), Krisitine Thomsen (Denmark), Ina Wagner (Austria) and John Drumm (Ireland).

of view of content) and their position on the labour market.The different backgrounds of the delegates meant that a wide variety of subjects and points of view came up for discussion. Matters on which there was general agreement included:

* women have major interest in new technological developments, because computerization is often introduced precisely in those fields of work where women are the predominant force, and because newly-created jobs will to a lesser or greater degree require knowledge of computers;
* women need t o adopt an active approach to computerization so that they can influence developments, because if they fail to do so they will find themselves at a disadvantage on the labour market, the quality of their work will be impaired and jobs are likely to disappear.

Broadly speaking, the papers presented at the conference related to three main issues: training, system analysis and design, and the position of women on the labour market. In this report these three issues are dealt with first. Subsequent, the main points of the conference are summarizes, followed by some comments by the authors.

2.1 Training

In the papers on this subject, considerably importance was attached to the fact that it is usually women instructors who teach other women how to use computers. It was emphasized that allowance must also be made in the design and content of these courses for the specific backgrounds and learning strategies of women.
The British duo Mary Jennings and Ruth Wallsgrove, of the Microsyster consultancy bureau, put forward some particularly useful ideas as regards the transfer of knowledge to women. Microsyster was established as a non-commercial bureau, subsidized by the city of London, to provide guidance and advice to women on computer-related mattes. They assists women's groups in London in the procurement and use of microcomputers and software packets. They said that had noticed in their word that women are far more reluctant that men to experiment themselves with technical devices. Women need a teaching method which is geared to what they must do in everyday situations.
More information on training and guidance provided to women was given by Marja Vehviläinen from Finland. She noted that due to the nature of their jobs, women are usually unable to exert much influence on the way their working life is organized and on the role which computer systems play in their work. Vehviläinen gave details of a method enabling women to acquire a deeper insight into computerized systems an to get to grips with them. This was achieved in one case in Finland by the formation of a 'study-circle' comprising ten employees with secretarial duties (nine women and one man) who met for two hours a week for a period of six months. At these meetings, papers were given, the participants practiced on computer systems, their individual working situations were analyzed, and the use of computers was discussed. The result was that at the end of the six months each of the participants felt more at ease with computerized systems.
The paper of Ylva Ericsson of the Swedish Ministry of Social Affairs and Employment dealt with the way the government in her country is endeavouring to encourage the employment of women and girls in occupations of a technical nature. In the Swedish town of Trolhättan, for example, there was a great shortage of technicians and engineers. A project was therefore initiated whereby unemployed women were given the opportunity to take a course especially designed to enable them to get a highly-qualified technical job. The project was financed by the Swedish government and was implemented by the regional employment office in cooperation with the Institute for Labour Market Education and the Saab-Scania concern. Saab-Scania guarantied the women who successfully completed the course a job in the concern's production or maintenance department. A large number of women applied to take the corse, which comprised one year of theory, one year of on-the-job training, and, finally, another

year of theory. The course focused primarily on electronics and computer technology. At the start of the project the relevant departmental heads of Saab-Scania were invited to meet the trainees and each women was assigned a contact person within the company as a source of information. The project proved to be very successful: 31 women completed the course and now work at Saab-Scania. Since the course was so successful, a second project was started.
The International Labour Organization (ILO) is also encouraging the employment of women in technical professions. To this end the ILO is conducting research into successful projects in Third World countries, so as to include the results of such projects in training programmes. Heidrun Kaiser of the Federal Republic of Germany gave the conference details of a case study in the Philippines. Due to the brain drain, there is a great shortage of computer experts in that country. For that reason, large numbers of men and women are receiving government-sponsored training for this field of work.

System analysis and design

During this part of the conference, subjects related to knowledge of work organization, user participation and new methods of system development were discussed. Various speakers asserted that female system analysts and designers were better placed that their male colleagues to analyze the work performed by women. It was also claimed that they can more easily create a basis for mutual trust and confidence that male analysts and designers; in this context mention was made of the social skills possessed by women.
Ingela Josefsson (Sweden), one of the first speakers at the conference, dealt with specific features of work for which no allowance is made in the present methods of computerization. She referred to the importance of 'tacit knowledge', i.e. knowledge based partly on experience and partly on intuition which is used implicitly when a situation has to be assessed. Referring to Watzlawick's communication theories and other sources, she raised the very pertinent question whether such knowledge can be "automated".
Gro Bjerknes and Tone Bratteteig, two system designers attached to the University of Oslo, gave the conference details of their efforts to develop a system for nurses at two hospitals. The purpose of the system is to provide a summary of the attention a patient has received in a given period and an overview of what remains to be done. The aim is to facilitate the transfer of information to the next shift of nurses or to the doctors. They said it was difficult with the customary methods of system analysis and design to produce a system that nurses themselves found practicable and that made allowance for the often instinctive judgements which are of crucial importance in the nursing profession. In order to develop a better system, the two designers are now employing different methods, such as participative observation, in-depth interviews and production of various prototypes which can be discussed with the nurses so as to arrive at the best solution.
The Swedish cartographer Cecilia Tygesjö, who works for the city of Malmö, is participating in a project called MBK90 whereby maps are produced by a computer. The aim of the project is not only to develop new equipment, but also to create new kinds of working activities so as to improve the work of all concerned.
Another delegate from Sweden, Jarmo Sandberg, a system analyst, emphasized the importance of involving users in efforts to improve the quality of computer systems and the work of those who use them, As a system designer he co-operated in the development of a system., for issue by the Swedish department of housing and its 24 regional offices, for the granting of subsidies.
The paper given by Riita Kalmi of Finland contained an idea for a new method of system analysis and design: the Human Scale Information System Methodology. She noted that problems experienced in the use of information systems can usually be trac-

ked back to the gap which exists between information technology and human beings. Too often technology is regarded as an unassailable fact of life, while the social aspects are seen as inevitable consequences of its existence. Kalmi put forward a different line of approach, based on the user's situation. The general idea is to relieve large integrated information systems of their anonymity and take away their 'black box' image by describing them in terms of Human Scale Information Systems (HIS). By means of this HIS approach, users learn to look upon themselves as an important link in the chain of communication processes which take place in an organization. The user performs a certain task and the computer is merely an aid. In the Knowledge and Work Project, the HIS approach is being developed further and it is being tested by means of four empiric studies. Kalmi said it had been established that the new approach will enable improvement of the quality of work and of user situations, while it will also help users in their everyday work with information systems. In the longer term, it may also facilitate the introduction of other computerized systems (e.g. personal computing). Since women account for a major proportion of office staff, this approach should be given serious consideration by women who work with computerized systems.

2.3 The position of women on the labour market

The papers devoted little attention top the consequences f computerization for the position of women on the labour market. Newspaper cuttings gave delegates some insight into the situation in Ireland. At the opening of the conference, Ireland's Minister for Social Affairs, Gemma Hussey, outlined the computer training available to women in her country. She noted that these facilities had only got off the ground after a women had become Ireland's Minister for Education. Women were encouraged to take existing courses and new ones were being launched. While an increasing number of women are taking such courses, however, their fields of work seem to be largely restricted to secretarial duties, programmers and/or analysis. Women are hardly ever appointed project managers.

Ina Wagner from Austria also looked at the working situation if women. On the basis of research results from seven case-studies and an investigation conducted among 1,245 employees (male and female) involved in office automation she concluded that there is not one dominant organizational concept by means of which employers introduce computerization. Women have a variety of different experiences as regards to automation, depending on their occupation, on their position in the office hierarchy and on the automation strategy adopted by management. It emerged that the position of secretaries improves as a result of computerization, because more information is available, because they can answer questions faster and because they can provide technical support to managers. Typists continue to perform a very limited task. The position of female clerks who perform routine work is threatened most by computerization. On the other hand, automation offers new opportunities to the more experienced office staff, whether male or female, particularly if they know something about computers. The upshot of this situation is that the costs and benefits of computerization are divided unequally over the different categories of employees. Wagner claimed this was a conscious choice by management as a strategy to increase control over the labour process. Different interests make it difficult for women to undertake joint action. On the other hand, there are also some common experiences with computerization: the pressure and pace of work is increasing, there are health problems and more and more work is being done by computers.

The paper given by Kea Tijdens of the Netherlands dealt with the job-segregation of the sexes in the banking world in the Netherlands. She noted that the negative consequences of computerization are greater for female employees than for their male colleagues. In the 1970's, an increasing number of young women were recruited as bank clerks and receptionists. As a result the number of female employees is now close to the number of male employees. However, their positions differ markedly from the point of view of salary and career prospects. In recent years the gap between men and women has widened rather than narrowed. Part-time work among women is increasing, more and more women are working on the basis of temporary contracts or employment, clerical jobs are disappearing due to computerization and so on. The conclusion drawn by Tijdens was that bank executives had managed to create a flexible, female segment of the work force.

3. EVALUATION OF THE CONFERENCE

3.1 Conclusions

The conference chairwomen, Boel Carlsson, emphasized in her opening address that the aim should be to develop strategies for women which make it possible to utilize computerization in a way which improves the quality of the posts held by women and their positions on the labour market. The conference made a contribution towards achievement of this objective in the sense that information emerged on:

* experiments with training for women focused on non-traditional occupations (Sweden, Italy, the Netherlands, Federal Republic of Germany, ILO);
* experiments with recruitment and training of women for technical professions (Sweden);
* experiments with different approaches to computerization (Norway, Sweden, Finland and the United Kingdom);
* the effects of office automation on the quality of work performed by both men and women;
* the effects of working with visual display units (VDU's) on the health of women (Stockholm conference);
* strategies enabling women and trade unions to exert influence (Italy, Austria, ILO);
* the role played by governments and unions in other countries as regards "informatics for women";
* intuitive experience which cannot be substituted by means of the present methods of computerization.

The papers and discussions made is clear that knowledge of the way work is organized within a certain company or organization is of crucial importance if a well-designed system is to be produced. Knowledge acquired through user participation results in a more practical design for users and an improvement in the quality of their work. Government-backed projects in the fields of awareness-enhancement, guidance and education have also had positive effects for women: they are now more widely represented in computer-related occupations. The dismantlement of job segregation in these occupations - where men are still the predominance force - is therefore of great importance if women are to improve their position on the labour market.

Attention was also devoted to the reinforcement of the power bases already established by women. The importance of having networks of women was emphasized. One of the firmest suggestions to be put forward was for a more permanent status within IFIP, with regular conferences to look at the question of women and computers. This gave rise to an interesting exchange of views with the IFIP. representatives who were present: one of these representatives endorsed the importance of the issue, but urged a less "feministic" approach because this might deter "people". Another IFIP representative felt obliged in a closing speech to express the view that the issue was too important to be left solely to women. At the next conference, which will possibly be held in the Netherlands, we therefore look forward to seeing a large number of male IFIP members who are interested in the subject of women and computers.

3.2 Comments.

An air of optimism prevailed at the conference in view of the successful participation of female users in computer-related matters, training opportunities for women, and systems designed by and for women; in short, efforts to give women a better footing in the world of computerization. This is a rightful endeavour when one considers the minor role of women currently play in computerization. However, this optimism left little room for discussion of the still far-from-rosy position of women on the labour market. It remains a fact that men with the loss of large numbers of jobs and/or relegation to routine work behind the keyboard of a terminal. In some of the papers it was pointed out that women suffer far more adverse consequences of computerization than their male colleagues, but in the discussions this fact was touched on only in passing. Similarly, there was a regrettable lack of developments in the desired way with a strategy which is aimed at increasing the number of female system analysts/designers and at getting more women to participate in training schemes? There is a need to consider how successful this strategy has been so far. The women trade unionists at the conference, in particular, would have appreciated an exchange of views on the different kinds of strategies which are feasible.
Many of the delegates based their arguments for strengthening the position of women in computer- related occupations on the specific qualities possessed by women. Among other things, women were said to have grater social skills which made them particularly suitable for the analysis and design of systems in cases where participative methods were required. Women were also said to be guided more by their intuition and for that reason had a greater insight into the work performed by women. Among the other points mentioned was that women were better placed to provide training to other women. The delegates from the United Kingdom and the Netherlands expressed doubts as to the validity of these arguments. It would be interesting to see whether such suppositions can be substantiated by research. To our way of thinking, more discussion based on the social sciences, such as insights gathered from labour market theories, organizational theory, social dynamics and cognitive psychology, would have been a fruitful addition to the proceedings.
Attention was drawn at the conference at the field of tension: computerization experts refuse to accept that the social sciences can provide an added value to their work. They make haphazard use of socio-scientific knowledge and when they make an accidental "discovery" it turns out to be established knowledge in the social sciences.
The papers and speeches provided information on a series of interesting experiences with computerization projects. Many of these projects were in progress, however, which meant that few conclusions could be drawn as yet. There was an over-representation of contributions of the Scandinavian countries. A contribution from the Netherlands on women's vocational schools for occupations in the field of informatics wold have been most welcome. As regards Third World countries, the only substantial contribution came from the ILO, and this too concerned a project in progress. The inventory of successful stimulation programmes for women in those countries might also yield useful information for the industrialized countries. There was a widespread feeling of optimism at the conference.Most of the women delegates looked forward to the future in a fighting spirit, convinced that major investments on the part of women will lead to positive results. The general mood was that women were going to conquer the computer.

Amsterdam/The Hague
December 1986

Published by PTT, The Hague, 1986

Women, Work and Computerization: Forming New Alliances
K. Tijdens, M. Jennings, I. Wagner, and M. Weggelaar (Editors)
Elsevier Science Publishers B.V. (North-Holland)

APPENDIX 2

THE PROGRAM OF THE CONFERENCE WOMEN WORK AND COMPUTERIZATION IN THE NETHERLANDS, 1988

Program on Wednesday, April 27, 1988	
9.00 - 10.00	Registration and coffee
10.00	**Opening and welcome** by the conference chairperson Clio Presvelou (Netherlands), professor on Home Economics at the University of Wageningen
10.10	**Strategies and education** by Saskia van Hoek (Netherlands), director of the Alida de Jongschool
11.00	Coffee break
11.30	**Trade union strategies** by Judith Gregory (United States), researcher at Department of Communications of the University of California
12.20-13.30	Lunch
13.30	**Computer System Development-Alternative Design Strategies** by Kari Thoresen (Norway), researcher at the Norwegian Computing Center
14.20	**Quality of labour, in countries of the first and the third world** by Ursula Huws (United Kingdom), director of Empirica London
15.10-15.40	Tea Break
15.40	**Sexual division of technology: is computing technology the same again or different?** by Cynthia Cockburn (United Kingdom), senior researcher at the City University of London
16.30	**Conclusions** by Boel Carlsson (Sweden), one of the organizers of the first and the second conference on Women, Work and Computerization

Workshop on Thursday April 28, 1988

Theme A: Challenges of organizational change

A1. New technology in industrial organizations

chairwoman and report: Ciska de Koning (The Netherlands)
referee and report: Birgitta Frejhagen (Sweden)
referee: Rosalinda Pineda-Ofreneo (Filipines)
papers:
New technology and the employment position of women in the confectionery industry
Erik Poutsma and Dorien Trommel (The Netherlands)
Impact of the introduction of new technologies on women's work in South West France: Women's critical assessment and suggested strategies
Marie-Claude Bergouignan, Marie-José Cameleyre and Marie-Claire Rouyer (France)

A4a. Organizational strategies

chairwoman: Boel Carlsson (Sweden)
referee: Ewa Gunnarson (Sweden)
papers:
Polarized patterns: Effects of computerization on women's career opportunities
Angelika Volst and Ina Wagner (Austria)
New technology and gendered division of labour
Päivi Korvajärvi (Finland)

A4b. Organizational strategies

chairwoman: Clio Presvelou (The Netherlands)
referee: Katalin Nagy (Hungary)
referee: Kea Tijdens (The Netherlands)
report: Marja Oomens (The Netherlands)
papers:
Polarized patterns: Effects of computerization on women's career opportunities
Angelika Volst and Ina Wagner (Austria)
New technology and gendered division of labour
Päivi Korvajärvi (Finland)

Theme B: Patterns of access and opportunities in the labour market

B1. New technology and women's position on the labour market

chairwoman: Monique Chalude (Belgium)
referee: Birgitta Söderholm (Sweden)
referee: Pilar Vasquez (Mexico)
papers:
Women, work and technology
Mary Ann Madison and Vary T. Coates (USA)
Women's labour market position from the perspective of technological change
Anna-Maija Letho (Finland)

Theme C: Strategies for system development

C2. Computerization and women's knowledge

chairwoman and report: Tone Bratteteig and Gro Bjerknes (Norway)
referee: Paula Goossens (The Netherlands)
papers:
Computerization and women's knowledge
Lucy Suchman and Brigitte Jordan (USA)

C3. Working style of system designers from a gender perspective

chairwoman and report: Mary Jennings (United Kingdom)
referee: Elin Kvande (Norway)
referee: Bente Rasmussen (Norway)
referee: Anneke Steffens (The Netherlands)
papers:
A feeling for systems development work. Design of the ROSA-project
Susanne Bødker and Joan Greenbaum (Denmark)

Theme D: Training: new opportunities for women?

D2. Training and acces to the labour market

chairwoman: Jose Siersema and Marja Visser (The Netherlands)
referee: Els Rienstra (The Netherlands)
referee: Cydney Pullman (United States)
reports: Cynthia Cockburn and Diane Whitehouse (United Kingdom)
papers:
Realizing training programmes for women on informatics
J. Verhoef and H. Voorhaar (The Netherlands)
Technical culture, training and female employment
Ghislaine Doniol-Shaw (France)
Women, recurrent education and new technologies
S. Gensior, L. Hohenberger and H. Schicke (Federal Republic of Germany)

Theme E: Empowering strategies for women

E2. Office automation and positive action

chairman: Frank Boddendijk (The Netherlands)
referee: Sara Ellen Goodman (Sweden)
report:
papers:
Consonance or clash
Gien Tuender-de Haan (The Netherlands)
Using the computers in the interest of women: an equal opportunity information system
Paola M. Manacorda and Clelia Bertello (Italy)
An experience of positive action at Italtel
Bruna Osimo (Italy)

Workshop on friday april 29, 1988

Theme A: Challenges of organizational change

A2. Home based telework

chairwoman and report: Thea Weijers (The Netherlands)
referee: Hester de Vries (The Netherlands)
papers:
Bringing it all back home: a study of new technology homework in Australia
J. Wajcman, B. Robert and R. Tanter (Australia)
Home based telework - selected highlights from a case study in the Federal Republic of Germany
Monika Jaeckel (Federal Republic of Germany)

A3. Office automation and clerical work

chairwoman: Margaret Weggelaar (The Netherlands)
referee: Cecelia Bergström (Sweden)
papers:
Strategies of introduction of office automation in mixed and traditionally female jobs within two work units of the french electricity and gas public utility companies
Marta Ormos and Maya Blameble (France)
Influencing the content of women's work in automated offices
Juliet Webster (United Kingdom)
Not over our heads - women and computers in the office
Microsyster (United Kingdom)

Theme B: Patterns of access and opportunities in the labour market

B2. Opportunities for women in male dominated computer jobs

chairwoman: Ellen van Oost (The Netherlands)
referee: Martha Kirejczyk (The Netherlands)
referee: Cynthia Cockburn (United Kingdom)
papers:
The times, are they a-changin'?
Frida de Jong and Loes Steensma (The Netherlands)
Women's participation in the computer industry in Singapore
Lim Gek Kheng (Singapore)

Theme C: Strategies for system development

C1. Models for system-analysis from a women-perspective

chairwoman: Gro Bjerknes and Tone Bratteteig (Norway)
referee: Guri Verne (Norway)
papers:
A framework for understanding (women's) work and its computerization
F. De Cindio and C. Simone (Italy)
System design in social organization and methodological background - problems and chances for women's participation
Margit Falck (German Democratic Republic)

C4. Can trade unions influence the design process for the benefit of women?

chairwoman: Judith Gregory (USA)
referee: Titia Bos (The Netherlands)
report: Gien Tuender (The Netherlands)
papers:
Decentralized system development: advantageous for women?
Leikny Øgrim (Norway)
Technology Agreements and strategies for women workers
Anne P. O'Brien and Monique Ramioul (Belgium)

Theme D: Training: new opportunities for women?

D1a. Are women successful in technical training programmes?

chairwoman and report: Thea de Poel (The Netherlands)
referee: Jos Spetter (The Netherlands)
referee: Jasminka Novak (Yugoslavia)
papers:
Myth and reality about women and technology
Bente Elkjaer (Denmark)
A training for women in the field of computer technics
R. Knoops, H.S. Helsma and K.J.A. Wynands-Stab (The Netherlands)
Gender and learning, do women learn differently?
Deborah Brecher (USA)
The impact of office technology on clerical worker skills in the banking, insurance and legal industries in New York city: implications for training
Cydney Pullman and Sharon Szymanski (USA)

Theme E: Empowering strategies for women.

E1. Technical change and women's jobs in banking

chairwoman: Iske ter Haar (The Netherlands)
referee: Diane Whitehouse (United Kingdom)
report: Sally Wyatt (United Kingdom)
papers:
Strategies of female workers to improve their positions within the process of automation on banking
Marleen Defever and Kea Tijdens (The Netherlands)
Technological change, internal labour market and women's jobs: The case of the banking sector
Diane G. Tremblay (Canada)

METEORITE HUNTING

HOW TO FIND TREASURE FROM SPACE

Geoffrey Notkin

Host of Science Channel's *Meteorite Men*

With an introduction by Christopher Cokinos, author of
The Fallen Sky: An Intimate History of Shooting Stars

Edited by Dr. Larry and Nancy Lebofsky

STANEGATE PRESS

Photography

Suzanne Morrison
Leigh Anne DelRay-Crowell
Pablo del Rio Larrain
Steve Arnold
Qynne Arnold
Lisa Marie Morrison
Stu Jenks
Caroline Palmer
Sonya Gay Bourn
John Kashuba
Libby Egleson
Geoffrey Notkin

Illustrations

Timothy Arbon

Certain parts of this book originally appeared in a somewhat different form in Geoffrey Notkin's science column *Meteorwritings* on Geology.com. That material is reproduced here with the gracious permission of publisher Dr. Hobart King, for which we are extremely grateful.

Design and art direction by Stanegate Studios, Tucson, AZ
Published by Stanegate Press, LLC
www.stanegate.com

SECOND EDITION
ISBN 978-0-9847548-0-9

Printed in the United States of America

For my brilliant father, Sam Notkin,
who first showed me the wonders of the night sky

PREFACE

Meteorites are the most remarkable things on our planet, and they are not even from this planet. They are pieces of extraterrestrial rock and metal that have fallen onto the surface of the earth. Meteorites are a tangible link with an existence outside of the world we know and are, literally, visitors from outer space. They are important to researchers because within them are clues to the origin of suns, planets, asteroids and, possibly, even life itself. They are valued by enthusiasts because a space rock is, perhaps, the most intriguing of all collectibles. The rarity of meteorites and the fact that they are the only way in which most of us will ever be able to touch a piece of an alien world makes them of great interest to an ever-expanding network of devotees. Collecting and hunting for meteorites are exciting hobbies that are rapidly growing in popularity.

Meteorites first appeared in my life when I was about six years old. I grew up in Greater London and was an avid rock hound from an early age. My fascinating parents encouraged my interest in science: my father sharing his love of astronomy and telescopes, and my mother accompanying me both on frequent trips to science museums and fossil hunting missions into the chalk quarries of southern England.

It all came together for me when, as a little boy, I first walked into the Hall of Meteorites in London's Geological Museum. In a moment of revelation, I realized that by recovering and studying meteorites, we were looking deep into the history, mystery, and makeup of other worlds. For me, meteorites became the locus where astronomy and geology became one. That flash of insight was the spark that ignited a lifelong passion for space rocks.

I found my first meteorite in 1994. Three years later I met Steve Arnold, an experienced meteorite hunter, and joined him on a sometimes grueling, sometimes thrilling expedition to the Atacama Desert in Chile on the hunt for space rocks. My life would never be the same.

Steve and I became close friends, expedition partners, and—many years later—co-hosts of the Science Channel's award-winning television series *Meteorite Men*.

In my career I have found literally thousands of space rocks. Some were deeply buried and had been hidden underground for tens of thousands of years. Others were lying right on the surface when I spotted them, only two or three days after falling from the sky. The smallest meteorite I ever recovered weighed only 0.2 grams—or 1/140th of an ounce—and largest was a staggering 124 kilograms (273 pounds).

If the idea of exploring strange landscapes in the hope of finding ancient visitors from space lights a spark inside you, then you are the person for whom I wrote this book. My journey has been long and fascinating and, after hundreds of emails and phone calls from *Meteorite Men* fans, I came to the realization that there was a void within the world of meteorites. Many people want to go on an adventure and find their own space rock, but there has never been a guide book to help them along the way—until now. In *Meteorite Hunting: How to Find Treasure from Space* I share my personal knowledge and experience—gathered the hard way over scores of expeditions, hundreds of thousands of miles, many countries and many years—with those who dream of finding their own meteorites.

After all, what could be more exciting and amazing than looking for lost pieces of the universe that have fallen to earth?

Geoffrey Notkin

A 9.1-gram stone meteorite as found. This Ash Creek L6 ordinary chondrite fell on February 15, 2009, in McLennan County, Texas, after a spectacular daytime fireball that could be seen nearly 100 miles away. Note the extremely fresh blue-black fusion crust.

ACKNOWLEDGMENTS

The people listed here are not only those who helped directly with the book, but also those who helped me gain the knowledge and experience that I would need in order to write it.

Any list of acknowledgments that relates to my career as a meteorite specialist must begin with my expedition partner and *Meteorite Men* co-host Steve Arnold—not least for putting up with my antics during countless expeditions for almost a decade and a half. Steve helped show me the ropes when I first started out and, since 1998, has been featured in many of my published articles. Without Steve there would have been fewer expeditions and far fewer laughs. Steve made *Meteorite Men* the TV series possible, as well as many of the discoveries and unique experiences that form the backbone of this book. Not only is Steve one of the greatest meteorite hunters of all time, but he is also a friend and comrade.

Suzanne Morrison, Director of Operations for my company Aerolite Meteorites LLC, took many of the excellent photos included in this work, ran the business while I was on the road filming *Meteorite Men*, took on extra duties which allowed me the time needed to write this book, and also gave me valuable input every step of the way. Timothy Arbon who created the original scientific illustrations for *Meteorite Hunting: How to Find Treasure from Space* is also a key member of my staff, as are Lisa Marie Morrison and Beth Carrillo. Their contributions have been many and are deeply appreciated. Leigh Anne DelRay-Crowell is a gifted photographer and former Aerolite employee, and several of her images grace these pages.

My true love, Libby Egleson, stuck with me through many journeys and challenges, emailed me daily, and kept the home fires burning while I traveled—sometimes for weeks on end—to harrowing locales in search of space rocks. She has brought me joy and inspiration.

The Fallen Sky: An Intimate History of Shooting Stars is one of the finest books about meteorites ever published, and its author, Christopher Cokinos, wrote the introduction to *Meteorite Hunting*. There is an old rock 'n' roll adage that it is a sometimes good, and sometimes risky idea to have a band better than your own perform as your opening act. A gamble of that sort makes the headline act work harder to try and keep up, and so it is when an author of Chris's stature introduces my writing. I am grateful for his friendship and support, and especially for the way in which he has used his beautiful prose to illuminate the wonders of meteorites.

Dr. Joel Schiff, Dr. Larry and Nancy Lebofsky worked closely with me over the years and became respected friends. They gave me the opportunity to see many of my works in print in *Meteorite* magazine during their notable careers as editors of that publication. Nancy and Larry brought their considerable talents to bear and served as the editors for this book. I thank them for making *Meteorite Hunting* the best that it could be.

Dr. Hobart King, the publisher of Geology.com—one of the world's top science websites—invited me to write a series of articles for him, which became the *Meteorwritings* column. Dr. King generously allowed me to reproduce some of that material in this book and I am extremely grateful to him.

Paul Harris and Jim Tobin of Meteorite.com, The Meteorite Exchange, and the online magazine *Meteorite Times* are cornerstones of the meteorite world. Paul and Jim have supported me and my work, and have helped to build and educate the community in so many ways, for so many years, that I could not even begin to write a list of everything that they have accomplished. They both rock!

Rob Matson is one of the most brilliant people I have ever met. His original work with strewnfield research, lake bed hunting, and Doppler radar is both groundbreaking and influential. Time and again, Rob has assisted us with technical advice for the TV series, local intelligence, and strewnfield data. When I grow up, if I am half as smart as Rob, I shall consider myself a genius.

Sonya Bourn, Senior Producer for *Meteorite Men*, not only steered three seasons of our television show through about 120,000 miles of travel and untold hazards and difficulties, but also became a dear friend, confidante, and champion of our work. Executive Producer Ruth Rivin first came up with the idea for *Meteorite Men*, and also became a valued friend, as did Kathy Williamson and Bob Melisso who, between them, directed the early episodes with great style and dedication. My thanks to Eric Schotz and his talented team at LMNO Productions for believing in Steve and myself, to Christo Doyle and his colleagues at Discovery, and especially to Science Channel's radiant General Manager Debbie Myers.

Anne Husick and legendary rock musician Lach are not exactly meteorite people, but they are among my oldest and best friends. Even though they have, from time to time, looked at me in a rather odd manner and teased me about my unusual profession, they always had my back. The same goes for my brother, Andrew Notkin.

Sonny Clary ("I know they're out there Geoff. They just *gotta* be!") appeared in two episodes of *Meteorite Men*, provided a roof over my head, shared his ATVs, his knowledge and experience, and is the finest of expedition companions. As a fireman and EMT, he has dedicated his life to saving others. In his spare time he takes off into the wilds with his dog, Brix, and is an incredibly successful meteorite hunter.

I have shared many adventures with Ruben Garcia, a fellow rock 'n' roller and accomplished hunter. Ruben and Hopper, the Meteorite Dog, guest-starred in the Ash Creek episode of *Meteorite Men,* and I have benefited greatly from his friendship and expertise.

Terry Boudreaux once initiated a conversation about a television show that had great and far-reaching consequences. Thank you for lighting a small fire that turned into a big one.

Many luminaries in meteoritics, astronomy, and impact studies have provided friendship, encouragement, and ideas. They include Mike Jensen of Jensen Meteorites and his brother Bill, Dr. Art Ehlmann and Teresa Moss of the Oscar E. Monnig Meteorite Collection, Dr. Laurence Garvie (the "Third Meteorite Man") and Dr. Carleton Moore of the Center for Meteorite Studies at ASU, author and adventurer Professor Roy Gallant, my great pal Geoff Cintron, the mighty Bob Haag who is and will always be the original "Meteorite Man," Matt Morgan of Mile High Meteorites who has exactly the same sense of humor that I do, fellow hunter and favorite neighbor Twink Monrad, Darryl Pitt of the Macovich Collection, David Eicher of *Astronomy* magazine, my brilliant researcher Katherine Rambo, genius inventor Bill Mason, Andrzej Pilski publisher of *Meteoryt* (Poland), Dr. Alain Carion whose published works and research have been an inspiration to me, and George Winters of the AAPS.

Special thanks also to friends and colleagues Blaine and Blake Reed, Qynne Arnold, Bob Holmes, Michael Blood and Angel Hayes, Bob Falls, José Guggiari, Martin Altmann, Jeff Kuyken, Maria Haas, Mike Martinez, Art Jones, Martin Horejsi, Derek Yoost, Bill Kroth, Mike Miller, Professor John Warme, Thomas and Carin Österburg, Dr. Andy Tomkins, the late O. Richard Norton, Dorothy Norton, Eduardo Jawerbaum, Hans Koser, Rob Wesel, Mike Bandli, Jason Philips, Charles Elias, Brad Kern, Sarah Kennedy, Rob Reisner, Marlin Cilz, Paul Swartz, Warren Lazar, and my much-missed mentor, Professor Jim Kriegh.

Al Mitterling once sat with me at the El Charro restaurant in Tucson and—over a few beers—encouraged me to write a book about meteorites. I never forgot that conversation.

Al and Iris Lang of R. A. Langheinrich Meteorites and the Langheinrich Fossil Preserve have been admired friends and colleagues for many years, and Al is a giant in both meteoritics and paleontology. Anne Black is President of the International Meteorite Collectors Association, owner of Impactika Meteorites, and my elegant showroom partner during the Tucson gem and mineral shows. Al, Iris, and Anne's contributions to the field, and to my personal journey, have been enormous.

Guy Rovella of Aardvark Printing brought the book *Meteorite Hunting* to life, Norma Morrison assisted greatly with proofreading and copy editing, and Stu Jenks took the rockin' back cover photo. Thank you all! I have amazing friends.

TABLE OF CONTENTS

INTRODUCTION

What's the most interesting thing within 50 or 60 miles of your house? A strip mall? An art museum? A national park?

What about outer space?

Sure, outer space—if your Buick could achieve escape velocity. Think of it. Space is that close and that far away. After all, your Buick can't fly at Mach 34, and only about 500 people have ever flown in space.

So if you're interested in the cosmos, you have to be content with NASA TV, *2001* reruns and summer stargazing, right?

Not if you become a collector of rocks from space. Or, better yet, a person who actually hunts for space rocks, who scours deserts and parking lots, farms and mountainsides for meteorites, those primordial bits of the asteroid belt, the moon or Mars that are just waiting to be discovered.

Meteorites are often called "the poor man's space probe" precisely because we don't have to go to them. They come to us. In fact, on every square mile of the planet every year will fall about eight new meteorites of a gram or more. A gram is pretty light, but there are enough new falls and old finds of heftier meteorites to sustain the passion of meteorite hunters like Geoff Notkin. And perhaps, after reading his book, meteorite hunters like you.

Meteorite hunting is a pursuit rich in connections to the stars and adventures for one's life. I wrote in my memoir and history of meteorite-hunting, *The Fallen Sky*, that "the untrained eye can ... mistake some meteorites for chunks of concrete, as if a cosmic road crew had jackhammered a solar highway and sent its congeries spilling down, briefly lit, to land among soybeans or ferns. Those who quest for meteorites, however, recognize them amidst the average rubble of the earth the way a birder hears rare song untangling itself from a forest full of sound. Passion does that: It sharpens one's senses, it changes the world."

Of course it takes more than passion to be good at something. It takes knowledge. Then it takes experience. To be good at meteorite hunting, you first need this book. Here you'll learn about everything from how to identify meteorites in the field to the legalities of hunting meteorites on public and private property. You'll learn about metal detectors, fireball chases and the best places to hunt for space rocks. These and other points of information are important—crucial, in fact—but like any good teacher, Geoff Notkin not only gives you the details, he does so after enticing you with the big picture: "What could be more exciting and amazing," he asks, "than looking for lost pieces of the universe that have fallen to earth?" The answer: not much.

Geoff should know. He has revitalized and revolutionized meteorite hunting through his enthusiastic use of social media, the web and, most importantly, co-hosting the entertaining and educational television show, *Meteorite Men*. He's the perfect guide to the novice and veteran meteorite hunter. You'll find that out in this excellent book.

So save the Buick for road trips to strewnfields. Don't know what those are? Read on.

Christopher Cokinos

WHAT ARE METEORITES AND WHERE DO THEY COME FROM?

Most meteorites were once part of large asteroids that broke up, or perhaps never fully formed more than four billion years ago. A very few are known to have originated on Mars and our own moon. These fragments of alien worlds wandered in the cold void of space for long periods of time before crossing paths with our own planet. Their tremendous terminal velocity can result in an encounter with the air above us at a staggering 17,000 miles per hour, or more, resulting in a short and fiery journey through our atmosphere.

Meteor is the scientific name for the atmospheric effect caused by a piece of extraterrestrial matter burning in our atmosphere, and should not be confused with the word meteorite, which refers to any material that actually lands. A bright and typically very short-lived meteor is caused by atmospheric drag and friction acting on an incoming body that becomes so hot it literally incandesces, as does the air around it. Most meteors glow for only a few seconds or less, and that brief period of intense heat is part of what makes any surviving meteorites so very unique and fascinating. The extremely high temperatures cause surfaces to melt and flow, creating remarkable features that are entirely unique to meteorites such as *regmaglypts*, *fusion crust*, *orientation*, and *rollover lips* (see Chapter 2).

METEOR SHOWERS

There are a number of periodic meteor showers ever year; the Perseids in August and the Leonids in November usually being the most interesting to observe. Annual meteor showers are the result of our planet passing through debris trails left by comets. The meteors we see during those annual displays are caused by small pieces of ice that rapidly burn up in the atmosphere and never make it to the surface of our planet. Every year, without fail, we receive calls or emails from people who saw the Perseid meteor shower in August, and believe that, as a result, meteorites fell in their yard or driveway. We tend not to investigate such reports because the periodic meteor showers do not produce meteorites.

Artist's impression of cometary fragments on course for impact on a planet.

METEOROIDS AND METEORITES

We use the term *meteoroid* to describe a potential meteorite. It is the physical body that travels through space, encounters our atmosphere and creates the phenomenon known as a meteor. To clarify: A body traveling in space towards our planet or entering our atmosphere is a meteoroid; the visual effect of superheated gas glowing around a meteoroid is described as a meteor; and anything that survives and lands on our planet's surface is a meteorite.

Meteorites are among the rarest materials found on earth. Some of them pre-date the solar system, which makes them the oldest things any human has ever touched. In order to find meteorites it is first important to understand what they are made of, and how they differ from terrestrial rocks.

THE THREE MAIN TYPES OF METEORITES

Although there are a large number of sub classes, meteorites are divided into three main groups: irons, stones and stony-irons. Almost all meteorites contain extraterrestrial nickel and iron, and those that contain no iron at all are extremely rare.

When the average person imagines what a space rock might look like, he or she often thinks of iron meteorites. It is easy to see why. Iron meteorites are dense, very heavy for their size, and have often been forged into unusual and sometimes spectacular shapes as they plummet, melting, through our planet's atmosphere. Although irons may be synonymous with most people's perception of a typical space rock's appearance, they are only one of three main meteorite types, and rather uncommon compared to stone meteorites, especially the most abundant stone meteorite group—the ordinary chondrites.

IRON METEORITES

When I give lectures and slide shows about meteorites to museums, schools, and *Meteorite Men* fans, I always enjoy beginning the presentation by passing around a softball-sized iron meteorite. Most people have never held a space rock in their hands and when they do pick up an iron meteorite for the first time, their faces light up and their reaction is, almost without fail, to exclaim: "Wow, it's so heavy!"

Iron meteorites may have originated in the molten cores of long-vanished asteroids. They are among the densest materials on earth and will stick very strongly to a powerful magnet. Iron meteorites feel heavier than most earth rocks—if you have ever lifted up a cannon ball or an iron dumbell, you will get the idea.

[left] Weathered iron meteorite fragments collected at the Monturaqui Crater in Chile. [above] One of our finds from the Muonionalusta expedition in 2010. This iron meteorite weighs 10 kg (22 lbs) and was photographed immediately after it came out of the ground. Note the unusual colors caused by oxidation. The rounded surface was caused by tumbling in an ancient glacier.

[top] A Campo del Cielo iron meteorite found in Argentina. This specimen has been cleaned and resembles the way the meteorite would have looked at the time of fall. [above] A Canyon Diablo iron from Meteor Crater, Arizona, in as-found condition, with hand-painted collection number. Note the presence of regmaglypts on both pieces.

[top] An 870-gram Mundrabilla iron exactly as I found it in Western Australia. Exposure to the elements for many thousands of years has caused the iron to acquire an attractive patina. [above] An iron meteorite fragment from the Whitecourt Crater in Alberta, Canada. The oxidation is typical of an iron that has been buried in damp ground.

STONE METEORITES

The largest group of meteorites is the stones. The majority of stone meteorites are believed to have once formed part of the outer crust of a planet or asteroid. Many stone meteorites—particularly those that have been on the surface of our planet for an extended period of time—frequently look much like terrestrial rocks, and it can take a skilled eye to spot them when a person is meteorite hunting in the field. Freshly fallen stones will exhibit a black fusion crust, created as the surface is heated during flight. The vast majority of stone meteorites contain enough iron for them to easily adhere to a powerful magnet.

[below] A 21.5-gram Buzzard Coulee H4 stone meteorite displaying excellent black fusion crust and sitting where it fell among canola stalks in rural Saskatchewan.

Stone Meteorites: Chondrites

The *chondrites* are the most abundant group of meteorites. They take their name from the *chondrules* of which they are comprised—small, colorful, grain-like inclusions which originated in the solar nebula and, therefore, pre-date the formation of our planet and the rest of the solar system, making them the oldest known matter available to us for study. The presence of chondrules is a key factor in identifying many stone meteorites (see Chapter 2). Iron content is used to differentiate between the three types of ordinary chondrites designated H, L, and LL. Enstatite chondrites contain unusual sulfide minerals and take the classifications EH and EL. Carbonaceous chondrites are highly oxidized, contain organic compounds, and are divided into five classes (CM, CR, CO, CV, CK). The R chondrites, a rare group, are high in iron-rich olivine and are named after the Rumuruti stone, which was seen to fall in Kenya on January 28, 1934.

[above] The fusion-crusted, oriented stone we found while hunting the San Juan strewnfield in Chile. The original meteorite weighed 870 grams. After removing some material for classification, the remaining mass weighs 639.4 grams.

[left] Detail of densely-packed, colorful chondrules in the LL3.8 chondrite Northwest Africa 5793.

Stone Meteorites: Achondrites

Achondrites are similar to volcanic rocks from earth and are "without chondrules," hence their name. Igneous activity within the parent bodies where they formed caused melting and recrystallization that eradicated all traces of ancient chondrules. Most achondrites contain little or no iron, making them much more difficult to find than other meteorite types, although fresh specimens often display a remarkable glossy fusion crust which looks almost like enamel paint. Several achondrites (the howardites, eucrites, and diogenites) are believed to have originated from within the asteroid Vesta.

[above] Achondrites, such as this Camel Donga from Australia, are very rare. They do not contain chondrules and have little, if any, metal. They are lighter in weight than most other meteorites and often exhibit a bubbly, shiny fusion crust. Magnets and metal detectors are of little use when searching for achondrites.

Stone Meteorites: Moon And Mars Rocks

Do we really find lunar and martian rocks on the surface of our own planet? The answer is yes, but they are extremely rare. Only about 100 different lunar meteorites (lunaites) and approximately 40 martian meteorites have been discovered on earth, and they all belong to the achondrite group. Impacts on the lunar and martian surfaces by other meteorites fired fragments into space and some of those fragments eventually fell on earth.

In financial terms, lunar and martian specimens are among the most valuable meteorites, sometimes selling on the collectors' market for as much as $1,000 per gram, making them worth many times their weight in gold. The identification of lunar and martian meteorites requires a great deal of expertise. Washington University in St. Louis has an excellent and entertaining website about the identification of lunar meteorites. It was written by specialist Dr. Randy Korotev and is a valuable resource if you are interested in this specialized subject.

STONY-IRON METEORITES

The rarest of the three main types, the stony-irons account for less than 2% of all known meteorites. They are comprised of roughly equal amounts of nickel-iron and silicates and are divided into two groups: *mesosiderites* and *pallasites*.

Stony-Iron Meteorites: Mesosiderites

The mesosiderites are the smaller of the two stony-iron groups. They contain both nickel-iron and silicates and usually show an attractive, high-contrast, silver and black matrix when cut and polished. The seemingly random mixture of inclusions can lead to some very striking features. The word mesosiderite is derived from the Greek for "half" and "iron," and they are very rare. Of the thousands of officially cataloged meteorites, less than 100 are mesosiderites. During *Meteorite Men* Season Two, we found a remarkable 3.3-kilogram mesosiderite at the Vaca Muerta strewnfield in Chile.

[below] The spectacular Vaca Muerta mesosiderite that we found on the 2010 Chile expedition while filming *Meteorite Men.* Note the high contrast between the iron-nickel matrix (silver) and the silicate inclusions (black) which is a characteristic typical of mesosiderites.

Stony-Iron Meteorites: Pallasites

Perhaps the most alluring of all meteorites, and certainly of great interest to private collectors, pallasites consist of a nickel-iron matrix packed with olivine crystals. When olivine crystals are of sufficient purity and display an emerald-green color, they are known as the gemstone peridot. Pallasites take their name from a German zoologist and explorer, Peter Pallas, who described the Russian meteorite Krasnojarsk, found near the Siberian capital of the same name in the eighteenth century. When cut and polished into thin slabs, the crystals in pallasites become translucent giving them a remarkable otherworldly beauty. If you have watched our show, *Meteorite Men*, you may have seen us find giant pallasites at Brenham, Kansas, and our secret Alpha Site, as well as smaller examples of the Imilac pallasite in Chile.

[above left] A translucent slice of the Imilac pallasite. [top] A corroded specimen of the Brenham pallasite found by H.H. Nininger in the 1930s. Note the hand-painted collection number which identifies it as a Nininger piece. [above] The author and Steve Arnold with a pallasite weighing 124 kg (273 lbs) found at Brenham in 2008.

HOW TO IDENTIFY METEORITES IN THE LABORATORY AND IN THE FIELD

It takes knowledge, practice, and experience to tell the difference between a meteorite and an earth rock. This fact is demonstrated time and time again by poorly-researched newspaper interviews with people who think they have found a space rock. The typical story goes something like this: A man is walking his dog or driving home at night and sees a bright fireball in the sky. He is convinced that it landed "just over there." The next day he goes out searching, finds a strange-looking rock in a field or driveway, and calls the local newspaper to report that he has found a meteorite. A few inaccurate articles are published and then, eventually, the rock gets taken to a specialist who determines that it is not a meteorite, but rather basalt, or hematite, or cement, or old iron slag from a smelter. The witness really did see a fireball, but he did not know how to recognize a meteorite on the ground, so he picked up the first unusual terrestrial rock that he saw, mistaking it for a meteorite.

It may seem obvious, but it needs stating: If you want to find meteorites, you need to know exactly what to look for. Most meteorites have a high density; the vast majority contain a great deal of iron and will easily adhere to a strong magnet. Meteorites generally look and feel different from the earth rocks around them. They do not contain the common mineral quartz, and, in general, do not contain *vesicles*—small pinprick holes in the surface of a rock caused when gas escapes from cooling molten material. The volcanic rock pumice, often used in skin care for the removal of callouses, contains vesicles which is one reason why it is very light in weight. If your suspected meteorite looks like a sponge with tiny holes on its surface, it is probably volcanic rock or slag of earthly origin.

METEORITE IDENTIFICATION: THE MAGNET TEST

Practically all meteorites contain a significant amount of extraterrestrial iron and nickel, so the first step in identifying a possible meteorite is the magnet test. Iron and stony-iron meteorites are especially rich in iron, and will stick to a powerful magnet so strongly that it can be difficult to separate them. Stone meteorites, for the most part, also have a high iron content (up to about 20%) and a good magnet will adhere to them as well. Many earth rocks will also be attracted to a magnet, so this is not a definitive test, but it is a step in the right direction. It is worth noting that meteorites will show a uniform attraction to a magnet, whereas terrestrial rocks will often demonstrate a greater attraction on one face.

Lunar and martian meteorites, and most achondrites (stone meteorites without chondrules), contain little or no iron and even a powerful magnet will generally have no effect upon them. These meteorite types, however, are so extremely rare that, as a general rule, we discount specimens that will not adhere to a magnet, especially if they do not exhibit fusion crust.

METEORITE IDENTIFICATION: WEIGHT AND DENSITY

Iron is heavy and most meteorites feel much weightier in the hand than an ordinary earth rock should. A softball-sized iron meteorite will likely weigh between six and eight pounds, making it seem unnaturally dense. Imagine holding a ball bearing as big as a grapefruit and you will get the idea.

[left] Excavating a buried iron meteorite in the Muonionalusta strewnfield, north of the Arctic Circle in Sweden. The glacial moraine in which the Muonionalusta irons are found also contains terrestrial iron oxides which will register with a metal detector. Every find was checked with a magnet while still in the ground.

[top] The vast majority of meteorites contain iron and will easily adhere to a good magnet. These weathered stone fragments that we found in Nevada had been sitting on a dry lake bed for tens of thousands of years, yet still had enough iron content to jump onto my magnet. [above left] Detail of fresh fusion crust on the carbonaceous chondrite Allende, which fell in Chihuahua, Mexico, on February 8, 1969. This specimen was picked up immediately after the fall. [above right] Fusion crust is fragile and will eventually erode. Note the weathering cracks clearly visible on a different piece of the same meteorite, picked up several years later.

VISUAL IDENTIFICATION: FUSION CRUST

When a meteoroid streaks through our atmosphere, tremendous heat is generated by atmospheric resistance. The surface of the rock melts and the air around it incandesces. As a result of this brief but intense heating, the surface forms a thin, dark rind called fusion crust. Meteorites literally began to burn up in our atmosphere, so they tend to appear darker than the terrestrial rocks around them. A freshly-fallen stone meteorite with intact fusion crust will look as black as a charcoal briquette. Desert varnish forms on the surface of some earth rocks, particularly in arid areas, and can easily be mistaken for fusion crust by an untrained eye. True fusion crust does not occur on earth rocks. It is delicate and will oxidize or weather away over time, so stone meteorites that have been on the earth for long periods of time often have a dark reddish color. Fusion crust also forms on irons and stony-irons, but it is fragile and tends to oxidize quickly.

[below] A fusion-crusted whole stone from the Ash Creek, Texas, fall of February 15, 2009. Note how the stone acquired light brown streaks of mud as it hit the ground.

VISUAL IDENTIFICATION: REGMAGLYPTS

Regmaglypts, also known as thumbprints, are oval depressions of varying sizes are found on the surface of many meteorites. These indentations resemble the marks a sculptor might make with his fingers on a wet lump of clay, hence their name. The size of regmaglypts on a particular meteorite is usually proportional to the size of the meteorite itself. A 100-gram iron may be covered with scores of tiny thumbprints, each just a few millimeters in length, while a 100-kilogram meteorite might display regmaglypts that are each several inches in length. Regmaglypts, created as the meteorite's outer layer melts during flight, are another feature unique to space rocks.

[above] Multiple fine, well-formed, overlapping regmaglypts on a 204-gram Sikhote-Alin individual. [left] Regmaglypts are most frequently seen on iron meteorites, but they can also occur on stone meteorites, as demonstrated by this excellent 828.6-gram example of the Juancheng meteorite which fell in China in 1997. [right] Details of regmaglypts on cleaned and uncleaned iron meteorites.

VISUAL IDENTIFICATION: FLOW LINES

As a meteoroid hurtles through our atmosphere, its surface may melt and flow in tiny rivulets known as flow lines. The patterns formed by flow lines can be minute, often thinner than a strand of human hair, and they are one of the most unique and intriguing surface characteristics of meteorites.

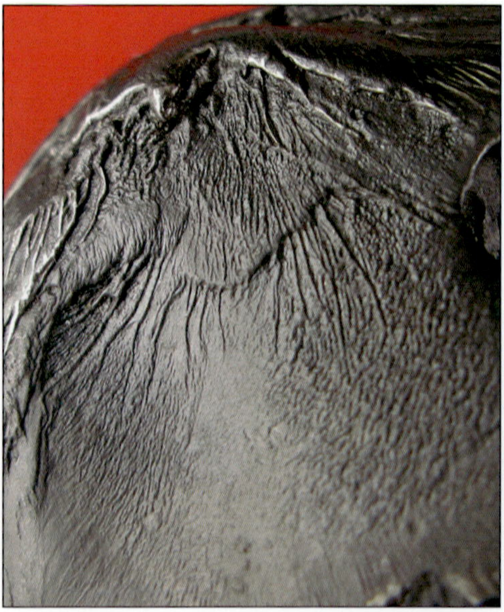

[top left] Detail of overlapping flow lines on the nose of a highly oriented Sikhote-Alin iron. [top right] Delicate flow lines on the leading edge of a Taza (NWA 859) iron. [above] Long, fine flow lines and shiny, black fusion crust characterize this Millbillillie eucrite which fell in Australia in 1960.

VISUAL IDENTIFICATION: ORIENTATION

Most meteoroids spin and tumble as they plummet through the atmosphere. Occasionally, one will maintain a fixed orientation towards the surface of our planet, causing the leading edge to *ablate* into a shield, nose cone, or bullet shape. When meteorites ablate, some of their mass is removed as a result of vaporization. Meteorites which display such features are quite rare and are described as *oriented.*

[top] Steve Arnold with the massive 3/4-ton meteorite that he unearthed in Kansas in 2005. It is the largest oriented pallasite in the world and the largest meteorite ever found in the Brenham strewnfield. Photograph © Arnold Meteorites [above left] This highly-oriented Sikhote-Alin iron acquired a shield shape as a result of ablation during flight. [above right] An oriented stone meteorite found in the Sahara Desert.

VISUAL IDENTIFICATION: ROLLOVER LIPS

A typical oriented meteorite will have a convex leading edge and a flat or concave trailing edge. Sometimes molten material will flow onto the trailing edge during flight, creating a fine, yet distinct feature known as a rollover lip. This intriguing characteristic is a glimpse into the processes that give meteorites their unique shapes and features.

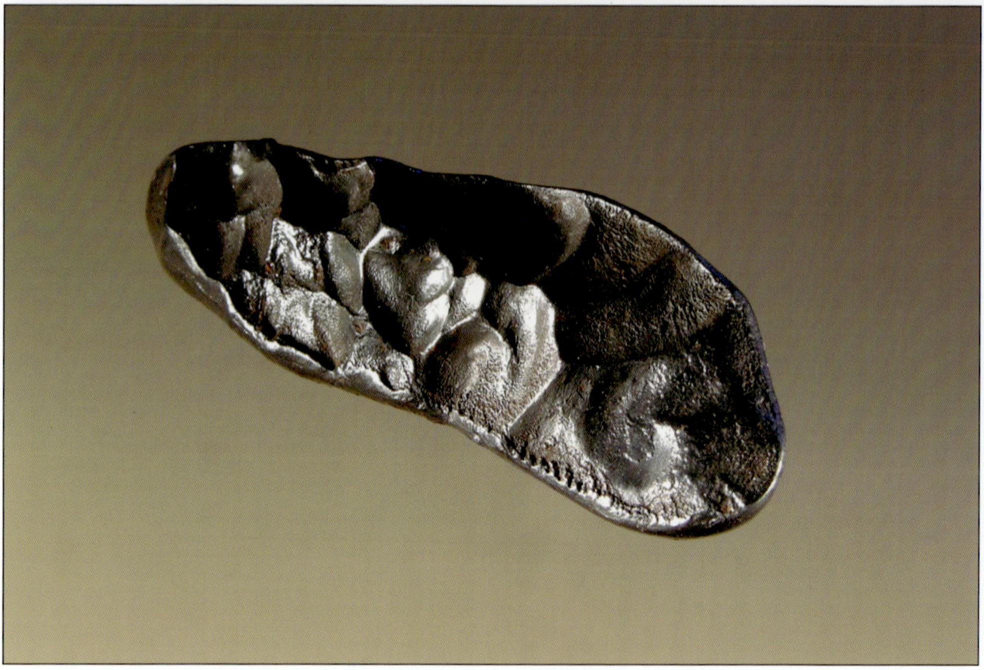

[top] Bassikounou is an H5 chondrite that fell near Hodh Ech in Mauritania on October 16, 2006. This 12.1-gram oriented individual shows a very distinct rollover lip. [above] An oriented 27-gram Sikhote-Alin iron with a 100% rollover lip on the trailing edge. Complete rollover lips such as this are a rare feature on meteorites. [above right] Abundant chondrules are typical of the L4–6 chondrite NWA 869. [far right] A small slice of the Xingyang chondrite that fell in China in 1977 exhibits shiny metallic flakes in a dark silicate matrix.

VISUAL IDENTIFICATION: CHONDRULES AND METAL FLAKES

Stone meteorites known as chondrites are the most abundant meteorite type. They are composed largely of chondrules, which are tiny, grain-like spheroids, often of widely differing colors. Chondrules are believed to have formed in the solar nebula disk before the planets in our solar system and are not present in earth rocks. Chondrites are usually also rich in metal flakes of iron-nickel. Shiny blobs of this extraterrestrial alloy are sometimes visible on their surfaces, although you may need a hand lens to see them. A simple identification test involves removing a small corner of a suspected stone meteorite with a file or bench grinder and examining the exposed face with a loupe. If the interior displays metal flakes and small, round, colorful inclusions, it may well be the real thing. Carrying a small metal file with you in the field can assist in chondrite identification. A small "window" in a suspected meteorite, and inspection with a loupe or magnifying lens, will make it easy to identify chondrules if they are present. Be careful when grinding a window as excessive cutting can damage a specimen, thereby decreasing its value and aesthetic appeal.

LAB TESTING OF METEORITES: NICKEL CONTENT

Nickel is rare on earth but almost always present in meteorites. If a suspected meteorite passes the magnet test and looks promising following a visual inspection, we may elect to conduct a test for nickel. Assay labs can perform an analysis of the nickel content for a modest fee, but it is necessary to cut off a small sample in order to perform such a test. Some labs and universities with meteoritics departments can perform more sophisticated tests without damaging a specimen. For example, the Ion Beam Analysis of Materials (IBeAM) laboratory at the Center for Meteorite Studies at Arizona State University uses accelerated ions to determine, with great accuracy, the composition of samples. In simple terms, that means we can discover the chemical makeup of a specimen without slicing it in half. The IBeAM can perform an analysis from a tiny cut sample, and results appear on a computer screen within a few seconds. A compositional analysis showing somewhere between three and ten percent nickel will almost certainly indicate an iron or stony-iron meteorite. Most stone meteorites also contain nickel, but the overall percentage is lower. A rare sub-class of iron meteorites, the *ataxites*, is characterized by high nickel content. The Tishomingo ataxite, discovered in Oklahoma in 1965, is extremely unusual in that it consists of 32.5% nickel.

HOW MUCH ARE METEORITES WORTH?

The scientific importance of meteorites cannot be over emphasized. They have great research value to academics, and financial value to collectors.

Meteorites may have spent millions of years traveling through the vacuum of space at temperatures approaching absolute zero. After landing on our planet, they are immediately exposed to oxygen and moisture, and the process of terrestrialization begins. Iron starts to oxidize and delicate fusion crust will begin to corrode. The sooner a fresh fall is delivered into the hands of a researcher, the more valuable it is to science. There is a real urgency to recover new falls and get them to a lab as quickly as possible, where measuring the decay of radioactive isotopes can provide important scientific information.

Much like other collectibles—comic books, baseball cards, stamps, coins, or fine art pieces—the financial value of meteorites is determined by a range of factors including rarity of type, condition, and aesthetic appeal. On the collectors' market, meteorites are generally sold by weight, and measured using the metric system. While there is a fairly well-established average price for some of the more widely available meteorite types, a specimen that has a remarkable story attached to it—such as the Peekskill stone that hit a parked car in New York state in 1992—or a piece that is visually pleasing to look at will command a higher value.

Meteorite prices vary just like any other collectible. A survey of the market in 2010 showed unclassified stone chondrites, picked up by nomads wandering in the Sahara Desert, readily available for about $0.80/gram. Attractive stones from a witnessed fall, such as Gao-Guenie (Burkina Faso, Africa, March 5, 1960) could be purchased for about $1.50/gram and a good quality one-kilogram specimen of the Campo del Cielo iron meteorite from Chaco Province, Argentina, might sell for about $500 ($0.50/gram).

The Russian iron Sikhote-Alin (fell February 12, 1947) is the largest single meteorite event in modern recorded history. Individuals—meteorite specimens which landed as one intact piece, rather than exploding on or near the ground—are typically well-preserved and demonstrate marvelous sculptural qualities and surface features. A premium Sikhote-Alin specimen will carry a price tag of $3 to $4/gram.

[left] A premium quality Sikhote-Alin iron meteorite such as the example pictured is worth several thousand dollars on the collectors' market. [above right] Meteorites for sale at the annual Tucson gem and mineral shows. Meteorite collecting is rapidly growing in popularity and space rocks can change hands for large sums of money.

Pallasites are packed with olivine (the gemstone peridot) and are particularly desirable when cut and polished because of the alluring color and translucency of the crystals they contain. Prepared slices of the most attractive pallasites such as Imilac (Chile), Glorieta Mountain (New Mexico), and Esquel (Argentina) are prized for their colorful gemstones and long-term stability and will fetch between \$20 and \$45/gram. Most meteorites are dense, so a quality pallasite slice about the size of a small dinner plate could be worth thousands of dollars.

At the high end of the pricing scale are unusual types such as the diogenite Tatahouine (fell June 27, 1931, Foum Tatahouine, Tunisia). A prime specimen will easily fetch \$50/gram while rare examples of lunar and martian meteorites have sold for \$1,000/gram—almost 20 times the current price of gold!

When valuing meteorites, the price per gram typically diminishes the larger a specimen is. In other words, an attractive 50-gram example of the Gibeon iron from Namibia would be worth \$50 or \$60 to a collector, or upwards of \$1,000/kg. A 30-kilogram mass of the same meteorite might sell for \$500/kilogram or half the price per gram of a smaller piece.

If you find a new meteorite, the most important thing to do is find out if it has scientific value. Once you have carried out your field tests with a magnet and magnifying lens, consider sending it—or a piece of it—to a university with a meteoritics department or an accredited lab. Once a find has been studied, classified, and named (see Chapter 15), it is actually worth more to the finder because it will have been included in the database of all known meteorites—The Meteoritical Bulletin Database.

Meteorites are worth money and a significant new discovery can provide unexpected income for the finder. A few years ago, a landowner in Kansas contacted me. He noticed a large unusual rock, weighing about 36 kilograms (79 pounds), on his property and thought it might be a meteorite. After looking at some photos, I was reasonably sure that he had found a real space rock. At my request, he sent me a small sample, and the minute I saw it, I knew it was a genuine meteorite. A cut face displayed abundant chondrules, and it also adhered strongly to a magnet. I then sent the sample on to a highly respected meteoriticist who performed a detailed lab analysis for us. The new find was classified as an H5 chondrite, and given the official name of Coffeyville. I assisted the finder in having his find cut into slices in order to display its interesting interior, and the resulting sales were in the thousands of dollars. Sometimes even an accidental find can generate good revenue.

METEORWRONGS, HOT ROCKS, AND HOW THEY CAN HELP YOU BECOME A BETTER HUNTER

Years ago I added a page to my website, www.aerolite.org, with tips on how to identify meteorites. It includes photographs of earth rocks and genuine meteorites, along with some simple tests that finders can carry out at home. Our "Aerolite Guide to Meteorite Identification" proved so popular that people began sending us suspected meteorites, and a number of them turned out to be the real thing. In some cases the finder was just curious and wanted to keep the meteorite in his or her collection. In other instances we helped the finder get the new space rock classified, named, inducted into the official literature and, eventually, sold for good money. We have received an large assortment of "meteorwrongs"—earth rocks that look like meteorites but are not. Some of those rocks were intriguing and we mailed them back to the finders at their request. We were instructed to discard others which had no value, and they now happily populate my rather unusual-looking rock garden.

We have been sent flint, chalk, basalt, granite, limestone, lead, copper ore, iron slag, and even man-made glass by people who thought they had found meteorites. The terrestrial rocks most commonly mistaken for meteorites are the iron oxides magnetite and hematite. Magnetite can be particularly tricky because it is usually metallic-looking, heavy for its size, and typically adheres well to a magnet. It is not quite dense enough to be a typical meteorite, however, and with a practiced eye the attentive hunter or enthusiast will quickly learn to distinguish magnetite from extraterrestrial material.

The single most popular "Is this a meteorite?" we receive is *botryoidal* hematite. This intriguing stone displays a reddish tint that is quite similar in appearance to oxidized meteorites, sometimes appears burned or melted, and often has a rounded, almost bubbly surface that looks—to the untrained observer—as if it must have fallen, burning, from the sky. Nothing could be further from the truth. The term botryoidal, derived from the Greek, means, loosely, "like a bunch of grapes." Botryoidal hematite is actually a collection of iron oxide spheres formed by the radial deposition of mineral material around nuclei. This commonly-misidentified rock typically has little or no attraction to a magnet, so remember to field test your finds. All but a very few extremely rare meteorite types will instantly react to a powerful magnet.

If you want to learn to really understand meteorites, buy a couple of inexpensive examples for study. Canyon Diablo (Arizona) is an excellent example of a somewhat terrestrialized iron meteorite, and a modest-sized piece can be easily obtained for less than \$50. Unclassified stone meteorites from Northwest Africa—often called NWA XXX—are also affordable and relatively abundant on the collectors' market. If you are serious about finding your own space rocks, buy a stone meteorite and an iron meteorite from a reputable dealer, and familiarize yourself with how they look and feel, and listen to what they sound like under your metal detector.

METEORWRONGS ARE, LITERALLY, EVERYWHERE

If you are going to start hunting for meteorites, the one thing I can guarantee is you are going to find meteorwrongs. Your metal detector can only tell you so much; you need to do the rest.

Metal detectors that are used for hunting meteorites are calibrated to return a signal when they find iron. Sophisticated modern hand-held detectors such as the Fisher F75, a personal favorite of mine, have the capability to distinguish between different types of metal.

[left] An unexploded artillery shell, discovered while meteorite hunting in Chile's Atacama Desert.
[overleaf] A World War II-era 50-caliber machine gun cartridge on the salt flats in Utah.

The F75 employs a digital readout—from 1 to 100—that can help you identify, to some degree, the makeup of a buried target. Iron reads between 4 and 15 on the display panel, but iron does not always mean "meteorite." When I was hunting at the Henbury crater field in Australia, I quickly learned that a 13 on my control panel meant I had the real thing. One larger mass close to the surface read a 7, so when you are learning it is a good idea to dig and examine all ferrous targets. A very low reading on the scale, such as 3 or 4, is usually man-made refined iron, most likely nails or bolts. At the Brenham, Kansas, site the iron-rich pallasites that we excavated read 11 on the meter, while stone meteorites from Buzzard Coulee gave us a 14, so there can be a significant variance in readings from one meteorite type to another.

Even though I have hunted for meteorites in some of the most remote places on earth, it seems that, most of the time, someone has been there before me and carelessly left their trash. Wherever people lived, worked, farmed, or fought, they left things behind, and many of those things contain iron. While hunting at the Alpha Site in Kansas, Steve and I were, at times, right on top of the old Santa Fe Trail, and that area is just full of lost and discarded items from wagon trains—all of them buried. We did a lot of digging before we found a real meteorite.

Most man-made items are found in the top foot of soil, so the deeper you go, the better the chances that your target is not something that was fabricated by a human being. That is not always the case; sometimes trash is buried on purpose, and deeply. While filming the *Meteorite Men* pilot in Brenham, Kansas, we received a massive return signal from Steve's giant detector. We were inside the strewnfield and quite close to areas where large meteorites had been found in the past, so we were elated. We dug down a couple of feet, but did not uncover the target. Reasonably sure that we had a meteorite, we brought in a backhoe, only to find half an hour later that our "new find" was a 55-gallon oil drum that had, for some inexplicable reason, been buried deep in a field.

In 2009 we received very special permission to hunt inside the small craters that surround the main structure at the Odessa meteorite site in Ector County, Texas. Those craters are actually the remains of dig holes where large meteorites were uncovered during the 1940s, and were not formed by the meteorites themselves. We thought it likely that some meteoritic material remained in the pits and, after getting a strong signal at the bottom of one of the pits, we brought in a mechanical excavator. Our excitement turned to dismay when we found that the old dig holes had been used as a dump and, instead of a meteorite, we unearthed a mass of old beer cans. They were all empty, which was no fun at all.

Items that I have found while looking for meteorites (most of them underground) include soda cans, sardine tins, horseshoes, Frontier-era wagon wheels, plow blades, hub caps, screwdrivers, animal traps, bolts, nails, hammers, wrenches, axe heads, Civil War Minié balls, batteries, a full clip of live ammunition for a .32 pistol, logging equipment, chains, a Wild West-era revolver, aircraft parts, tractor parts, a large crowbar, an uncapped oil well, shoes with iron heels, brass 50-caliber machine gun cartridges, sunglasses, keys, money, several unexploded bombs, artillery shells, and two very large missiles. It is worth noting that if you are digging targets near a military test range, or in areas that saw heavy combat in the modern era, such as northern Europe, there is always a chance that you will come across live ammunition or unexploded military ordnance. This has happened to me on at least five separate occasions, so use extra caution if you are working in such areas.

Picking up a meteorwrong from the surface is one thing; digging one up from several feet underground is another. It can be tiring and discouraging to spend half an hour toiling through hard ground with a muddy shovel only to reveal a foot of rusty pipe. That is why, when people ask me what you need in order to be a successful meteorite hunter, I say: "Determination." Be prepared to dig up a lot of trash on your way to finding meteorites, especially in areas that have been farmed, mined, or were once settled.

HOT ROCKS

A "hot rock" is a terrestrial stone that sets off a metal detector. It is an old gold prospector's phrase that has been adopted by meteorite hunters. It is very important to remember that there are many different types of earth rocks that contain iron, so if your target turns out to be stone, rather than man-made iron trash, do not automatically assume that it is a meteorite. This is when your ability to identify meteorites in the field will be most valuable: Does your find stick strongly to a magnet? Does it exhibit any of the visual characteristics we might expect to see, such as remnant fusion crust, regmaglypts, exposed chondrules, or metal flecks? The successful meteorite hunter is not only adept at finding and excavating targets, but also knows how to look for the tell-tale features that are unique to space rocks.

Some areas are rife with hot rocks; others will have none. The Mundrabilla strewnfield in Australia contained no hot rocks at all—only thousands of square miles of limestone, and limestone will not generate a return signal to a metal detector. If I received a signal, I knew it was either a meteorite or a piece of metallic trash, and since we were hunting in a remote location there was very little man-made trash present. In contrast, Arizona's Gold Basin strewnfield is so rich in hot rocks that—on my early hunts there, when I was relatively inexperienced—I dug up ten or twenty hot rocks for every meteorite.

[above] The common terrestrial iron oxide hematite is often mistaken for meteorites by novices. Hematite typically does not show any attraction to a magnet and will usually leave a red streak on a white ceramic tile, while an iron meteorite will leave little or no streak. The surface features on some hematite specimens also have a visual resemblance to regmaglypts.
[right] Unexploded ordnance, large and small, discovered while searching for space rocks.

A LEARNING EXPERIENCE

I view finding and digging up meteorwrongs as doing extra credit in metal detector school. Every time you find a target and dig it, you become a better detectorist. In time you will be able to identify nails, buried barbed wire, and most hot rocks just by the unique sound they make through the detector's speaker.

The more hours you spend in the field, the better a hunter you will become. Finding meteorites takes time and you often have to dig up a lot of "wrongs" before you locate a "right." Try to view every expedition and every dig—even if you only find cans and bullets—as a learning experience. My friend Sonny Clary, a dedicated and amazingly successful meteorite hunter, once said to me: "Geoff, every hour I spend in the field not finding anything brings me one hour closer to my next new meteorite."

[top] When I found this stone meteorite, it had only been on the ground for a few days. Because it was part of a witnessed fireball, it is described as a "fall." [above] The Gold Basin L4 ordinary chondrite has been on the earth for an estimated 20,000 years and is described as a "find." It is weathered and resembles some terrestrial rocks in appearance, but it still displays remnant fusion crust, flow lines, and faint regmaglypts.

METEORITE FALLS VERSUS METEORITE FINDS

The novice meteorite hunter or collector might, understandably, be confused by the following exchange. A colleague calls me on the telephone and says: "I have a new stone meteorite for sale. Are you interested?" to which I reply: "Is it a fall or a find?"

Clearly, in order to recover a meteorite you have to determine where it landed. All meteorites fell out of the sky at some point, but "fall" and "find" have specific meanings in the world of meteorites.

A fall refers to a *witnessed fall*—one or more meteorites that were actually seen in flight by credible observers, and they are usually associated with fireballs. A fireball is a term used to describe a particularly large meteor.

A *find* is a meteorite that has been discovered on the ground, but is not part of a witnessed event. Falls are typically recent, while finds have often been on the earth for long periods of time. Witnessed falls tend to carry more financial value than finds, as they are forever associated with a particular time and place; may be documented by fireball footage; and sometimes have intriguing stories attached, such as the Claxton meteorite which hit a mailbox in Georgia in 1984. Finds could conceivably be recovered anywhere, whereas recovered falls are usually the result of ground searches following a documented fireball event.

Over the years, a number of finders have come to me with a single meteorite that they discovered while hiking, gold prospecting, or rock hounding. They almost always ask me if they should go back and look for more. They want to know if the meteorite they found was the only one, or if others fell at the same time. The answer is we usually do not know. Sometimes only a single meteorite will fall, sometimes hundreds or even thousands of fragments rain down over a finite area. The Lorton, Virginia, meteorite is a perfect example of a solitary fall. On January 20, 2010, a single stone punched through the roof of a dentist's office and broke into several pieces. A thorough search of the surrounding areas failed to turn up any additional pieces. The Sikhote-Alin fall (February 12, 1947, Siberia) scattered literally thousands of pieces over a forest in eastern Siberia, and meteorites continued to be found at the site for decades.

The only way to discover if your meteorite was part of a shower, or if it was just a single piece, is to thoroughly hunt the area surrounding the initial find. Meteorites appear to sometimes fall in small clusters, and Gold Basin (Arizona) is an excellent example of this phenomenon. I have hunted at the site for hours, or even days, and not found a single meteorite. When I did eventually make a find, I carefully searched the immediate area, and often found several more pieces in close proximity within only a few minutes.

If you find a meteorite that clearly shows a broken face, or is an angular fragment rather than a complete individual with a smooth, rounded surface, it is possible that the other pieces fell somewhere, perhaps nearby. Bear in mind that meteorites may fragment here on earth long after a fall due to terrestrial weathering. Pieces can also be transported some distance away from their original fall site by natural processes such as floods, landslides, and other forms of erosion.

During their final stages of flight, after experiencing atmospheric braking, a meteorite might be traveling at 300 miles per hour. That is equivalent to about 440 feet per second, so if a meteorite explodes several miles above the ground, and the fragments begin traveling along only very slightly different flight paths, they could be widely dispersed by the time they reach the surface.

As a general rule, if I find a meteorite on the ground I always hope and assume that it is just one of many. I recommend marking the exact find location with a small flag or GPS unit, and then searching outwards in all directions, in the hope of finding more.

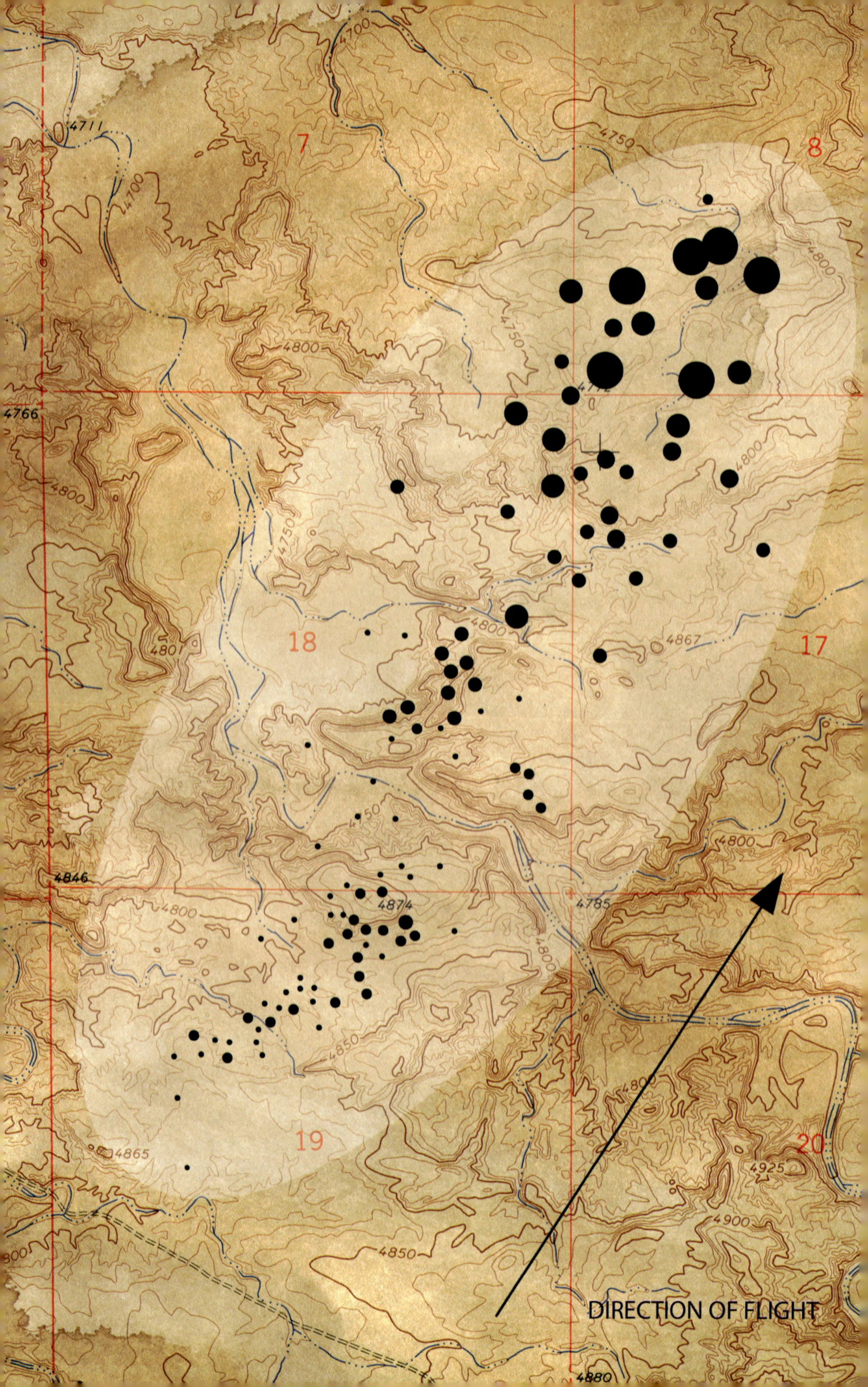

7
8
18
17
19
20
4711
4700
4750
4800
4766
4801
4772
4867
4846
4874
4785
4850
4865
4925
4900
4880
DIRECTION OF FLIGHT

UNDERSTANDING STREWNFIELDS

When a meteoroid enters the atmosphere at high speed, it rapidly heats up, and if it is of sufficient size, will turn into a fireball. Consider this: The meteoroid has been traveling in space, nearly at absolute zero, for perhaps millions of years. When its surface begins to heat up, the temperature of the mass's exterior rises several thousands of degrees in just a few seconds. When metal is heated, it expands. Combine that expansion with the tremendous atmospheric pressure the expanding mass suddenly encounters, and the result can be a very large explosion.

The great Russian meteoriticist E.L. Krinov suggested that the thousands of meteorites that hit the ground following the massive Sikhote-Alin meteorite fall (February 12, 1947, Siberia) may all have originally been part of a single, very large mass. As the mass fragmented, it dropped pieces ranging from one gram or less to hundreds of kilograms within a relatively small area. That zone, where multiple pieces from the same meteorite land at the same time, is known as a strewnfield. Understanding strewnfields is one of the most important things for the successful meteorite hunter.

Once a fireball has gone out, it continues to travel in what is known as *dark flight*. Atmospheric braking has slowed down the incoming material from its original entry velocity, and surviving fragments are thought to hit the earth at between 200 and 300 miles per hour. Typically, the smallest pieces fall first, while the greater mass and inertia of the larger pieces carries them a greater distance. In a textbook strewnfield we would find many small pieces at the "small end," less abundant but larger pieces in the middle, and a few very large masses at the "big end." If a fall zone is mapped and all the finds logged, it is often possible to draw a lazy circle around the fall zone and this is known as the strewnfield ellipse. Although it is a commonly used term in meteoritics, in our experience the distribution of fragments is usually more akin to the shape of an arrowhead.

If you have a rough idea of the boundaries of an existing strewnfield, you can plan where best to hunt. It is usually easier to find pieces at the small end, because there are likely to be more of them. However, some hunters prefer to gamble in the hopes of finding one large mass towards the big end, and larger masses are easier to see if they are on the surface, and easier to detect if they are buried. Bear in mind that the distribution of pieces based on size is only a model and there are always exceptions and variations. It is possibly or even likely that a few small pieces will be found near the large end, and some medium-sized fragments will have fallen closer to the small end. By noting the size and location of all finds, you can begin to build a map of the strewnfield.

Some strewnfields are quite small, and some are vast. You can walk from the smallest crater at Henbury, Australia, to the largest in about ten minutes, while the Gibeon strewnfield in Namibia covers hundreds of square miles. The size of a strewnfield is often related to the angle of entry taken by an incoming meteoroid. A steep angle of entry will result in most or all fragments falling in a relatively small zone, while a long shallow angle will give the meteoroid more time to burn and fragment, all the while scattering material along its flight path.

The success that Steve and I had while hunting the Park Forest fall (May 29, 2003, Illinois) is a perfect example of how understanding the dynamics of a strewnfield helped us recover multiple meteorites. Several large masses were seen to hit the roofs of houses in the Chicago suburb of Park Forest. We assumed these were the largest stones that had fallen, and we also knew the approximate angle of flight taken by the incoming meteoroid. By following that flight path upstream (away from the big end and toward the point of origin), we increased our chances of finding other meteorites that were smaller than the house smashers, and which would have fallen a few seconds earlier.

[left] Diagram illustrating a typical strewnfield ellipse and fall distribution.
The black dots show where meteorites fell and also represent their varying sizes.

After locating several mid-sized stones, we traveled further upstream and, as we did, the meteorites we found became smaller and smaller, until I noticed a 0.2-gram (0.007 of an ounce) specimen in a parking lot, many miles from the big end of the strewnfield. We chose to call that point the boundary of the small end of the strewnfield, as we probably would not be able to spot any smaller fragments.

In the following scenario I am hunting an imaginary strewnfield in which I believe numerous meteorites have fallen. A fireball is seen near the fictitious town of Smithsburg and I receive a report that a farmer found a single meteorite of 100 grams (3.5 ounces) on his property. I assume that the 100-gram stone—being of modest size—might be in the middle of the fall zone. After speaking to eyewitnesses I have a good idea of the direction of flight. I travel upstream, following the flight path, towards the small end. I find a 50-gram stone, and two hours later a 20-gram stone, so I know I am still within the strewnfield and moving towards the small end. After another hour of hunting, I find a 5-gram stone and after that nothing, meaning I have probably moved past the small end of the zone and have entered an area where no meteorites fell. At that point I may backtrack and travel towards the large end where I hope to find fewer, but larger, masses.

If I then mark all of my finds on a map, and draw a line from the smallest to the largest, I am able to estimate, more accurately, the actual direction of flight. The more finds that are made, the more detailed and comprehensive the strewnfield map becomes.

Every strewnfield is different and you will need to employ a variety of tactics on each hunt, but understanding the extent and direction of a strewnfield will greatly improve your chances of finding meteorites within it.

[top left] Team members grid a field during a strewnfield hunt. Gridding refers to the systematic covering of an area using an organized search pattern. [left] Meteorites can fall anywhere and hunting conditions are not always ideal. My friends Rob Reisener and Sonny Clary accompanied me on this expedition into marshy woodlands in the American South. We knew meteorites had fallen there, but the thick vegetation made hunting extremely difficult, even hazardous, as venomous snakes are common in the area. [above] Sometimes meteorites are discovered in the strangest places. I found this 21.5-gram Park Forest stone sitting in a parking lot in a Chicago suburb.

METEORITE CRATERS AND IMPACT PITS

Craters are formed when large meteorites traveling with sufficient velocity hit the earth, resulting in an explosive event. Most meteorites do not form craters, either because they were not large enough or were not traveling quickly enough. Small meteorites often come to rest on the surface of the ground without even making an indentation. Meteorites that land in soft ground sometimes punch a hole into the soil and become entirely hidden; we witnessed this phenomenon with both the Park Forest and the Ash Creek (February 15, 2009, West, Texas) falls. Larger meteorites, especially if they land in sand or dry soil, may remain where they fell, sitting in a shallow, concave indentation known as an impact pit. I have seen very distinctive impact pits—some almost the size of a child's wading pool—in the Imilac and Vaca Muerta strewnfields in Chile.

If you look at the surface of the moon through a telescope, you will notice that it is covered with impact craters. Our planet has also been pummeled by meteorites—large and small—for billions of years, but the protective curtain of our atmosphere both retards and ablates potential meteorites, meaning less material will make it to the ground. The air that we breathe slows down incoming meteoroids and also reduces their mass. Our planet is also a dynamic entity with moving tectonic plates, oceans, wind, and rain; the moon is not. The geologic processes at work on the earth continually change the shape of the landscape, meaning that very old meteorite craters will be eroded over time, and this is very well demonstrated by some ancient impact structures such as Vredefort (South Africa) and the Alamo Breccia (Nevada). As we have already discussed, most meteorites are rich in iron and will likely decay over the millennia in our oxygen-rich, moist atmosphere. Ancient meteorite craters, therefore, should not be expected to produce meteorites, as they may have completely decomposed. The 100,000-year-old Monturaqui Crater in Chile's northern Atacama Desert is an excellent example of what happens to crater-forming iron meteorites over time. While *impactites*—terrestrial rocks that have been melted or otherwise altered by the tremendous heat and pressure of a large meteorite impact—are relatively abundant and well-preserved around the crater, all that remains of the original impactor is weathered fragments, most of which are less than 2 grams in weight.

[left] Hidden deep in Canadian woods, the Whitecourt Crater in Alberta is only about 1,100 years old and is a protected site. [above] The spectacular 100,000-year-old Monturaqui Crater in northern Chile.

There are less than 200 positively identified separate meteorite crater sites worldwide (the Sikhote-Alin strewnfield in Siberia comprises about 100 distinct craters but is only listed as one site). Most of them are very old, and only about 15 have produced actual meteorite specimens. The Earth Impact Database, managed by the Planetary and Space Science Centre (PASSC) at the University of New Brunswick, is an excellent resource; it provides a list of all known craters and crater fields.

The first significant impact structure to be widely studied was the famous, and erroneously named, Meteor Crater (craters are formed by meteorites, not meteors) in northern Arizona. Daniel Moreau Barringer, a mining engineer and entrepreneur, was convinced that a gigantic meteorite lay buried beneath the crater floor. During the early 1900s he spent several years and a substantial fortune drilling and trying to locate the hidden mass. We now know that meteorites are not to be found buried at the bottom of large craters, except for some small fragments that may have fallen back inside after the impact event. Meteorite craters are formed by an impact and explosion of immense proportions. As a result, much of the meteorite is destroyed and any pieces that are not vaporized are thrown outside of the crater. In some instances, ejected fragments have been found miles from the actual point of impact. So, if you are searching for meteorites at a known crater site, the hunt should begin at the rim and move outwards, with the help of metal detectors.

Many of the world's most famous impact sites, such as Meteor Crater, Wolf Creek (Australia), and Odessa (Texas) are protected, and meteorite hunting is not allowed. One of the things that drew me to the Monturaqui Crater, and prompted a second complicated expedition to the hostile Atacama Desert in order to see it, was the fact that it is remote, undisturbed, and access is entirely unrestricted. My co-host, Steve Arnold, and I had the amazing experience of camping overnight on the floor of the actual crater—something that could be accomplished at only a handful of sites worldwide. We were careful to drive only existing dirt tracks so as not to damage the crater walls.

[left] Meteor Crater in northern Arizona is regarded as the best-preserved large meteorite impact site on earth, and is thought to be between 25,000 and 50,000 years old. The iron meteorites found around the crater are named Canyon Diablo after a nearby geological feature. [top] The author standing inside an impact pit formed by a large piece of the Imilac meteorite, Atacama Desert, Chile. [above] The largest of the 15 meteorite craters inside the Henbury Crater Preserve in Australia's Northern Territories. The impact is estimated to have occurred about 5,000 years ago and the event may have been witnessed by indigenous peoples.

[top] Sometimes extreme measures are required to reach the most inaccessible sites. During a 1999 expedition to Siberia's Popigai Crater, led by Professor Roy Gallant, the only way in or out was by a cargo helicopter that landed us *inside* the crater. The cliff faces behind us are composed of impactites and were once part of the floor of the ancient astrobleme. [above] The Rassokha River flows though Popigai. During our nine days inside the crater, we traveled from one site to another using inflatable rafts.

People frequently contact me, certain in their minds that they have discovered a new meteorite crater. New discoveries are rare—perhaps one new meteorite crater comes to light every year or so. Impact specialists actively search for new craters using satellite imaging and mapping software. Even with this advanced technology the chances of finding a new crater are slim, but it does happen. In 2007 a new impact site was discovered in northern Alberta in Canada, and it is now known as the Whitecourt Crater. Deep in the woods, the site had been known for generations by woodsmen and hunters and was regarded locally as a strange anomaly. In 2007 some enthusiasts took metal detectors to the site and, after a brief search, began finding rusty, angular, metallic fragments. Those pieces were taken to the University of Alberta and positively identified as a type IIIAB iron meteorite. I had the great pleasure of visiting Whitecourt in the fall of 2009, while filming Season One of *Meteorite Men*. We received express permission from the university to search there, and we found a number of excellent pieces. Whitecourt, like many other craters, is a national preserve, and there is a $25,000 fine for hunting in the restricted area. So, if you plan on hunting at a crater site, be sure it is legal to do so.

There are many natural and man-made features that have been mistaken for meteorite craters, and these include bomb craters, sink holes, man-made stock tanks, dried-up watering holes and ponds, and even buffalo wallows. If you think you have discovered a new meteorite crater, start by examining historical records of the area. Was mining carried out at the site? Was it once a farm or military test range? If the answer to one of these questions is yes, then any crater-like formations may have been caused by earlier human activities.

If it is the real thing search with metal detectors around the rim of the feature may turn up meteorite fragments. If there are no meteorite fragments to be found, the positive identification of an impact site can be complex. Verification may require the expertise of impact specialists who may search for shocked quartz, the presence of the element iridium (rare on earth but more abundant in meteorites), and impactites. Some sites that appear to be ancient craters, such as the Upheaval Dome in Utah, are still a source of debate among some academics, so there is not always an easy answer to be found regarding the origin of strange circular depressions in the ground.

Very large impact sites are sometimes called *astroblemes* (star wounds). The Popigai Crater in northern Siberia is one such site. Its great age of 35.7 million years and phenomenal size—100 kilometers (62.5 miles) in diameter—make it one of the five largest known astroblemes on earth. I had the great privilege of joining the very first international team to visit Popigai on a 1999 expedition led by the great astronomer, author, and adventurer Professor Roy Gallant. He chronicled our three-week expedition in his excellent book *Meteorite Hunter*. Even though any meteorites that survived the cataclysmic Popigai event had long since vanished, it was one of the most exciting adventures of my career, and I returned to the United States with a marvelous collection of impactites. Some of those specimens contain microdiamonds formed by the intense heat and pressure of that ancient impact.

Meteorite craters are the point at which asteroids and our planet intersect, with sometimes apocalyptic results. In addition to being promising sites for meteorite hunters, they are also places of extraordinary beauty and wonder. Well-preserved meteorite craters are rare and important historic sites, and should be treated with respect. The passage of tourists and cattle caused erosion to the walls of the Henbury Craters in Australia, and although visitors are welcome to walk around them, special permission is required to venture down inside.

The days I spent hiking around Henbury, our visit to Whitecourt, and our overnight sojourn at the bottom of the magnificent Monturaqui Crater, were some of the most memorable times of my life.

Even if you cannot get permission to hunt at a protected crater site, consider giving it a visit. The grandeur of these footprints of cosmic devastation are both inspiring and humbling—a reminder of how vulnerable our little planet really is.

The author [left] and Steve Arnold with the 101-kg pallasite they recovered while filming *Meteorite Men* Season Two, in Kansas

THE IMPORTANCE OF RESEARCH

Another question I am asked all the time is: "How do you know where to look for meteorites?" If we are chasing a witnessed fireball in an attempt to locate freshly-fallen rocks, then we usually have a fairly good idea of where to begin the hunt. But most of our work takes place at sites where meteorites have been found before. If numerous meteorites fell at a known spot, it is possible or even likely that not all of them have been recovered. The great meteorite hunter and researcher Harvey Harlow Nininger was fond of saying that a strewnfield is never hunted out, and we have proved him right time and again. Heavily-hunted and well-known strewnfields such as Brenham (Kansas) and Vaca Muerta (Chile) have delivered major finds for us even though we knew many people had hunted those areas before us.

There are several reasons why we can return to sites that have already been thoroughly searched and still find space rocks. One factor is that some strewnfields are very large, covering ten, twenty, one hundred square miles or more. It is virtually impossible to carefully search every square foot of a strewnfield that could be the size of hundreds of football fields, so a meticulous search will often produce meteorites that others have missed. Other considerations are improvements in equipment and improvement in your own abilities. We have, many times, visited well-documented strewnfields and found new material because we were using better detectors or because our skill and knowledge as hunters had increased.

Research becomes one of the most important tools in the meteorite hunter's arsenal.

Meteorites have fallen over the entire surface of our planet and—odds are—one probably landed somewhere near you. Read books about meteorites, look in newspaper archives for stories about historic witnessed falls, sift through scientific papers and see if you can find a clue that will help you know where to begin your hunt. The Meteoritical Bulletin Database is an excellent online resource, as it lists all officially classified meteorite falls and finds, many of them with coordinates. Bear in mind that not all published coordinates are precise. Sometimes only an approximate find location is given—such as the nearest town—and there have been cases where meteorite finders have deliberately provided information that is not strictly accurate in order to prevent others from hunting at their sites. Other find locations are simply not known. An example is the famous Turtle River iron meteorite from Minnesota. One mass of 22.39 kg (49 pounds) was reportedly plowed up by a farmer in Beltrami County during the 1950s, but it was not until 1968 that the iron was identified as an authentic meteorite by Glenn Huss of the American Meteorite Laboratory. During the intervening years, details of the exact find location were forgotten, and it remains a mystery to this day. There are many such stories, and many "lost" meteorites waiting to be rediscovered.

Sometimes I carry out my own research, and sometimes I hire a specialist to assist me. My personal researcher, Katherine Rambo, is an experienced meteorite collector, writer, and an expert on vintage maps, making her a perfect addition to my team. Research is a skill like anything else, so when I ran into problems while planning our Henbury Crater expedition to Australia (*Meteorite Men* Season Two, Episode Eight), I enlisted Katherine to help. After examining obscure papers, maps, and documents in various archives, she was able to provide me with valuable clues about old find locations that led to some significant meteorite discoveries during that trip.

[above] Immediately after recovery, the fall location of a Park Forest stone meteorite is recorded on a map.

OWNERSHIP, LANDOWNERS, AND PERMISSION TO HUNT

The ownership of meteorites is a potentially complicated issue, and something the would-be meteorite hunter should research before commencing a hunting trip. In the United States meteorites that are found on private land typically belong to the landowner. In other words, if a meteorite fell in my neighbor's yard and I found it, the rock would still belong to him. Meteorites recovered on federal land belong to the Smithsonian Institution and should be turned over for study. Regulations for state land and Bureau of Land Management (BLM) properties vary, and you should check with the appropriate authorities to determine if it is legal to hunt in your target area.

To further complicate matters, some BLM areas are open for private collecting, but finders are not allowed to sell material they recover, making such locations attractive for the collector, but not for the meteorite hunter who makes a living from finding and selling space rocks. Local BLM officers will be able to advise you on the specifics of areas they oversee, so be sure to do your homework and get accurate, up-to-date information before beginning a hunt. Most national monuments and national parks prohibit the removal of any material at all—meteorites, plants, rocks, even sand—and there are serious penalties for those who are foolish enough to ignore federal regulations. The use of metal detectors is illegal in most federal reserves and hefty fines and even jail time can be imposed on anyone caught poaching.

In some instances, such as the Gold Basin episode from the first season of *Meteorite Men*, we received express permission from the National Park Service to hunt in a restricted federal reserve. In order to do so, we first had to find an academic sponsor (in this case the Oscar E. Monnig Meteorite Collection at TCU, Fort Worth), and arrange for an experienced Park Ranger to accompany us on the hunt. Every find was carefully documented and turned over to the Park Service which then presented all of our meteorites from that expedition to the Monnig Collection on long-term loan. We knew, going in, that we would not be able to keep a single meteorite that we found, but it was worth it for the experience of hunting in an undisturbed area.

Due to the complexity of obtaining permission to hunt on some federal and state land, we often prefer to work with private landowners. After determining who owns the land on which we are interested in hunting, we will set up a meeting to ask for permission to work on that property. Some landowners are supportive and extremely interested in our work, and generous property owners have sometimes allowed us not only to hunt on their land, but also to keep everything that we find. In most cases we make a deal with the landowner and turn over a percentage of all finds, either in the form of money or material. In other cases we offer to pay a flat fee to hunt for a set period of time.

There are no set guidelines or fixed percentages for this type of operation, as each site is different, and each meteorite find is unique. Be respectful to landowners and be straightforward with them. Gaining access to someone's property is a privilege, not a right. While I have generally been welcomed by property owners, there have also been cases of would-be meteorite hunters getting chased off private land with a shotgun. Meteorites often fall in rural areas, and sometimes people choose to live in remote areas for a reason; they do not want to be bothered by others. If a landowner does not want you on his or her property, move on. Badgering property owners gives meteorite hunters a bad name and can get you into trouble. Remember that trespassing and removing material from someone's land without permission are crimes for which you can be prosecuted.

Some meteorite hunters have set an unfortunate precedent in which they offer to give up 50% of all finds to their host. While this may be viable in some cases, a 50/50 split is not always realistic. We may spend months, or even years, working at a particular site. An extended search will likely involve us bringing in expensive, heavy equipment, paying for hotel bills, gasoline, and other operating expenses. If a find is made, we then have to excavate it and invest more time in identifying, preparing, and possibly selling the material we have recovered. Although it is true that the landowner is the legal owner of meteorites on his property, our expertise and expenditure of time, money, and effort are worth a great deal to the partnership between owner and hunter.

In April of 2009 I journeyed to western Saskatchewan, Canada, with my great friend and sometime hunting partner, Lisa Marie Morrison. We were hoping to find pieces of the Buzzard Coulee stone meteorite which had fallen over many square miles of rolling farmland, following a spectacular fireball in November of the previous year. One of our colleagues had already provided us with valuable local intelligence, so our first mission was to meet with the owner of our prime hunting location, and we quickly came to an arrangement with him. As we were hunting on foot, armed only with magnet canes, and did not require the assistance of heavy equipment, in this case we happily agreed to a 50/50 split. The landowner was trusting and gentlemanly, and granted us full access to his extensive property without any supervision. During a three-day hunt we found over 100 excellent stone meteorites. At the end of the expedition we returned to the landowner's office, showed him our finds, and he graciously allowed us to choose which pieces we wanted to keep for ourselves. Our Buzzard Coulee hunt was the perfect example of how meteorite hunting can be a beneficial experience for all parties involved: Lisa and I were able to conduct a thorough search—with full permission—we found numerous fine pieces for our collections, and the landowner received a significant quantity of valuable meteorites, which he later sold.

[above] This meteorite find weighed well over 100 kg and had to be removed from the ground using a backhoe. This type of large-scale operation requires special permission and cooperation from landowners. [above right] While hunting in Texas we maintained a detailed map showing where we had express permission to search.

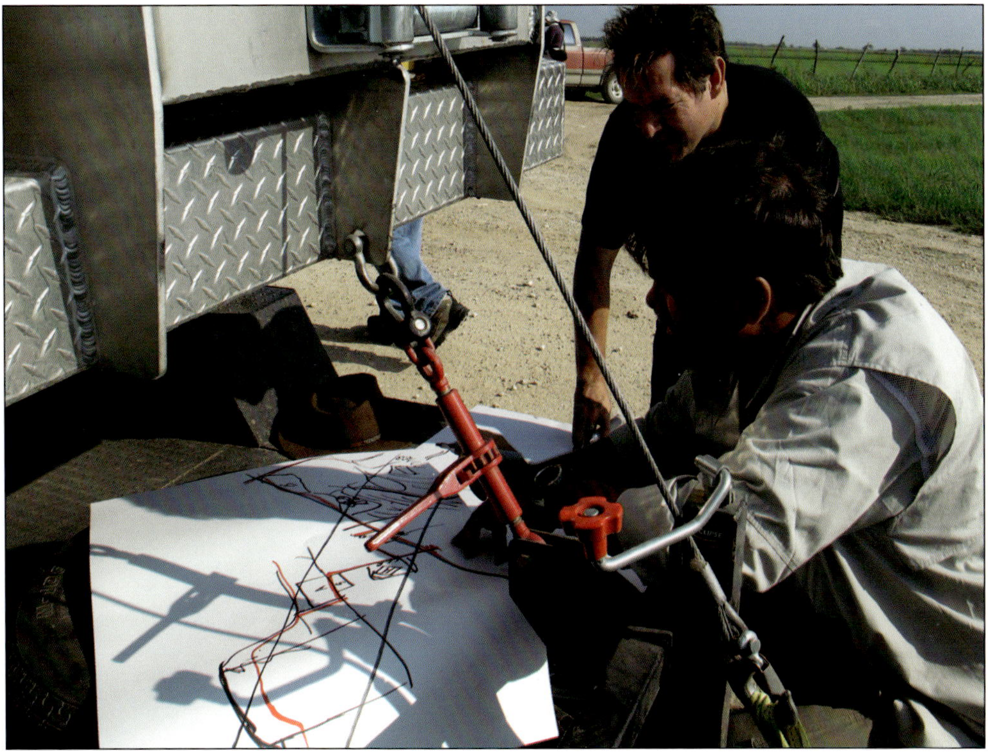

The Buzzard expedition also serves to illustrate another important point: Even though we were extremely successful and found a large number of attractive stones, we did not take them home with us at the end of the trip. Meteorites found in Canada cannot leave the country without an official export permit. This regulation ensures that Canadian academics have the opportunity—if they wish—to purchase, at fair market value, finds made in Canada, thereby ensuring that important meteorites remain in the country for study. Before we left Canada we mailed all of our meteorites to a colleague. We went home successful, but also—at least temporarily—empty handed. In the case of the Buzzard Coulee fall, a number of fine specimens had already been made available to academia for research purposes, and so the Canadian authorities eventually released our finds. When our specimens arrived in the United States, it was an exciting day. We got to see our finds again, safe in the knowledge that we had followed established protocol and were now in possession of legally exported meteorites.

In Australia, all meteorite finds belong to the Commonwealth and cannot be removed from the country without written approval from the authorities. Export permits are sometimes granted but, again, this takes time and must be done through proper channels. While filming Season Two of *Meteorite Men*, I spent 3 1/2 weeks hunting in Australia at the famous Henbury and Mundrabilla sites. I found many small meteorites and several exceptional larger specimens. It was heartbreaking to leave those beautiful pieces behind, but I was eventually granted export permits and—some months later—my finds made their way safely, and legally, to the United States where they found a welcome home in my personal collection.

Argentina, home of the giant Campo del Cielo iron meteorite strewnfield, no longer allows the exportation of any meteorites; neither does Namibia, the African nation which encompasses the great Gibeon strewnfield. So, if you are planning on hunting overseas, be very sure to check legalities first. Winding up in a South American prison would not a pleasant end to an expedition.

HUNTING STRATEGIES: HISTORIC METEORITES AND KNOWN STREWNFIELDS

Nothing gets my *Meteorite Men* co-host Steve Arnold more charged up than a good fireball chase. My friend Sonny Clary prefers the challenge and solitude of long-range desert hunts for cold finds. I love research, and history, and the history of meteorites; for me, there is nothing more satisfying than visiting classic locations such as Odessa (Texas), Holbrook (Arizona), or Mundrabilla (Australia) and—with a lot of dedication and a bit of luck—proving, once again, that the old strewnfields are not hunted out.

Steve's incredible success at the Brenham strewnfield is a reminder to not always believe what you are told by meteorite experts. Meteorites were first found at the site in the 1880s. Harvey Nininger recovered hundreds of pounds during his 1933 expedition, and H.O. Stockwell found a mass weighing 454 kilograms (1,000 pounds) in 1949. Since then, almost every serious meteorite hunter in the country has been over the site; if you asked anyone about Brenham, he or she would say: "It's all hunted out. Don't waste your time there." Steve did not agree.

In 2005 he went to the strewnfield with a new hunting strategy and pulled hundreds of pounds of Brenhams out of the ground, including the largest ever found, which weighs a staggering 650 kilograms (1,430 pounds).

There are many strewnfields in this country and overseas where there are still meteorites in the ground. This is not my opinion, it is a fact. Gold Basin (Arizona) is still producing meteorites and it was first discovered in 1995. Since then, scores of meteorite enthusiasts have been to the site, some of them many times. A few gold prospectors camped there for months, and found literally hundreds of pieces. The Gold Basin strewnfield is very well known, and relatively easy to get to. There are still meteorites to be found, which gives you an idea of the tremendous amount of material that must have fallen there about 20,000 years ago.

One of our viewers wrote to me recently and said he had watched our Season One Gold Basin episode and was taking his son meteorite hunting at the site. It was the first time they had ever been out looking for space rocks, and on their second day they made a find. It so happened that the gentleman lived in the Washington, D.C. area, and when Steve and I were guest speakers at the 2010 USA Science and Engineering Festival, he brought their find to show us. Not only was it a real meteorite, but it was a lovely piece. It was the first meteorite they had ever recovered, and I thought it most impressive that they had managed such a thing on only their second day.

I have personally found meteorites in at least ten historic strewnfields that had already been heavily hunted and which were thought not to warrant additional search time. I was successful because I put in the time, used good equipment, tried to think outside of the box, and did my research. Meteorites can fall anywhere. Chances are there is an old strewnfield in your home state or county, maybe even close to your home. The United States is rich in documented meteorite falls. In 1807 an H4 chondrite fell near Weston, Connecticut. Over 150 kilograms (330 pounds) of material was recovered. There are probably other pieces still out there. More than 220 kilograms (484 pounds) of stones have been collected from the 1912 fall near Holbrook, Arizona. Norton County (Kansas) is a rare aubrite meteorite and a staggering 1.1 metric tons have been collected since the fall in 1948. These are just a very few examples.

[left] Excavating a complete 870-gram stone meteorite in the Gold Basin strewnfield. While the exposed surfaces had weathered considerably, fusion crust and regmaglypts were present on the buried sections. It turned out to be the best-preserved fusion crust we had ever seen on a Gold Basin specimen.

A good way to get your feet wet is to compile a list of meteorite falls and finds from your home state. Do some research on that type of meteorite: Is it a stone or an iron? How much is the meteorite likely to have deteriorated in the intervening years since its initial discovery? Are there any scientific papers or newspaper articles available about the meteorite?

Hunting in old strewnfields can be difficult for several reasons: Other people know about the site, so the easy-to-find pieces have probably already been picked up; over time, meteorites will become hidden under leaves, grass, or soil and, in some cases, new housing developments obliterate old fall zones; meteorites will oxidize and eventually decay in moist environments. These are obstacles that all meteorite hunters face. My co-host Steve Arnold is fond of saying: "If it was easy, everyone would be doing it."

You may not find anything on your first hunt, or your first ten hunts, but if you spend some time exploring old strewnfields you will begin to hone your skills. Your map reading abilities will improve; you will become more familiar with your GPS, metal detector, and digging tools; you will gain valuable experience in talking to landowners and initiating your own research. Meteorites are out there, it is just a question of figuring out exactly where.

As my friend Mike Miller has reminded me when we have been out hunting: "You have to walk right over them to find them."

[below] Abundant olivine crystals are present in this slice taken from one of our pallasites unearthed at the Brenham, Kansas, site.

HUNTING STRATEGIES: FIREBALL CHASING

A few times each year a large fireball is seen in the skies over our planet. The term *bolide* is often used to describe a fireball that is seen to break up, or explode, during flight. Some fireballs are old satellites or rocket fuel tanks burning up, but the most impressive fireballs are caused by cosmic debris entering our atmosphere. Some fireballs are completely consumed; others fragment and send meteorite pieces raining down upon the earth. When such an event occurs, enthusiasts sometime race to the scene, try to determine where pieces landed, and recover them if possible. This exciting and extremely difficult type of meteorite hunting is known as fireball chasing.

EYEWITNESS REPORTS

We often start on a new fireball chase by interviewing eyewitnesses. In these cases it is vital to take into account the ways in which untrained observers may react to the shock and surprise of seeing such an event.

Due to their great brightness large meteors and fireballs often create a remarkable optical illusion in which it appears that they have hit the ground somewhere nearby, especially if they are observed at night. The glowing fireballs we see in the sky are caused by atmospheric resistance and friction, but meteors stop ablating while they are at least seven miles high. If you are lucky enough to witness a bright fireball, and it begins dark flight while approaching you or directly overhead, it is possible that meteorites will land nearby. When a fireball apparently lands in the vicinity what we are usually seeing is it arcing away over the horizon, still high up in the atmosphere. Due to the curvature of the earth, the fireball may seem to hit the ground, but has in fact just moved out of our field of view and gone beyond the horizon. Because of their extreme brightness, fireballs can appear—to our human eyes—to be much closer than they really are. It is something I, myself, have been fortunate enough to witness a couple of times, which can be frustrating because it does look as if meteorites landed "just over there." If anything made it to the ground, however, it probably landed hundreds of miles away.

Such optical illusions often make eyewitness accounts of fireballs unreliable. Most fireball events last for only a few seconds, and most witnesses are not trained astronomical observers. Sometimes interviewees will insist that the fireball they saw was much closer than it possibly could have been. In 2009 I spoke with a number of people in Tucson who had excellent views of a June 23 fireball which caused a small number of meteorites to fall near the Whetstone Mountains in southern Arizona. One eyewitness was walking to his vehicle in a parking lot when the fireball streaked by. He insisted it was so close that he could feel heat on his face! This was clearly impossible, as the fireball was still many miles up at the time.

When interviewing witnesses who give confusing or clearly inaccurate reports, the best course of action is to be gracious, thank them for their time, and move on. There is no point standing in a parking lot arguing the laws of physics with someone who *knows* what he saw.

SONIC BOOMS

By collecting reliable eyewitness reports we can begin to plot a course for the fireball. Reports from people who also heard sonic booms or other associated sounds are the most valuable. When a body such as a jet aircraft or a meteoroid travels through the air faster than the speed of sound, a sonic boom is created. The explosion of a meteoroid during flight can also generate significant sound waves. A bright fireball can sometimes be seen from hundreds of miles away, especially at night. Sound waves do not travel as far, and they also travel much more slowly than light. Spectators who heard sonic booms, in addition to seeing the fireball are, therefore, usually within, or close to, the flight path. While investigating the Ash Creek fireball in Texas, we visited one landowner who showed us a window that had been cracked by the sonic boom. His house was directly under the line of flight.

VIDEO SURVEILLANCE FOOTAGE: THE MOST ACCURATE WITNESS

Our modern world is full of video cameras: security cameras outside of private homes, banks, and warehouses; small units mounted on police cars, and so on. Fireballs do not last for long, so the chances of an individual happening to be filming with a handycam during those few seconds are pretty slim, but it has happened. The Peekskill event of 1992 is an excellent example. Its spectacular flight was captured by several different individuals who were filming their kids in high school football games when the fireball rocketed overhead.

Video surveillance footage is highly prized by meteorite hunters. It is typically more reliable than eyewitness accounts, and it can be watched over and over again in an attempt to determine velocity and angle of flight. If video footage can be collected from more than one camera, and those cameras were pointed at different angles, we can sometimes calculate—with some degree of accuracy—where meteorites might have fallen.

DOPPLER RADAR

In recent years meteorite hunters have begun to appreciate the usefulness of Doppler radar, and how it can aid in tracking fireballs. Doppler is used in aviation and aerospace, and for weather forecasting as it tracks objects in the sky such as aircraft and cloud banks. If a large incoming meteoroid breaks up at an altitude of, for example, eight miles, a debris cloud of fragments will begin to fall and may take several minutes to reach the ground. Doppler radar systems used for weather prediction employ a periodic circular sweep, and if that sweep is in the right place at the right time, it may produce an image of plummeting meteorite fragments. Doppler data was used successfully for the first time during the Ash Creek fireball hunt in 2009. Data from the National Weather Center in Oxnard, California gave searchers on the ground a very good idea of where to look for meteoritic material. Steve and I also used Doppler data on our Wisconsin (April, 2010) and Utah (November, 2010) fireball chases, and it is one of the most promising new technologies available to the meteorite hunter.

BOOTS ON THE GROUND

Once we have narrowed the search area down to a manageable size, it is time to get boots on the ground. More than once, we have been out scouting a possible fall zone and have found a meteorite just lying in the road. If you consider how much asphalt there is in the world, driving all the roads in a suspected fall area allows you to cover a lot of ground quickly.

Much of the populated world is still covered by farmland and it seems that two out of every three fireball chases result in us getting to know some rural farmers. When we have arrived at our search area the most immediate matter at hand is getting permission to hunt (see Chapter 9).

Pretty much every square foot of land in the United States is owned by someone, whether it be private, state, or federal. Hunting on private property without permission is unethical and also illegal. Asking for permission from property owners is the right thing to do and can often add useful intelligence to the mix. If I am hunting within the fireball's path, it is likely that homeowners in the area heard or saw something that can help me with the search. Once in a while a local resident will amaze me by reporting that he heard something hit his roof—hard—shortly after the fireball, and such news can turn a chase into a successful recovery. Meteorites that have hit man-made objects, such as cars or buildings, are informally called hammer stones and are coveted by certain collectors.

When hunting in farmland it is important to be respectful of the farmer's property and that can include livestock and crops. More often that not that means walking on foot, with no equipment except for magnet canes.

If my team can make even one meteorite find, then the game is really afoot. By plotting the location of that find in relation to the fireball's flight path, we can begin to map the strewnfield and determine where other pieces might have fallen (see Chapter 6).

THE REALITIES

An average strewnfield might encompass 15 square miles or more, and that may not seem like such a big area until you start trying to cover it by foot. Freshly-fallen meteorites are highly sought after by researchers and collectors, and even a single light rain can cause some meteorite types to rust, thereby losing scientific and financial value.

Sometimes there is an immediate urgency to recover pieces; bad weather might be expected, or farmers may be planting or harvesting crops. Tractors can easily bury small meteorites, causing them to be lost forever. You may find yourself competing in the field with others who are searching for space rocks—professionals, academics, amateurs, or curious locals—and the excitement of the hunt sometimes brings out the worst in people. We have, unfortunately, witnessed several instances where unscrupulous meteorite hunters have lied to landowners, bad-mouthed rival teams, or brazenly pretended to be scientists representing a university research department in order to gain access! One part-time meteorite hunter promised landowners a share of everything he found, then later sneaked out of town without giving the property owner a penny. Do not be one of those people. Such antics give all meteorite hunters a bad name, and unethical behavior has a way of coming back to bite you later. The meteorite community is small, and bad news travels fast.

Fireball chasing is hard work. Statistically, the majority of meteorites that land are small, and you may spend a great deal of time hiking through rough ground, trying to spot tiny black rocks in tall grass, and wondering what ever possessed you to want to be a meteorite hunter in the first place. To make matters worse, sometimes there is just nothing there. For example, when Steve and I were searching for pieces of the Mifflin, Wisconsin, fireball we compared notes with other hunters. After averaging our colleagues' time in the field, the results were discouraging: One meteorite was found for approximately every 65 man-hours of hunting time—and these were experienced hunters who knew exactly what to look for.

Many meteoroids burn up completely in the atmosphere and no meteoritic material survives to be found. The only way to know for sure is to look, and the payoff can be great. There are few things as exciting as picking up a pristine meteorite that—only hours or days earlier—was still part of the cosmos.

[above] Persistence pays off. In 2009 I rushed to the Ash Creek (Texas) strewnfield a few days after the fireball of February 15. I found 13 meteorites in less than a week, while competing with other meteorite hunters. Seven months later I returned and found this 36.8-gram stone in plain sight, in an area that had already been hunted by numerous other people.

HUNTING STRATEGIES: COLD FINDS, HOT DESERT HUNTING, AND DRY LAKE BEDS

When a new meteorite is discovered that was previously unknown to science and is determined to be unrelated to a fireball or witnessed fall, it is described as a cold find. Making an entirely new discovery is one of the greatest achievements that a meteorite hunter can aspire to. Not only has the finder made an important contribution to the science of meteoritics, but his or her name will likely go down in the literature, in perpetuity, as the original discoverer.

Meteorites that have been on our planet for many years will, in most cases, have started to oxidize and, over time, may begin to look more and more like earth rocks. Fusion crust decays or weathers away, and iron meteorites rust or acquire a patina that is often reminiscent of old bronze. In other words, old meteorites can be more difficult to spot than freshly-fallen ones because they are camouflaged.

If old meteorites are sometimes hard to spot, then we want to look for them in places where they will stand out. Hunting for cold finds in a heavily wooded forest, for example, or in zones that are rich in terrestrial rocks, or in areas with tall grass will likely not be successful. Any meteorites that may have fallen there will be difficult to recognize among trees and boulders and will probably have been covered by falling leaves or decaying plants.

Old meteorites will still appear darker in color than most terrestrial rocks, so if we want to locate them visually we should look in areas where the odds are in our favor. Dry lake beds are a favorite search area for experienced meteorite hunters because they are wide open spaces, mostly devoid of vegetation, and generally have far fewer terrestrial rocks than other environments. If you spot a dark-colored rock on the surface of a lake bed, it may have fallen out of the sky.

[left] Dry lake beds are typically devoid of vegetation and can make for ideal meteorite hunting locations. [above] A new Nevada stone meteorite lying exactly where it was found. Note the relatively fresh, black fusion crust, indicating a fairly recent fall.

STRANDING SURFACES AND INCIDENCE OF FALLS

The majority of cold finds have been made in hot deserts in Northwest Africa, the American Southwest, the Australian Outback, and in Antarctica.

For the sake of argument let us use a simplified model which suggests that a single meteorite, weighing 10 grams or more, falls on each square mile of the earth every 10,000 years. In a region that experiences significant annual rainfall—northern Europe for example—we would expect that meteorite to entirely decompose before the next statistically predicted fall 10,000 years in the future. In an arid environment, however, such as the deserts of Northwest Africa, the first meteorite would likely survive the 10,000-year period, so when a second meteorite falls on the same square mile 10,000 years later, we now have two potentially recoverable space rocks in a relatively small area. If we then extrapolate and look at a surface that has remained undisturbed for 100,000 years we could, theoretically, have ten recoverable meteorites per square mile.

Our theoretical model may be used to explain how over 30 different meteorites have been found in the San Juan strewnfield in Chile in an area of only a single square mile. The conditions in that part of the Atacama are ideal for the long-term preservation of meteorites, and the area is also very remote so existing meteorites may not have been disturbed by wildlife, road building, farming, or other activities. There are few places in the world which are so perfect, but the find statistics at the San Juan site clearly demonstrate why we are particularly interested in old surfaces which are also dry and low in native vegetation. These areas where meteorites may have collected over great periods of time are sometimes known as stranding surfaces.

[above] Using an all-terrain vehicle to scout a dry lake bed for meteorites. [above right] How good are your eyes? This Mundrabilla iron meteorite was found on the surface, where it had doubtless been lying for thousands of years. Its color is only slightly different from the surrounding soil.

The Nullarbor Plain in Western Australia is another remarkable example of how old surfaces can produce abundant new meteorites. The Nullarbor is famous among meteorite enthusiasts because it encompasses the Mundrabilla iron meteorite strewnfield, where we filmed Season Two, Episode Seven of *Meteorite Men*. Since 1911 when Mundrabilla was first discovered, thousands of small irons have been found in the 30-mile long strewnfield, and they were all part of the same fall. A more intriguing statistic is that 600 different meteorites, unrelated to Mundrabilla, have also been found on the Nullarbor. Our science specialist for the Mundrabilla episode, Dr. Andy Tomkins, told me an amusing story about one of his meteorite hunting expeditions to the Nullarbor. On the way to a potential site, he stopped his truck for a few minutes to have a look around. As he opened the door to his vehicle and got out, he almost stepped on a meteorite. After analysis in the lab, it was determined to be a completely new cold find.

Although Antarctica is covered in ice, it is still regarded as a desert because of its lack of rainfall, and great numbers of meteorites have been found there. Conditions in Antarctica are unique. Unlike the Arctic, it is an ice-bound continent with mountains, valleys, and indigenous rocks. Meteorites have been falling on the continent for thousands of years. They become embedded in the ice and are, therefore, well preserved. Ice sheets travel, much like slow-moving rivers, so the imprisoned meteorites can sometimes be carried considerable distances. As ice sheets ride up against Antarctic mountain ranges, wind causes them to erode, sometimes revealing frozen meteorites. Although the circumstances of meteorite finds in Antarctica are unlike those in any other areas, the successful recovery expeditions there demonstrate just how many meteorites can fall in a finite area over long periods of time.

My friends Rob Matson and Sonny Clary have had great success making cold finds on dry lake beds in the United States. They have trained their eyes to look for out-of-place material in flat, desolate areas where very few earth rocks occur. While Rob does most of his hunting on foot, Sonny prefers to cover large areas of ground quickly on an ATV. Carrying a magnet cane with you (see Chapter 13) is an excellent idea. It allows you to quickly and easily conduct a magnet test on any interesting-looking rocks.

If you find the right type of environment, spend sufficient time on the ground, and familiarize yourself with what weathered meteorites look like, there is a chance of making a new cold find for yourself.

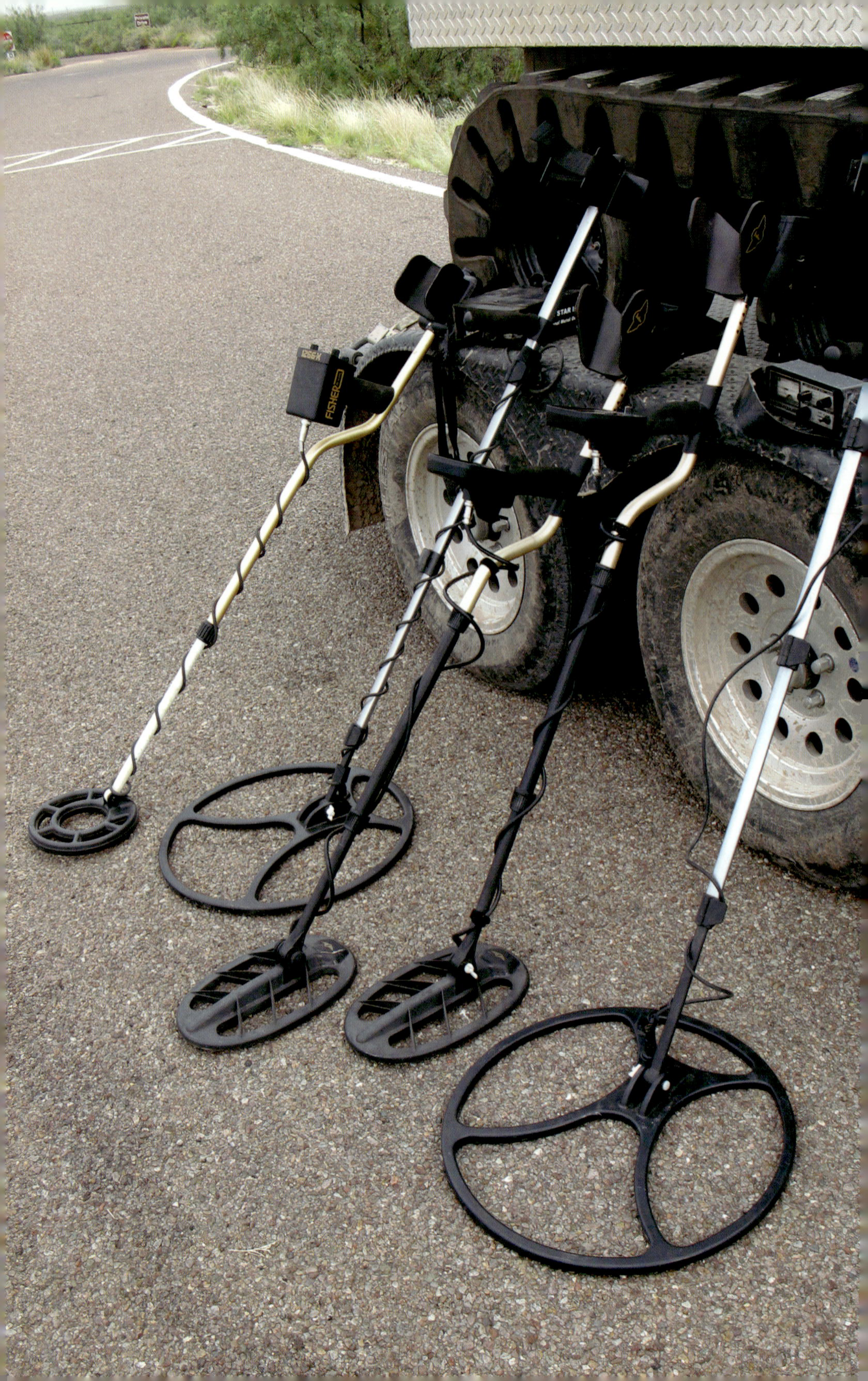
FISHER

TOOLS OF THE TRADE

METAL DETECTORS

We have already discussed the fact that almost all meteorites contain a significant amount of ferrous material, typically about 20% in most stone meteorites, roughly 50% in stony-irons, and 90% in irons. We have also discussed how the majority of meteorites are buried at varying depths, and so a good metal detector becomes a crucial tool for the meteorite hunter. While there are many different makes, models, and sizes, all metal detectors basically work on the same principle and most consist of the same basic components: control box, cable, coil (consisting of a transmitter coil and shielded receiver coil), and frame. The control box transmits an electromagnetic signal down the wire to the transmitter coil, which then sends that signal into the ground. When the electromagnetic field encounters a buried metallic object, a return signal is sent back to the receiver coil, then the control box, and is finally broadcast on a speaker, or through headphones to the user. A rigid plastic frame holds these components together, separates the metal circuits in the control box from the coil, provides a handgrip for the user, and space for batteries.

HAND-HELD DETECTORS (VLF)

The most widely-used detectors are hand-held units that employ VLF (very low frequency) technology. They are used for gold prospecting and relic hunting, among other things, and are popular with hobbyists as they are lightweight and easy to use. If you have ever seen a retiree walking the beaches looking for lost watches or jewelry, he or she is almost certainly using a VLF detector. These hand-held models typically have a coil diameter of about 10 or 12 inches. They are used for locating shallow targets and most have a maximum depth of about 18 inches. VLF units are lightweight, easy to transport, readily available from many different manufacturers, and relatively inexpensive. Their one drawback is limited range, which is a function of their small coils. VLF detectors can be used to hunt for any type of meteorite that contains iron, and are the best choice when hunting for stone meteorites, as their high degree of sensitivity can locate very small amounts of metal.

HAND-HELD DETECTORS (PI)

Pulse induction (PI) technology operates—in the broadest sense—in a similar way to VLF, but PI detectors are less susceptible to interference from mineralized ground, and can "see" through any type of layers: soil, sand, even underground shelves of rock. The electromagnetic field generated by any form of detector decays quickly as it travels down into the ground, and PI technology allows for the construction of larger coils. In simple terms, therefore, the larger your coil, the deeper it can "look" into the ground. The control box (which we sometimes call "the brain") of pulse induction units can be connected to a variety of different coils. When hunting for iron or stony-iron meteorites which I suspect are buried at a moderate depth, I use a larger 18-inch coil attached to a pulse induction control box, such as the Pulse Star II Pro.

PI detectors are ideal for sizable targets that are rich in iron, but most PI units will receive no signal at all from a stone meteorite. The control box and battery pack on a typical PI detector are larger and heavier than a VLF unit, so I wear mine on a strap around my neck. The Pulse Star's 18-inch coil has a maximum depth range of approximately 1 meter. It is much heavier than a conventional VLF unit, so bear in mind that your effective hunting time will be limited when employing a larger detector. Swinging an oversize coil for hours puts a significant strain on your arms and back, but also allows you to cover more ground.

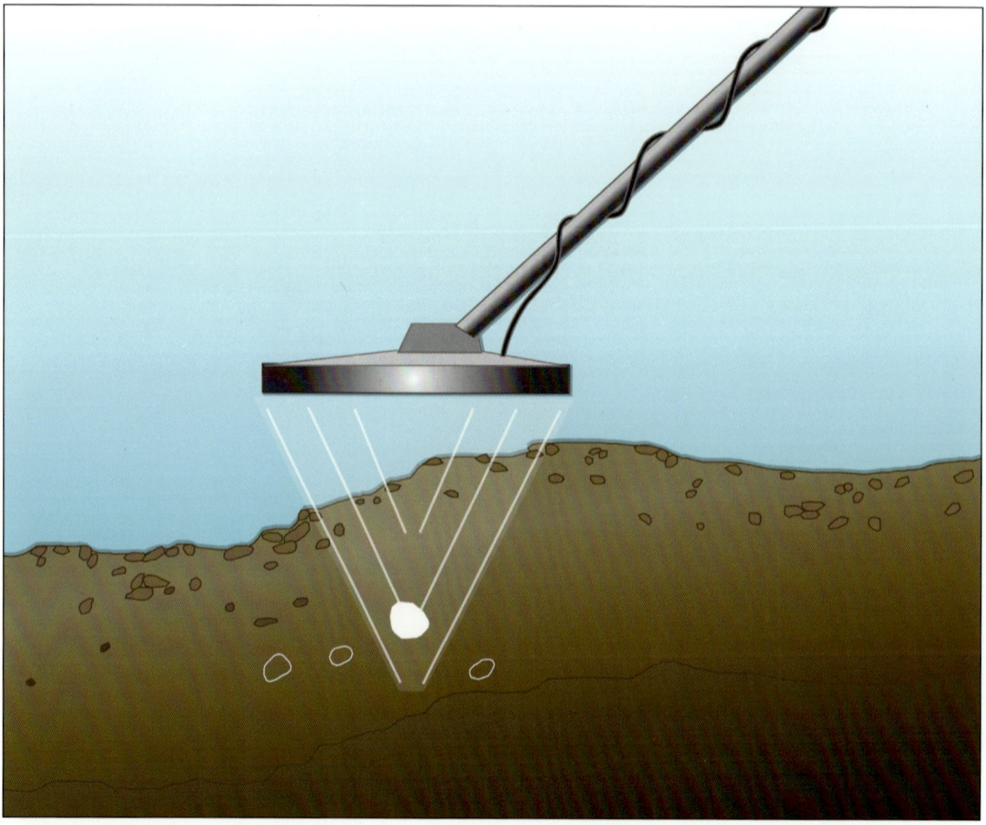

[above] This diagram demonstrates how the farther a detector's signal travels, the less area it covers. The diminishing "cone" of the detector's electromagnetic field can miss deeper targets. [right] Steve Arnold assembling his giant meteorite detector before the hunt begins. Since the coil will register any metallic targets, the entire frame must be constructed without using any metal bolts, bearings, or washers.

GIANT DETECTORS (PI)

If you have watched my TV series *Meteorite Men*, you have probably seen our giant meteorite detector in action. "The Big Rig," as we call it, was designed and built by my co-host Steve Arnold and has gone through many developmental stages over the years. The Big Rig consists of a giant coil suspended on a PVC sled and connected to a pulse induction control box.

As we have already discussed, the larger the coil, the greater its effective range, so the largest version of The Big Rig can locate a buried target of significant size at 3 meters or more, which is an extraordinary depth for any detector. The drawbacks in using a giant rig are the expense of building it and the difficulty in transporting the sled and coil. A working unit of that size costs in excess of $5,000 to construct and the act of towing it over rough ground often results in damage and additional repair costs. Building and operating a large rig is complicated and expensive, and requires a great deal of practice. It is not something recommended for the beginner, and should only be used in areas where the hunter expects to encounter very large, deeply-buried targets that contain at least 50% iron.

When hunting the Muonionalusta strewnfield, north of the Arctic Circle in Sweden, we were in a protected area where motor vehicles were prohibited. As the meteorites at that site are deeply buried, we still needed to use an extra-large detector. We used a pulse induction "brain" to power a portable coil that measured 2 meters on each side. It was large enough to find deep targets, but still maneuverable enough to use in a densely-wooded forest.

METAL DETECTOR TIPS AND TRICKS

Metal detectors do not use a great deal of power, and a set of fresh batteries will provide many hours of use with most units. As with any quality piece of electronic equipment, use good quality batteries, and do not mix different brands, or new and old batteries. It is a good idea to keep a spare set of fresh batteries in your backpack or equipment bag, and remove batteries from the detector if you will not be using it for extended periods.

TAKING CARE OF YOUR DETECTOR

Your detector is a powerful and sensitive piece of scientific equipment. Treat it as you would a good stereo, camcorder, or expensive camera. Try not to drop it or bang it into objects while searching. It is a good idea to invest in a carrying case if you will be transporting your detector in a vehicle or on an airplane. Most detector frames can easily be stripped down into smaller sections, and when I am going on an international hunt I break mine down and pack it carefully in a hard shell Pelican case. Most detectors will work in the rain, but they might not like it very much. Try to keep your detector dry and do not submerge it in streams or lakes unless it is a waterproof model. Try to keep the control box clean and free from dust (not always an easy task in the desert) and do not leave it lying unattended in the hot sun for extended periods.

CARRYING SPARE EQUIPMENT

Professional meteorite hunters usually carry more that one detector, but that is not usually necessary for the beginner. Find a detector that you like, and if you plan on embarking upon extended trips consider purchasing an additional unit as a backup. There are many inexpensive but reliable used detectors on the market.

If I am traveling overseas, I always take at least two detectors, along with spare parts. When I visited the Imilac strewnfield in Chile's Atacama Desert, it took about four days of travel in each direction to get to the hunt site. I never want to take the chance of getting to a difficult-to-reach site only to discover that my detector has malfunctioned. Spares for modern detectors are not readily available in many other countries and, even if they are, I do not want to waste valuable field time trying to repair damaged equipment.

[left] The author with two different metal detectors in the Imilac strewnfield in Chile.
[above] When my detector first found this buried object, I thought it was trash.
I dug it anyway and it turned out to be an excellent Henbury iron meteorite.

GET TO KNOW YOUR EQUIPMENT

Much as a musician may search for years for the perfect guitar, you should find the detector that suits you best. Different detectors have different controls, varying operating parameters, and varying weights. I am a Fisher devotee and they are my detectors of choice, but there are many other good companies, such as White's, Garrett, Pulse Star, and Minelab. The most important consideration is that you purchase a detector that is sensitive to iron. Although this may seem obvious at first, be aware that most detectors are designed to search for gold and other precious metals. Many detectors include a discrimination circuit that can be used to filter out trash (usually iron, steel, and aluminum) if the user is looking for gold nuggets. When hunting for meteorites, be sure that discrimination is set to "off," or you may miss ferrous targets which could be meteorites. Spend some time reading the manual, practicing, and getting to know your equipment before you take it out in the field. It is very frustrating to get to your hunting site and discover that you do not fully understand how your detector works.

DIG EVERY TARGET

The experienced meteorite hunter can often tell—just by the sound emanating from the detector—whether or not a buried target is likely to be a meteorite. Attuning your ears to the variance in sounds produced by different targets might take months or years of practice, and even the professional hunter can easily be fooled. When I was hunting the Henbury Craters strewnfield in Australia during 2010, I received an extremely strong signal and was convinced I had passed my detector over a tin can hidden in high grass. I dug the target anyway, and it turned out to be a beautiful 468-gram iron meteorite worth about $1,700. So, when you locate a good target be sure you find out what it is. Yes, you will dig up a lot of trash, but you will also know that you have not left a possible meteorite in the ground.

PACK OUT TRASH AND FILL IN YOUR HOLES

My policy is to always leave hunting sites a little cleaner than I found them. I always pack out my own trash as well as meteorwrongs. If I go to the trouble of digging up a horseshoe, a pile of nails, or a tin can, I put them in my backpack and dispose of them properly. It is also extremely important to fill in your dig holes. Some meteorite sites have been closed permanently because landowners became annoyed at irresponsible hunters who left a mess behind. Cows and horses can stumble in a hole and break a leg, and such accidents do not make for happy ranchers. Respect the land and the landowner, close gates behind you, clean up after yourself, and you will be invited back to hunt another day.

METAL

It is obvious that metal detectors find metal, but it is surprising how many detectorists load themselves up with metallic objects such as rock hammers, magnets, steel canteens, and so on. A sensitive detector will pick up metallic objects on your body, so position your gear accordingly. My Fisher F75 is so sensitive it will even register the small metal eyelets in my hiking boots, thereby giving a false signal. If your footwear contains metal—and most boots do—be sure to swing your detector away from your feet, and keep other metal objects as far away from the coil as possible.

HEADPHONES: YES OR NO?

Headphones do, in a sense, increase the hunter's ability to find deeper targets, or targets that only contain a modest amount of iron. Wearing headphones allows you to hear fainter signals, get a better idea of the tone of the signal, and they also minimize distracting outside noise. Using headphones can also extend the battery life of your detector. As headphones can obscure exterior sounds, they can also be a potential hazard: You may not hear the warning of a rattlesnake, or may miss your hunting buddy calling for assistance. If you decide to hunt with headphones, keep the volume at a reasonable level and be extra vigilant for potential dangers.

[below left] An assortment of meteorwrongs found while hunting, including: Shrapnel from an artillery shell, Civil War bullets, riding tack, aircraft parts, 50-caliber machine gun cartridges, nails, bolts, and the cylinder from a Wild West revolver. [above] A GPS unit is used to record the latitude and longitude of a new meteorite find.

GLOBAL POSITIONING SYSTEM

Global Positioning System (GPS) units are devices that use a signal to communicate with orbiting satellites and then display the user's location in latitude and longitude. They are accurate to within a few feet and also typically show your compass heading and altitude above sea level. The seasoned meteorite hunter uses a GPS to guide her/him to a particular site with known coordinates, mark the location of new finds, and also record travel and distance covered. Many times, we have made a find, logged it into a GPS unit, then returned days or—in some cases—years later to continue the hunt. Logging the location of each find is a useful record for the hunter, and find data are welcomed by meteoriticists who are attempting to plot all known meteorites from a particular find or fall. Such information is useful to researchers who work on mapping strewnfields. At the end of the day, a GPS can help you navigate directly back to your starting point and also let you know how many miles your tired feet have covered.

Most GPS units are small and compact and easily fit in a vest pocket, backpack, or on your belt. More sophisticated models include color screens, topographical maps, and even satellite photos onto which the hunter or hiker can superimpose a route or hunting path. I never leave my truck without my GPS—in fact, I usually carry two different models—and I log the exact location of every find. A basic model can be purchased from outdoor outfitters or camping stores for about $100.

MAP AND COMPASS

While a GPS device is an invaluable tool, in my opinion nothing beats an old school map and compass. As a Royal Air Force cadet back in the United Kingdom, I studied advanced orienteering (map reading). As a visual person, I like to be able to see exactly where I am and where I am headed on a printed map. If my GPS gets lost, damaged, or—in rare cases—is unable to lock on to a satellite, a map and compass might end up being my only option for finding a safe route home, and you never have to replace the batteries in a map.

MAGNET CANES AND RARE EARTH MAGNETS

Since meteorites almost always contain iron, a good magnet is an essential item for your field kit. When we find a suspected meteorite on the surface, or when we excavate one from underground, the first thing we do is test it with a magnet. Most meteorites will jump to a powerful magnet and adhere strongly.

We use rare earth magnets on every expedition. They are lightweight, though somewhat fragile, and much more powerful than conventional magnets. I generally use a 29-pound pull magnet, which is strong enough so that even an iron-poor meteorite will adhere to it. Steve and I have also constructed magnet rakes—devices that carry an array of magnets and which can be pulled by hand, or behind a vehicle. On our rakes we have used magnets with an 80-pound pull, but they can be extremely dangerous, and should be used with great care. I always wear gloves when handling them. A strong magnet can erase a computer hard drive, credit cards, hotel room keys, iPods, and similar devices, so be careful where you store them. I wrap my magnets in a woolen sock and transport them inside my hiking boots. Magnets will also jump towards each other at blinding speed if placed too close together. If your finger gets between them when that happens, it is an unpleasant experience you will not forget.

When hunting for freshly-fallen meteorites on the surface, we test all interesting looking rocks with a magnet cane. A magnet cane is easy to construct for yourself. We have used walking sticks, metal poles, and even wooden dowel rods as a cane, and attached a magnet to one end using duct tape. Touching your magnet cane to possible meteorites can save you from bending over repeatedly to look at suspect rocks.

The Holbrook strewnfield in northern Arizona has produced thousands of pea-sized stone meteorites since the witnessed fall of 1912, and the Holbrook dunes are also home to a sizeable population of wild desert hares. Dried rabbit droppings can look remarkably like small chondrites, and meteorite hunter Suzanne Morrison, who has been very successful at Holbrook, is fond of saying: “If it doesn’t stick to a magnet, don’t pick it up!”

DIGGING TOOLS

When I head out into the field, I throw a couple of good digging shovels into the back of my pickup truck. You never know when you might find a target that is out of reach for small hand tools. When I travel internationally I carry a half-size shovel in my backpack. A folding military surplus entrenching tool is also a good option: It is lightweight and folds down for easy carrying.

The one indispensable tool for me is my rock hammer. I use a pickaxe-style tool with a steel blade and a durable plastic handle. I put one or two rare earth magnets on the flat surface of the blade, allowing me to both dig and test suspected meteorites with the same tool. A rock hammer can easily be carried in, or on, a backpack or belt mount. If you have a magnet affixed to your rock hammer, do not forget it is there. I have, more than once, leaned up against my truck, heard a loud clunk!, and discovered that my magnet and hammer had jumped out of their belt mount and securely affixed themselves to a vehicle door or side panel.

Another useful tool is a plastic gardening trowel, which weighs only a few ounces, costs a couple of dollars and fits easily in a backpack. When working at Brenham, Kansas, and at the secret Alpha Site, Steve and I discovered numerous large pallasites. The surfaces of these rare and beautiful meteorites are often encrusted with clusters of delicate olivine (peridot) crystals. An accidental blow from a steel shovel could destroy or damage beautiful and delicate crystals that formed billions of years ago. So, after removing the top layer of soil and exposing part of a buried meteorite, we will often dispense with steel shovels and complete the final excavation with plastic trowels that are much less likely to damage a potentially valuable meteorite.

[left] The author with a new find: A lovely fusion-crusted meteorite from the Buzzard Coulee fall. This specimen was found using just a simple magnet cane. [right] Pallasites are rich in olivine, and clusters of crystals are often found on the surface. In order to avoid damaging any of the delicate crystals, this pallasite was carefully excavated using a plastic trowel.

VEHICLES

Sometimes just getting to a site is half the battle. If you are going to be traveling on dirt tracks or exploring remote locations, be sure to select a suitable vehicle that can handle difficult terrain and which has plenty of ground clearance. I drive a modified 4WD pickup truck with top quality off-road tires and shocks. A trick I learned from my late friend Jim Kriegh—discoverer of the Gold Basin strewnfield—is to always keep a spare battery and tools with me in the vehicle. If you are working at an isolated location and discover that your truck has a dead battery, there may be nobody around to assist, and you could be facing a very long walk to get help.

Sometimes even the sturdiest 4WD truck is not sufficient to get where you want to be. When we were filming at the Whitecourt Crater in Canada during the fall of 2009, the ground was covered with snow and even a powerful off-road truck could not navigate the narrow woodland tracks that led to the site. We had to rent a convoy of ATVs to get us, our equipment, and the film crew to the hunt location. We have also used ATVs to tow Steve's giant detector sled and to search by eye for recent falls. We have employed a wide variety of other transport, including the Hydratrek tracked vehicle, our Orange County Choppers Meteorite Bike, and even mountain bikes to get to a prospective hunt site.

Whichever vehicle you choose to work with, make sure it is in good running order before you set out and carry spare food and water at all times. When exploring Chile's Atacama Desert we were so far from civilization each truck had to carry several spare jerry cans filled with fuel. Gas stations were sometimes hundreds of miles apart, so we made sure every vehicle was stocked with extra supplies.

KC
KC
KC

EXCAVATING METEORITES AND DOCUMENTING FINDS

A small meteorite buried a few inches below the surface may come out of the ground quickly and easily and become an instant addition to your collection. For shallow targets, I use my rock hammer and it is the perfect tool for the job.

Uncovering a large, buried meteorite is, for me, perhaps the single most exciting part of my work. The process is reminiscent of an archaeological dig, but the treasure we are excavating is much older than even the most ancient man-made artifacts. It can be an arduous and complicated procedure and may take several hours or more. As the soil, sand, or clay is removed, I always take a few minutes to photograph the emerging space rock. In situ photos are wonderful reminders of a new find and also important for hunting records. Take your time and record the details. Remember that a buried meteorite may have started to fragment while buried in wet ground. Once the find has been removed from the ground, check the hole with your metal detector. In several cases we have found several pounds of small but attractive fragments in the bottom of a hole after a large meteorite has been pulled out. Such material should be placed in plastic specimen bags and labeled for future reference.

I always carry extra specimen bags and a waterproof felt pen in my field bag or backpack, and write a field number, GPS location, or brief note on each bag that contains a meteorite. As time goes by it is easy to forget the details of a particular expedition. Spending a few seconds recording find numbers and locations in the moment will provide you with accurate data that you can refer back to at a later date.

Sometimes a meteorite is too deeply buried, the soil too dense, or our remaining time on site is too short to allow for a conventional excavation by hand. In those instances we might bring in a Bobcat or backhoe to speed up the process. We make arrangements with a local owner/operator, and even though a mechanized digger saves a great deal of time, we must be extra careful not to bash a buried meteorite with a quarter-ton steel excavator bucket. Finding a meteorite that is so big it requires the services of a tractor to retrieve it from the ground is a great problem to have, so if it happens to you, you will find a way.

When Steve and I were north of the Arctic Circle in Sweden hunting the Muonionalusta strewnfield, we found a beautiful 30.4-kilogram (67-pound) iron at a depth of about 1 1/2 meters. The meteorite had been transported by a long-vanished glacier and had been deposited in the terminal moraine—unsorted debris dropped as an ice sheet melts and recedes. The iron was securely wedged under a boulder that had also been dropped by the glacier. This find proved that our pulse induction detector could, indeed, see right though large rocks. It also proved that sometimes there is no substitute for hard manual labor. Our permit to hunt at the site specified that we were not allowed to use mechanized vehicles in the forest, so we had to dig the iron by hand. The combined efforts of our four-person team were not sufficient to shift the boulder, which easily weighed several hundred pounds. Sometimes meteorite recovery is all about determination, and there was no way we were leaving that marvelous specimen in the ground. After several strategy discussions, and several hours of experimentation, we greatly expanded the size of hole, dug around the boulder, and under it, until we were able to dislodge the trapped meteorite by having our friend and colleague Carin Österburg jump up and down on it until it worked loose.

Every expedition is different and every challenge requires a new solution. Persistence pays off and, once in a while, brute force wins out.

[above] Our 30.4-kg Muonionalusta meteorite after excavation and cleaning. [right] Examining a thin section under a microscope, and with polarized light, enables a specialist to determine a meteorite's composition. The specimen pictured is the CK4 carbonaceous chondrite NWA 4657. Photograph © John Kashuba.

AFTER THE FIND: NAMING AND CLASSIFICATION

Any number of experienced meteorite professionals can tell you whether or not your find is the real thing. My company, Aerolite Meteorites, offers a meteorite identification service for a nominal fee, as do several other prominent dealers and collectors. A positive identification, however, is only the first step when it comes to a completely new find.

A series of procedures must be followed in order to have a new meteorite formally recognized by the scientific community and that process can be fairly complicated. The first step is to have the meteorite classified. This requires a sophisticated laboratory analysis by a recognized meteoritics department and is typically not the sort of thing that the average geology or astronomy department is equipped to carry out. The UCLA Department of Earth and Space Sciences, NASA's Johnson Space Center, and the Center for Meteorite Studies at Arizona State University are examples of the small number of accredited research institutions that can perform meteorite classification.

If the new find is a stone meteorite, a thin section is made. This involves removing a small piece from the specimen and grinding part of it down until it is so thin that light can pass through. The section is affixed to a glass slide and examined under a powerful microscope using polarized and normal light. Examination under polarized light enables a specialist to determine which minerals are present in the rock. The majority of meteorite finds are ordinary chondrites and the classification process assigns them a letter such as H (high iron content), L (low iron content), or LL (low-low iron content). A number follows the letter and that indicates the amount of altering (by heat or water) that the chondrules experienced while still part of the parent body. Ordinary chondrites with a low number, 3.8 for example, have undergone very little alteration, while a number 6 denotes a high degree of alteration. So, an L4 chondrite such as Gold Basin will have a comparatively low iron content, along with chondrules that have undergone only a modest amount of deformation.

The classification of iron meteorites is dependent on a number of factors including the ratio of iron to nickel and the size and structure of kamacite and taenite bands (nickel-iron alloys). Iron meteorite classification will typically involve removing a piece for grinding and etching, so the *Widmanstätten pattern* can be examined and measured. Non-destructive bombardment by an ion beam can determine the exact percentage of nickel and other elements. The presence of nickel is a key factor in verifying almost all meteorites.

Laboratory time is expensive and in demand so there is often a significant delay in getting a classification done, as many meteorites are found each year, and there are only a few labs that have the experience and equipment to carry out this type of specialized work. In exchange for classification work, the finder donates a minimum of 20 grams, or 20% of the find (whichever is less), to the institution doing the work. The donated piece is referred to as the *type specimen* and will remain with the lab or university as part of their reference collection, enabling further study and comparison with future finds.

Once the classification has been completed, the details are presented to the Nomenclature Committee of the Meteoritical Society (NomComm), a group of professional meteoriticists who vote on the proposed names of new meteorites. A new fall or find is typically named after the nearest town to the point of discovery. If the find was made in a remote area, far from civilization, the nearest geographical feature is sometimes used, making for some very colorful names such as Starvation Flat, Cat Mountain, Crab Orchard, and Camel Donga. In order to finalize a new find and give it an official name, the NomComm also requires that the finder furnish them with the exact find location and the total known weight of all related recoveries. If a new find turns out to be a particularly rare or scientifically interesting meteorite, this process is sometimes fast tracked—as is the examination of new witnessed falls—but classification and naming of, for example, an ordinary chondrite can sometimes take two years or more.

Once the NomComm has approved a new meteorite, the final step is to have the classification and other details published in the *Meteoritical Bulletin*—a scientific publication that lists all verified meteorites.

Although getting a new meteorite officially recognized can be a rather long and time consuming process, I encourage you to go through the steps and follow proper procedure. New meteorite finds are important to science, and if you are the discoverer of one, your name will likely be included in the scientific literature of meteoritics for all time, and that is a significant accomplishment. Meteorites that have been properly named and classified are also worth more, in financial terms, to collectors. A recognized meteorite has a pedigree and a history and is usually more valuable than an unclassified find.

When a meteorite is found inside the boundaries of a known strewnfield, a specialist can usually carry out a simple visual comparison with other specimens that have already been recovered from the same site. If the new find shares common characteristics with other meteorites from that strewnfield, there is little point in getting an official classification done. For example: I will, from time to time, take a trip to the Holbrook strewnfield in northern Arizona. In 1912, a shower of thousands of stone meteorites rained down over the railroad town of Holbrook. There were many witnesses to the event and, to this day, small meteorites can still be found in the area if you are willing to put in the time. Most Holbrook meteorites are fusion-crusted individuals about the size and shape of a pea. So, if I am out walking the strewnfield with my magnet cane, and find a small, somewhat weathered stone meteorite that weighs a couple of grams and has the typical surface features of a Holbrook stone, I am not going to take up valuable lab time with a classification request. It is obvious to my trained eye that I have found a survivor from the 1912 fall. If, on the other hand, I am hunting in an area that does not encompass a known strewnfield and I discover a meteorite, it is likely new to science, and I would turn over a portion of that find for study and classification. Every new find adds valuable information to the literature of meteoritics.

[above left] Historic meteorites with museum labels and collection numbers. The five small, black stones are from the Holbrook, Arizona, fall of 1912. [above] When polished and etched with nitric acid, most iron meteorites display a lattice-like structure known as the Widmanstätten pattern. Intergrowth of the nickel-iron alloys kamacite and taenite creates this feature which is unique to meteorites. Photograph © Suzanne Morrison.

CRATER CAMP
ALTITUDE 9,045 FEET EST. 2010 POP. 2

IN THE FIELD: FUN AND SAFETY

Although meteorites are, and have always been, the great interest of my life, there was no particular day when I woke up and said: "I am now going to be a professional meteorite hunter." Recovering, collecting, and studying meteorites began for me as a hobby, grew into a passion, and ultimately became my profession.

Hunting for meteorites is not the best way to make a living; it is a calling. If you think you are going to head out into the field and strike it rich on the first day, or in the first year, think again. Meteorites are rare, and finding them requires patience, skill, dedication, knowledge and a lot of hard work. You might hunt in a strewnfield for an entire week, and find one walnut-sized space rock that is worth $50—or nothing at all. Financially, meteorite hunting does not often provide a good return on the investment of time, equipment, and expenses. But if, like me, you can appreciate the wonder, delight, and excitement that comes with being the first person—ever, in the history of the world—to see and hold a visitor from space, then meteorite hunting may be for you. If, on the other hand, your aim in life is a comfortable existence with a big house and expensive cars, then you may do better embarking upon a different career path. Although I do make my living from meteorites, it is the joy and amazement of finding them that keeps me going.

Meteorite hunting is time consuming and can be strenuous, even dangerous. If you are heading out into the field, make the same preparations that you would for camping or for a long hike. Always carry extra water and food. Maps, compass, and GPS units can be invaluable, not just to help you get to your destination, but to bring you back safely to your vehicle or camp site. It is easy to get lost in deserts or forests, so always let someone know where you are going and when you expect to be back. I have, several times, lost my way, even with a good map and my normally excellent sense of direction. So, when I begin a hunt, the first thing I do is log the GPS location of my truck. There have been a number of instances on desert hunts, miles from anywhere, the sun setting, when I discovered that I had become disoriented or turned around, and was not where I thought I was. At those moments, I was very glad that my GPS was able to lead me back to my vehicle.

Much of my work revolves around desert hunting. Deserts are dry and conditions often allow meteorites to be preserved for centuries or even longer. Some deserts are also home to snakes, scorpions, poisonous spiders, abandoned mine shafts, and other hazards. Good hiking boots are essential, and in areas that are likely to be populated by reptiles I always wear snake boots or snake gators. Carrying a small snake bite kit with you or keeping a first aid kit in your vehicle are sensible ideas, as is traveling with a cell phone. In remote areas that do not get cell phone reception, we sometimes take a satellite phone with us. Sat phones are expensive to use, but can be life savers.

Take sunscreen and a wide-brimmed hat with you on desert hunts. Be prepared for flies, mosquitos, poison ivy, and other annoyances in different environments. This is, by no means, a comprehensive list of potential hazards, so use common sense and always be prepared. Extra batteries or an extra canteen of water could literally save your life. For safety reasons, I prefer to have someone out there with me. If you encounter dangerous wild animals there is safety in numbers, and I have friends who, while hunting alone, broke a leg or suffered other serious injuries and had to literally drag themselves back to safety.

Some meteorite hunters enjoy the social aspect of expeditions and hunt in groups. Some, like Steve Arnold and myself, prefer to work as a small team. Others, like the great meteorite hunter Sonny Clary, are loners who head deep into the wilderness by themselves or with their dog. Meteorite hunting is, for me, a solitary and contemplative occupation. Steve and I have been working together for well over a decade and have embarked upon scores of expeditions together. We look out for each other, pitch in to move heavy gear, share ideas and strategies, but—despite all of that—when our boots first hit the ground we automatically take off in different directions and follow our own Muse.

Walking alone across a startling desert landscape gives me time to think, to enjoy the stillness, and to ponder the remarkable origin and history of the space rocks for which we are searching. Many times, for no good reason, an inner voice will say to me: "Geoff, look over there," or "The meteorites are just on the other side of that mesa!" I am a scientist and I do not really believe that such hunches or whims can empirically lead to anything concrete, but I follow them anyway. When the hunting day is done, and I return to the truck—exuberant because I have found something, or disappointed because my specimen bags are empty—it is good to know that my hunting buddy is somewhere nearby and has my back.

That is how I do it, but it is not the way you have to do it. This book is the world's first ever guide to meteorite hunting. I learned by doing, and so will you. Be careful, be smart, keep your eyes open, and let me know when you find your first space rock. Good hunting!

POSTSCRIPT

When I was first starting out in the meteorite world, I had the good fortune of dining with Marlin Cilz of the Montana Meteorite Lab—a highly respected meteorite expert and prep artist. He told me that we are only caretakers of the meteorites we find and collect; they will still be here long after we are gone.

Meteorites have survived an incredible journey of literally astronomical proportions and some carry within them clues to the answers of the great questions in science: How old is the universe? How did it form? What are the origins of life on our planet and is there other life, out there—somewhere? As such, meteorites deserve to be treated with care, perhaps even reverence. They are not a simple commodity like gold, diamonds, or copper. Part of being a responsible hunter and collector is acknowledging that a portion of important new finds should always be made available to academia for study; meteorites should be properly cared for, documented, labelled, and curated; and plans should be made for what will become of them after their human caretakers have passed on.

After all, they did travel a very long way to get here.

ABOUT THE AUTHOR

Geoffrey Notkin is the author of the book *Rock Star: Adventures of a Meteorite Man*, as well as more than 150 published articles on meteoritics, paleontology, adventure travel, technology, and the arts. He is a popular columnist, and writes *The Logical Lizard* for TucsonCitizen.com and *Meteorwritings* for Geology.com.

He hosts the hit television series, *Meteorite Men*, on Science Channel, and has made numerous other programs for the BBC, PBS, A&E, TLC, Discovery, Discovery UK, the National Geographic Channel, History Channel, and the Travel Channel.

Geoffrey is a widely published photographer and art director. His work has appeared in *Astronomy, Astronomy Now, Sky & Telescope, Wired, Reader's Digest, Robb Report, The Village Voice, Rock & Gem, Geotimes, Meteorite, Meteoryt* (Poland), *Seed, The Field Guide to Meteors and Meteorites, Arizona Daily Star, Tucson Weekly, Mechanical Engineering, American Theater Arts, Popsmear, Alternative Press, New York Press* and many other national and international publications. He is also an accomplished musician, and performed professionally for many years in England and the United States.

As owner of Aerolite Meteorites, LLC (www.aerolite.org), one of the foremost commercial meteorite companies, Geoffrey has worked with, and provided specimens to, many of the world's major institutions including, The American Museum of Natural History, New York; The Natural History Museum, London; The Vienna Museum of Natural History, Austria; The Center for Meteorite Studies at ASU, Tempe; and the Oscar E. Monnig Meteorite Collection in Fort Worth. He is a fellow of The Explorer's Club, and a member of the Electronic Frontier Foundation (EFF), the International Dark-Sky Association (IDA), the Association of Applied Paleontological Sciences (AAPS), and the International Meteorite Collectors' Association (IMCA). He received two Telly Awards for his work on *Meteorite Men,* and the minor planet 132904, discovered at Mount Palomar, was named after him in recognition of his contributions to science and education

Geoffrey was born on 14th street in Manhattan, and grew up in London, England. He studied geology, astronomy, photography, writing, and design in London, Boston and New York, and now resides in the Sonoran Desert of southern Arizona, with one eccentric cat, a flock of semi-tame quail, and no children.

LEARN MORE ABOUT METEORITES

AEROLITE METEORITES LLC

Geoffrey Notkin's company offers one of the world's finest selections of meteorites for sale, as well as science articles, meteorite photographs, expedition reports, news, and more.
www.aerolite.org
www.meteorites.co

CLUB SPACE ROCK

The world's largest meteorite community, Club Space Rock is a free, interactive online forum offering blogs, discussion groups, news, photography galleries, and the opportunity to meet some of the world's top meteorite specialists.
www.meteorites.ning.com

METEORITE TIMES

A free monthly online magazine presenting the best in meteorite news, research, and reports. *Meteorite Times* features exclusive columns by noted meteorite professionals.
www.meteorite-times.com

MEET THE METEORITE MEN

OFFICIAL WEBSITE

www.meteoritemen.com

SOCIAL MEDIA

www.facebook.com/meteoritemen
www.twitter.com/meteoritemen

www.aerolite.org